Computer Based

Numerical
and
Statistical
Techniques

Computer Based

Numerical
and
Statistical
Techniques

Kamlendra Kumar PhD

Assistant Professor and Faculty-in-Charge
Department of Mathematics
Shri Ram Murti Smarak College of Engineering and Technology
Bareilly, UP

Rakesh Kumar PhD

Associate Professor
Department of Mathematics
Hindu College
Moradabad, UP

CBS

CBS Publishers & Distributors Pvt Ltd

New Delhi • Bengaluru • Chennai • Kochi • Kolkata • Mumbai
Bhubaneswar • Hyderabad • Jharkhand • Nagpur • Patna • Pune • Uttarakhand

Computer Based
Numerical
and
Statistical
Techniques

ISBN: 978-93-87085-12-1

Copyright © Authors and Publisher

First Edition: 2018

Published by Satish Kumar Jain and produced by Varun Jain for

CBS Publishers & Distributors Pvt Ltd

4819/XI Prahlad Street, 24 Ansari Road, Daryaganj, New Delhi 110 002, India.

Ph: 23289259, 23266861, 23266867   Website: www.cbspd.com

Fax: 011-23243014   e-mail: delhi@cbspd.com; cbspubs@airtelmail.in.

Corporate Office: 204 FIE, Industrial Area, Patparganj, Delhi-110092

Ph: 4934 4934   Fax: 4934 4935   e-mail: publishing@cbspd.com; publicity@cbspd.com

Branches

- **Bengaluru:** Seema House 2975, 17th Cross, K.R. Road,
  Banasankari 2nd Stage, Bengaluru 560 070, Karnataka
  Ph: +91-80-26771678/79   Fax: +91-80-26771680   e-mail: bangalore@cbspd.com
- **Chennai:** 7, Subbaraya Street, Shenoy Nagar, Chennai 600 030, Tamil Nadu
  Ph: +91-44-26680620, 26681266   Fax: +91-44-42032115   e-mail: chennai@cbspd.com
- **Kochi:** Ashana House, No. 39/1904, AM Thomas Road, Valanjambalam,
  Ernakulam 682 018, Kochi, Kerala
  Ph: +91-484-4059061-65   Fax: +91-484-4059065   e-mail: kochi@cbspd.com
- **Kolkata:** 6/B, Ground Floor, Rameswar Shaw Road, Kolkata-700 014, West Bengal
  Ph: +91-33-22891126, 22891127, 22891128   e-mail: kolkata@cbspd.com
- **Mumbai:** 83-C, Dr E Moses Road, Worli, Mumbai-400018, Maharashtra
  Ph: +91-22-24902340/41   Fax: +91-22-24902342   e-mail: mumbai@cbspd.com

Representatives

- **Bhubaneswar** 0-9911037372   • **Hyderabad** 0-9885175004   • **Jharkhand** 0-9811541605   • **Nagpur** 0-9021734563
- **Patna** 0-9334159340   • **Pune** 0-9623451994   • **Uttarakhand** 0-9716462459

Printed at: Glorious Printers, Daryaganj, New Delhi

Preface

Numerical technique is the area of applied mathematics and computer science that creates, analyses, and implements algorithms for solving the problems of mathematics numerically. These problems arise throughout natural sciences, social sciences, engineering, medicine, and business. Growth in power and availability of digital computers has led to an increasing use of realistic mathematical models in science and engineering, and numerical analysis of increasing sophistication has been needed to solve these more detailed mathematical models of the world. The book entitled *Computer Based Numerical and Statistical Techniques* is designed to meet the requirements of BSc/MSc (mathematics and physics), BSc (computer science), engineering and MCA students of various Indian universities.

The subject matter has been discussed in such a simple way that students shall find no difficulty to understand it. The approach is to ensure conceptual understanding of numerical methods by relying on students' geometric instinct. Important definitions and derivations will help students develop a concrete understanding of concepts. There are a good number of solved examples with exercise to be worked out. Most of the questions have been selected from various university papers. Application-oriented problems also find appropriate place in the text. We are given preliminaries in the chapters where it is needed so that students can understand the topic without any difficulty.

For better understanding, the text is divided into twelve chapters which cover computer arithmetic and errors, roots of equations (modified Newton–Raphson, Birge–Vieta method for polynomials and Lin–Bairstow's method for quadratic factor), calculus of finite differences, interpolations (error in polynomial interpolation), piecewise and spline interpolation, approximation of functions (for continuous data, Chebyshev polynomial approximation, uniform (minimax) approximation, Gram–Schmidt orthogonalization process and Lanczos economization), numerical differentiation (Richardson's extrapolation method), numerical integration, solution of simultaneous linear algebraic equations (condition number, successive over relaxation (SOR) method and ill conditioning of system of equations), solution of ordinary differential equations, algebraic eigen values and eigen vectors of matrices. In this book, we have included algorithm of various methods as well. Gaussian integration methods, Gauss–Legendre integration, Radau integration method and Lobatto integration method are explained in detail in chapter 8. The error analysis of various methods is also given in the book. The last chapter is a part of statistical technique. Under this chapter, Z-test is given in detail. Appendix with short answer type questions is also included at the end of the book.

We express our heartiest gratitude to Shri Dev Murti, Chairman, Shri Ram Murti Smarak Trust, Bareilly, and Sahu Shankar Saran Kothiwal, President, Hindu College, Moradabad, for their support in bringing out this book. Our special thanks to Prof Prabhakar Gupta, Dean, SRMS College of Engineering and Technology, Bareilly. Our heartfelt thanks to Khalil Ahmad, former Professor, Department of Mathematics, Jamia Millia Islamia, New Delhi; Prof D Bahuguna, Department of Mathematics and Statistics, IIT Kanpur; Dr Brajesh Singh, Associate Professor, Department of Mathematics, HBTU, Kanpur; AK Mishra, Department of Mathematics, IIT (BHU); Dr Nagendra Kumar, Associate Professor, Department of Mathematics, MMH College, Ghaziabad; Dr Sanjeev Rajan, Head, Department of Mathematics, Hindu College, Moradabad, and Dr Tarun Kumar Garg, Associate Professor, Department of Mathematics, Satyawati College, Delhi. We are also thankful to our colleagues for their constructive suggestions.

We wish to put on records our sincere thanks to Mr Satish Kumar Jain, CMD, and Shri YN Arjuna, Senior Vice President Publishing, Editorial and Publicity, CBS Publishers & Distributors, New Delhi, as well as the editorial department for their kind cooperation at every stage. Constructive criticism and suggestions for improvement of the book are welcome. We hope that this book will receive a good response from students as well as teachers.

Last but not the least, the first author (Kamlendra Kumar) would like to thank his wife Mrs Pratima Gangwar and daughter Ms Lavanya Gangwar for their support. The second author (Rakesh Kumar) would like to express his heartfelt gratitude to his family members Nand Rani, Sunalika Singh and Krrish Kumar for their moral support, encouragement, inspiration and for bearing trouble knowingly or unknowingly during the writing of this book.

Kamlendra Kumar
Rakesh Kumar

Contents

Computer Arithmetic and Errors

1.1 INTRODUCTION

Various systems of computer arithmetic are available. They consist integer arithmetic, fixed point arithmetic, floating point arithmetic etc. The most common and popular arithmetic systems are integer arithmetic and floating point arithmetic. Here, we shall mainly discuss the later. Also, in the numerical solution of problems, we generally begin with some initial data and then compute, after some intermediate steps, the final results. The numerical data used are usually approximate. In addition, the methods which used may also be approximate and so the error in a computed result may be due to the error in the data, or the error in the method or both. In this chapter, we shall discuss some basic ideas regarding errors and their analyses.

1.2 FLOATING POINT REPRESENTATION OF NUMBERS

Two types of arithmetic operations are available in a computer. They are (i) integer arithmetic (ii) real or floating point arithmetic.

Integer arithmetic deals with integer operands, i.e. numbers without fractional parts. It is used mainly in counting and as subscripts.

Real or floating point arithmetic uses numbers with fractional parts as operands and is used in most computations. Scientific calculations are usually done in floating point arithmetic.

Due to economic considerations, computers are usually designed in such a way that each location (also called a word) in memory stores only a finite number of digits. As a result, all operands in arithmetic operations have only a finite number of digits.

To explain it, let us consider a hypothetical computer has a memory in which each location (or word) can store 6 digits and also has an arrangement to store one or more sign. One method of representing real number in such computer would be to suppose a fixed position for the decimal point and store all numbers after a suitable shifting if require with supposed decimal point (Fig. 1.1).

In this way, maximum and minimum possible numbers can be stored are 9999.99 and 0000.01 respectively. This range is utterly insufficient in practice so we take up a different way for representing real numbers. This way intend to save the maximum number of significant digits in

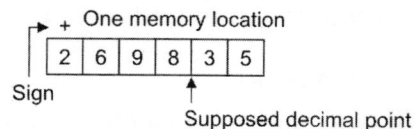

Fig. 1.1: A memory location storing the number 2698.35

a real number and also increase the range of values of real numbers stored. This way of representating and storing real numbers is called the *normalized floating point mode*. In this mode, a real number is signified as a combination of mantissa and exponent.

The mantissa should satisfy the following conditions:

 i. For positive numbers, it should be less than 1.0 and greater than or equal to 0.1
 ii. For negative number, it should be less than -1.0 and greater than or equal to -0.1

and the exponent is the power of 10 that multiplies the mantissa. We consider an example to clarify it.

The number 35.82×10^4 is expressed in this notation as $0.3582 \, E \, 6$, where $E \, 6$ is used to express 10^6. The mantissa is 0.3582 and the exponent is 6. The number is stored in a memory location as shown in Fig. 1.2.

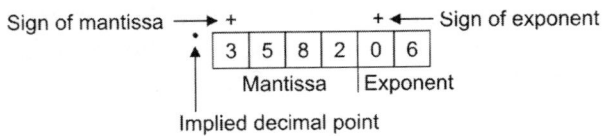

Fig. 1.2

1.2.1 Normalization

The shifting of mantissa to the left till its most significant digit is nonzero, is called *normalization*. The normalization is done to preserve the maximum number of useful (information carrying) digits. We consider an example to clarify it. Consider a number 0.0003485. In this number the leading zeros serve only to locate the decimal point and we can transfer this information to the exponent part of the number. So, the number stored as $0.3485 \, E \, 3$. We can see it in Fig. 1.3.

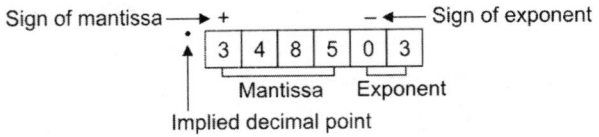

Fig. 1.3

When numbers are stored by this mode, the range of numbers that may be stored are 0.9999×10^{99} to 0.1000×10^{-99} which is clearly much larger than that used in the fixed decimal point notation. This increase in range has been found by reducing the number of significant digits in a number by two.

1.3 ARITHMETIC OPERATIONS WITH NORMALIZED FLOATING POINT NUMBERS

Arithmetic operations on floating point numbers consist of addition, subtraction, multiplication and division. We will take these operations one by one.

1.3.1 Addition

If two numbers are added and these numbers are represented in normalized floating point notation, then exponents of two numbers must be made equal and mantissa shifted **approximately**.

Example 1: Add the following floating point numbers:

   i. 0.3536 E 7 and 0.4323 E 7

  ii. 0.4232 E 6 and 0.2535 E 7

 iii. 0.2654 E 2 and 0.6255 E 4

 iv. 0.6824 E 99 and 0.3568 E 99

Ans. i. Here exponents are equal so mantissa are added

$$0.3536 \text{ E } 7$$
$$+\ 0.4323 \text{ E } 7$$
$$\text{Sum} = 0.7859 \text{ E } 7$$

  ii. Here exponents are different. The operand with the larger exponent is kept as it is

$$0.2535 \text{ E } 7$$
$$+\ 0.0423 \text{ E } 7 \qquad\qquad 0.4232 \text{ E } 6 = 0.0423 \text{ E } 7$$
$$\text{Sum} = 0.2958 \text{ E } 7$$

 iii. Here exponents are different. The operand with the larger exponent is kept as it is

$$0.6255 \text{ E } 4$$
$$+\ 0.0026 \text{ E } 4 \qquad\qquad 0.2654 \text{ E } 2 = 0.0026 \text{ E } 4$$
$$\text{Sum} = 0.6281 \text{ E } 4$$

 iv. Here exponent are equal so mantissa are added

$$0.6824 \text{ E } 99$$
$$+\ 0.3568 \text{ E } 99$$
$$\text{Sum} = 1.0392 \text{ E } 99$$

Here, sum of mantissa exceeds 1. Mantissa is shifted right and exponent is increased by 1. Then the sum is .1039 E100. But exponent part can not store more than two digits. This condition is called an overflow condition and the arithmetic unit will intimate an error condition.

Example 2: Add the following floating point numbers:

   i. 0.7432 E 5 and 0.4243 E 5

  ii. 0.5692 E 2 and 0.6293 E 3

Ans. i. Here exponents are equal so mantissa are added

$$0.7432 \text{ E } 5$$
$$+\ 0.4243 \text{ E } 5$$
$$\text{Sum} = 1.1675 \text{ E } 5$$

Here, sum of manissa exceeds 1. Mantissa is shifted right and exponent is increased by 1. The result is 0.1168 E 6.

  ii. Here exponent are different. The operand with the larger exponent is kept as it is.

$$0.6293 \text{ E } 3$$
$$+\ 0.0569 \text{ E } 3 \qquad\qquad 0.5692 \text{ E } 2 = 0.0569 \text{ E } 3$$
$$\text{Sum} = 0.6862 \text{ E } 3$$

1.3.2 Subtraction

The operation of subtraction of two numbers is nothing but adding a negative number. Thus the principle are the same as addition.

Example 3: Subtract the following floating point numbers:

   i. 0.6453 E 99 and 0.6432 E 99

  ii. 0.5422 E 3 and 0.5408 E 3

 iii. 0.8432 E 4 and 0.5411 E 3

 iv. 0.5432 E 3 and 0.3292 E 4

Ans.   i. The exponent are same so mantissa are substracted

$$\begin{array}{r} 0.6453\ E\,99 \\ -\ 0.6432\ E\,99 \\ \hline =\ 0.0021\ E\,99 \end{array}$$

In normalized floating point, the mantissa is ≥ 0.1. So, the result is 0.21 E-101. Exponent part can not store more than two digits. This condition is called *underflow condition* and the arithmatic unit will signal an error condition.

  ii. The exponent are same. So, mantissa are subtracted

$$\begin{array}{r} 0.5422\ E\,3 \\ -\ 0.5408\ E\,3 \\ \hline =\ 0.0014\ E\,3 \end{array}$$

In normalized floating point, the mantissa is ≥ 0.1. So, the result is 0.14 E 1.

 iii. The exponent are not same. The operand with the larger exponent is kept as it is

$$\begin{array}{r} 0.5411\ E\,3 \\ -\ 0.0843\ E\,3 \\ \hline =\ 0.4568\ E\,3 \end{array}$$

 iv.
$$\begin{array}{r} 0.5432\ E\,3 \\ -\ 0.0329\ E\,3 \\ \hline =\ 0.5103\ E\,3 \end{array}$$

Remark

Overflow and underflow condition

Any result larger than 0.9999 E 99 leads to an *overflow condition*. Similarly, if the result of an arithmetic operation gives a number less than 0.1000 E 99, then it is called an *underflow condition*.

1.3.3 Multiplication

Two numbers are multiplied in the normalized floating point mode by multiplying the mantissas and adding the exponents. After multiplication of the mantissas, the result mantissa is normalized as in addition/subtraction operation and the exponent appropriately adjusted.

Example 4: Multiply the following floating point numbers:

   i. + 0.5453 E 12 and 0.3111 E 15

  ii. + 0.2222 E 10 and 0.1234 E 15

 iii. + 0.2121 E 51 and 0.3334 E 50

 iv. + 0.1234 E 48 and 0.1111 E 55

Solution: i. $0.5453\,E\,12 \times 0.3111\,E\,15 = 0.1696\underline{4283}\,E\,3$

$$_{D^*}$$

$$= 0.1696\,E\,3$$

ii. $0.2222\,E\,10 \times 0.1234\,E\,15 = 0.02741948\,E\,25$

In normalized floating point, the mantissa is ≥ 0.1.

So, the result is $0.2741\underline{948}\,E\,24 = 0.2741\,E\,24$

$$_{D}$$

iii. $0.2121\,E\,51 \times 0.3334\,E\,50 = 0.07071414\,E\,101$

$$= 0.7071414\,E\,100$$

The result overflows.

iv. $0.1234\,E\text{-}48 \times 0.1111\,E\text{-}55 = 0.01370974\,E\text{-}103$

$$= 0.1370\underline{974}\,E\text{-}104$$

$$_{D}$$

The result overflows.

1.3.4 Division

In case of division, the mantissa of the numerator is divided by that of the denominator. The denominator exponent is subtracted from the numerator exponent. After the division of mantissas, the resulted mantissa is normalized as in addition or subtraction operation and the exponent appropriately adjusted.

Example 5: Solve the following floating point numbers:

   i. $0.8998\,E\,1 \div 0.1000\,E\,46$

  ii. $0.8989\,E\,5 \div 0.1000\,E\,97$

 iii. $0.1000\,E\,4 \div 0.9999\,E\,2$

 iv. $0.9432\,E\,2 \div 0.1000\,E\,98$

Solution:

   i. $0.8998\,E\,1 \div 0.1000\,E\,46 = 8.998\,E\,47$

$$= 0.8998\,E\,48$$

  ii. $0.8989\,E\,5 \div 0.1000\,E\,97 = 8.989\,E\,102$

$$= 0.8989\,E\,101$$

The result underflows.

 iii. $0.1000\,E\,4 \div 0.9999\,E\,2 = 0.1000\,E\,2$

 iv. $0.9432\,E\,2 \div 0.1000\,E\,98 = 9.432\,E\,100$

$$= 0.9432\,E\,101$$

The result overflows.

Example 6: Find the solution of the following equation using floating point arithmetic with 4-digit mantissa

$$x^2 - 1000x + 25 = 0$$

Give comments on the results so obtained.

Solution: We have

$$x^2 - 1000x + 25 = 0$$

$$x = \frac{10^3 \pm \sqrt{(10)^6 - (10)^2}}{2}$$

---

* D stands for discarded

Using floating point arithmatic with a 4-digit mantissa
$$10^6 = 0.1000 \, E7, \, 10^2 = 0.1000 \, E3$$

Thus $\quad\quad 10^6 - 10^2 = 0.1000 \, E7 - 0.1000 \, E3 = 0.1000 \, E7$

Thus $\quad\quad \sqrt{10^6 - 10^2} = 0.1000 \, E4$

The roots are thus $\left(\dfrac{0.1000 \, E4 + 0.1000 \, E4}{2} \right)$ and $\left(\dfrac{0.1000 \, E4 - 0.1000 \, E4}{2} \right)$ which are respectively $0.1000 \, E4$ and $0.0000 \, E4$.

One of roots becomes zero due to the limited precision allowed in calculation. Now, we reformulate the problem and remember that in the quadratic equation $ax^2 + bx + c = 0$, the product of the roots is given by c/a, then having determined the larger root, the smaller root may be obtained by dividing c/a by it.

So, $\quad\quad$ first root $= 0.1000 \, E4$

and $\quad\quad$ second root $= \dfrac{25}{0.1000 \, E4} = \dfrac{0.2500 \, E2}{0.1000 \, E4} = 0.2500 \, E\text{-}1$

Such a situation may be recognized in an algorithm by checking to see if $b^2 >> |4ac|$.

Example 7: For $x = 0.3845$ and $y = 0.3800$, calculate the value of $\dfrac{x^2 - y^2}{x + y}$ using normalized floating point arithmetic. Compare with the values of $(x - y)$.

Solution: $\quad\quad x + y = 0.3845 \, E0 + 0.3800 \, E0 = 0.7645 \, E0$

$$x^2 = (0.3845 \, E0) \times (0.3845 \, E0) = 0.1478 \, E0$$

$$y^2 = (0.3800 \, E0) \times (0.3800 \, E0) = 0.1444 \, E0$$

$$x^2 - y^2 = 0.1478 \, E0 - 0.1444 \, E0 = 0.0034 \, E0$$

Now, $\quad\quad \dfrac{x^2 - y^2}{x + y} = 0.0034 \, E0 \div 0.7645 \, E0 = 0.4447 \, E2$

Also, $\quad\quad x - y = 0.3845 \, E0 \div 0.3800 \, E0 = 0.0045 \, E2$

$$= 0.4500 \, E2$$

The value of $x - y$ obtained from $\dfrac{x^2 - y^2}{x + y}$ is different from direct value of $x - y$. Here

$$\text{error} = 0.4500 \, E\text{-}2 - 0.4447 \, E\text{-}2$$

$$= 0.5300 \, E\text{-}4$$

Example 8: Find the smaller root of the equation $x^2 - 500x + 1 = 0$ using four digit arithmetic.

Solution: Here $\quad\quad b^2 \gg |4ac|$

Roots of equation $ax^2 - bx + c = 0$ are $\dfrac{b + \sqrt{b^2 - 4ac}}{2a}$ and $\dfrac{b - \sqrt{b^2 - 4ac}}{2a}$

Product of roots are c/a

$\therefore$ smaller root is $\dfrac{c/a}{\left(\dfrac{b + \sqrt{b^2 - 4ac}}{2a} \right)}$ or $= \dfrac{2c}{b + \sqrt{b^2 - 4ac}}$

Here $a = 1 = 0.1000\,E1$, $b = 500 = 0.5000\,E3$, $c = 1 = 0.1000\,E1$

$$b^2 - 4ac = 0.2500\,E6 - 0.4000\,E1 = 0.2500\,E6 \text{ (upto four digit)}$$

$\therefore \qquad \sqrt{b^2 - 4ac} = 0.5000\,E3$

$\therefore \qquad$ Smaller root $= \dfrac{2 \times (0.1000\,E1)}{0.5000\,E3 + 0.5000\,E3} = \dfrac{0.2000\,E1}{0.1000\,E4}$

$$= 0.2000\,E2 = 0.0020$$

Example 9: Evaluate applying normalized floating point arithmetic for the following

$$1 - \cos x \text{ at } x = 0.1382 \text{ radian}$$

Assume $\quad \cos(0.1382) = 0.9905$

Compare it when evaluated $2\sin^2 x/2$. Assume $\sin 0.0638 = 0.6375\,E1$.

Solution: $\quad 1 - \cos(0.1382) = 0.1000\,E1 - 0.9905\,E0$

$$= 0.1000\,E1 - 0.0990\,E1 = 0.1000\,E1$$

Now $\qquad \sin x/2 = \sin(0.0638) = 0.6375\,E1$

$$2\sin^2 x/2 = (0.2000\,E1) \times (0.6375\,E1) \times (0.6375\,E1)$$

$$= 0.08128\,E1 = 0.8128\,E2$$

The value of $1 - \cos x$ obtained from alternate formula is differ by an error

$$= 0.1000\,E1 - 0.8128\,E2$$

$$= 0.1872\,E2$$

Example 10: For $e = 2.7183$, calculate the value of e^x when $x = 0.5250\,E1$. The expression for e^x is

$$e^x = 1 + x + \frac{x^2}{2!} + \frac{x^3}{3!}$$

Solution: $e^{0.5250\,E1} = e^5 \times e^{0.25}$

$$e^5 = (0.2718\,E1) \times (0.2718\,E1) \times (0.2718\,E1) \times (0.2718\,E1) \times (0.2718\,E1)$$

$$= 0.1484\,E3$$

Also, $\qquad e^{0.25} = 1 + (0.25) + \dfrac{(0.25)^2}{2!} + \dfrac{(0.25)^3}{3!}$

$$= 1 + 0.25 + 0.03125 + 0.002604 = 0.1284\,E1$$

Now $\quad e^{0.5250\,E1} = (0.1484\,E3) \times (0.1284\,E1) = 0.1905\,E3.$

Example 11: In case of normalized floating point representation, associative and distributive laws are not always valid. Give example to prove this statement. If the normalization on floating point is carried out at each stage, prove the following:

   i. $a(b - c) \neq ab - ac$, where $a = 0.4444\,E1$, $b = 0.3434\,E1$, $c = 0.3424\,E1$.

   ii. $(a + b) - c \neq (a - c) + b$, where $a = 0.4556\,E1$, $b = 0.4447\,E1$, $c = 0.4533\,E1$.

Solution: A consequence of the floating point representation is that the associative and the distributive laws of arithmetic are not always valid.

We consider the following example intentionally to illustrate the inaccuracies that may build up due to shifting and truncation of numbers in arithmetic operations.

Non-distributivity of arithmetic

Let $\qquad\qquad\qquad a = 0.4444\,E1$, $b = 0.3434\,E1$, $c = 0.3424\,E1$

Then
$$b - c = 0.0010 \text{ E}1 = 0.1000 \text{ E}1$$
$$a(b - c) = 0.4444 \text{ E}1 \times 0.1000 \text{ E}1$$
$$= 0.0444 \text{ E}0 = 0.4440 \text{ E}1$$
$$ab = 0.4444 \text{ E}1 \times 0.3434 \text{ E}1 = 0.1526 \text{ E}2$$
$$ac = 0.4444 \text{ E}1 \times 0.3424 \text{ E}1 = 0.1521 \text{ E}2$$
∴
$$ab - ac = 0.1526 \text{ E}2 - 0.1521 \text{ E}2 = 0.0005 \text{ E}2$$
$$= 0.5000 \text{ E}1$$

Thus, $a(b - c) \neq ab - ac$

which shows the nondistributivity of arithmetic.

Non-associativity of arithmetic

Again, let
$$a = 0.4556 \text{ E}1, b = 0.4447 \text{ E}1, c = 0.4533 \text{ E}1$$
$$a + b = 0.4556 \text{ E}1 + 0.4447 \text{ E}1 = 0.4556 \text{ E}1 + 0.0044 \text{ E}1$$
$$= 0.4600 \text{ E}1$$
$$(a + b) - c = 0.4600 \text{ E}1 - 0.4533 \text{ E}1 = 0.0067 \text{ E}1$$
$$= 0.6700 \text{ E}1$$

Again,
$$a - c = 0.4556 \text{ E}1 - 0.4553 \text{ E}1$$
$$= 0.0023 \text{ E}1 = 0.2300 \text{ E}1$$
$$(a - c) + b = 0.2300 \text{ E}1 + 0.4447 \text{ E}1 = 0.6747 \text{ E}1$$

Thus, $(a + b) - c \neq (a - c) + b$

which shows the non-associativity of arithmetic.

EXERCISE 1.1

1. Add the following floating point numbers:
   i. 0.4546 E5 and 0.5433 E5   [Ans. 0.9979 E5]
   ii. 0.6434 E99 and 0.4845 E99   [Ans. Overflow condition]
   iii. 0.4546 E3 and 0.5433 E7   [Ans. 0.5433 E7]
2. Subtract the following floating point numbers:
   i. 0.36143448 E7 – 0.36132346 E7   [Ans. 0.1110 E4]
   ii. 0.9432 E4 – 0.5452 E3   [Ans. 4509 E3]
   iii. (0.9432 E4) – (0.6353 E5)   [Ans. 0.8797 E4]
3. Represent 25.47×10^5 in normalized floating point mode.   [Ans. 0.2547 E7]
4. Explain underflow and overflow conditions in floating point's addition and subtraction.
5. Multiply the following floating point numbers:
   i. 0.1111 E51 and 0.4444 E50   [Ans. Overflow condition]
   ii. 0.1234 E49 and 0.1111 E54   [Ans. Underflow condtion]
   iii. 0.5334 E0 and 0.1132 E25   [Ans. 0.6038 E17]
   iv. 0.5543 E12 and 0.4111 E15   [Ans. 0.6038 E17]
6. Perform the following operations:
   i. 0.1000 E5 ÷ 0.9999 E3   [Ans. 0.1000 E2]
   ii. 0.9998 E5 ÷ 0.1000 E98   [Ans. Underflow condition]

7. Obtain a second degree polynomial approximation to $f(x) = (1 + x)^{1/2}$, $x = [0, 0.1]$ using Taylor's series expansion about $x = 0$. Use the expansion to approximate $f(0.05)$ and bound the truncation error.

$$\left[\text{Ans. } (1 + x)^{1/2} = 1 + \frac{x}{2} - \frac{x^2}{8} + \frac{1}{16} \frac{x^3}{[(1 + \eta)^{1/2}]^5}; 0 < \eta < 0.1, f(0.05) \right.$$

$$\left. = 0.10246875 \times 10^1, |T| = 0.625 \times 10^{-4} \right]$$

8. Compute a middle value of numbers $a = 4.568$ and $b = 6.762$ using the four digit arithmetic and compare the result by taking $c = a + \dfrac{(b - a)}{2}$.

9. Find the value of $\sin x \simeq x - \dfrac{x^3}{3!} + \dfrac{x^5}{5!}$ with an absolute error smaller than 0.005 for $x = 0.2000 \, E \, 0$ using normalized floating point arithmetic with 4 digit mantissa.

10. Find the value of $(1 + x)^2$ and $(x^2 + 2x) + 1$ when $x = 0.5999$ E-2.

11. Prove the following consequence of the normalized floating point representation of numbers by taking $x = 0.6667$.

$$6x \neq x + x + x + x + x + x$$

1.4 Error Analysis

Numerical analysis may be regarded as a process to develop and evaluate the methods for computing required numerical results from the given numerical data. It incorporate three broad steps in the process:

i. Given data, called the input information

ii. The algorithm which is based on finite set of rules giving a sequence of operations for solving a specific type of problem

iii. The result obtained, called the output information

Input information $\longrightarrow$ The algorithm $\longrightarrow$ Output information

Clearly, the error may be due to in data (input), or the errors in method (algorithm), or both. Our aim will be to minimize these errors so that we can find best possible results.

1.5 Numbers and Their Accuracy

We often come across two types of numbers, one exact and other approximate. Exact number are 2, 3, 6, 8, 3/2, 1/4 etc. while approximate numbers are those which represent the numbers to a certain degree of accuracy. The approximate numbers are $\frac{1}{3}, \sqrt{2}, \pi$ etc.

1.6 Significant Digits

The digits which are used to express a number are called significant digits or significant figure. The numbers 2.2415 and 0.0014 contains 5 and 2 significant digits respectively. Similarly the numbers 1.0145 and 2400 have 5 and 2 significant digits respectively. Since, the zeros serve only to fix the position of decimal point.

1.7 Source of Errors

The broad sources of errors in numerical analysis are given as:

 i. Input errors in data and modeling
 ii. Algorithmic errors
 iii. Computational errors.

1.8 Rounding Off

Amongst above errors, computational errors are due to rounding off a number is very frequent. The reason is that we frequently come across numbers with large number of digits, and in practice, it is desirable to limit such numbers to a manageable number of digits. The process is called rounding off.

Numbers are rounding off according to the following rule:

To round off a number to a significant digit, we shall discard all digits right of the nth digit. For this purpose, we have the following rules.

If this discarded number is:

 i. Less than 5 in $(n + 1)^{th}$ place, leave the nth digit unaltered, e.g. 4.563 to 4.56.
 ii. Greater than 5 in $(n + 1)^{th}$ place, increase the nth digit by unity, e.g. 5.8957 to 5.896.
 iii. Exactly 5 in $(n + 1)^{th}$ place, increase the nth digit by unity if it is odd, otherwise leave it unchanged, e.g. $10.485 \simeq 10.48$, $10.475 \simeq 10.48$.
   The number thus rounded off is said to be correct to n significant digits.

1.9 Chopping

In it, extra digits are dropped by truncation of numbers. Let us suppose that we are using a computer with a fixed word length of four digits then a number like 14.82463 will be stored as 14.82.

1.10 Types of Errors

In numerical analysis, we generally come across following types of errors:

 1. **Inherent errors:** Most of the numerical computation are not exact. This is because of either due to given data being approximate or due to the limitations of computer aids, such as calculator, mathematical tables and the digital computer. Because of this limitation, we have to round off the numbers and this is the cause of rounding off errors.

    Although such errors can be minimized by carrying the computations to more significant digits at each step of computation by retaining at each step of computation at least one more significant figure that is given in the data. Finally, perform the last operation and round off.

 2. **Truncation errors:** Truncation errors refers to the error in a method, which occurs due to the use of approximate formula in computation, e.g. when a function $f(x)$ is evaluated from an infinite series for x after truncating it at a certain number of terms.

    We can understand it by taking another example. In calculus of finite difference, we always construct polynomial approximation in a given interval. Further, we use only finite number of terms depending on the number of known observations. All this leads to error in the final result. This error is called truncation error. Truncation error is a kind of algorithm error. Also if

$$e^x = 1 + x + \frac{x^2}{2!} + \frac{x^3}{3!} + \frac{x^4}{4!} + \frac{x^5}{5!} + \ldots \infty = Y \text{ (say)}$$

is truncated to $1 + x + \frac{x^2}{2!} + \frac{x^3}{3!} + \frac{x^4}{4!} = Y'$ (say) then truncation error $= Y - Y'$.

3. **Absolute error:** Absolute error is the numerical difference between the true value of a quantity and its approximate value. Thus if x is the true value of a quantity and $\bar{x}$ is its approximate value, then the absolute error E_A is given by

$$E_A = |x - \bar{x}| = |\text{Error}|$$

4. **Relative error:** The relative error E_R is defined by

$$E_R = \frac{E_A}{x} = \frac{|\text{Error}|}{\text{True value}} = \left|\frac{x - \bar{x}}{x}\right|$$

5. **Percentage error:** The percentage error E_p is defined by

$$E_p = 100 E_R = 100 \left|\frac{x - \bar{x}}{x}\right|$$

Some Important Remarks:

i. If a number correct to m decimal place, then error $= \left(\frac{1}{2}\right) \times 10^{-m}$, e.g. if the number 2.71828 is correct to five decimal places, then error $= \left(\frac{1}{2}\right) \times 10^{-5} = 0.000005$.

ii. If the first significant digit of a number is k and the number is correct to n significant digits then, relative error $< \frac{1}{(k \times 10^{n-1})}$.

Example 12: Define absolute error and relative error. An approximate value of π is given by 3.1428571 and its true value is 3.1415926. Find absolute and relative errors.

Solution: True value = 3.1415926

Approximate value = 3.1428571

Error = True value – Approximate value

$= 3.1415926 - 3.1428571 = -0.0012645$

Absolute error $E_A = |\text{Error}| = 0.0012645$

Relative error $E_R = \dfrac{E_A}{\text{True value}} = \dfrac{0.0012645}{3.1415926}$

Example 13: Suppose 1.414 is used as an approximation to $\sqrt{2}$. Find the absolute and relative errors.

Solution: True value $= \sqrt{2} = 1.41421356$

Approximate value = 1.414

Error = True value – Approximate value

$= 1.41421356 - 1.414 = 0.00021356$

Absolute error $E_A = |\text{Error}| = 0.00021356$

$= 0.21356 \times 10^{-3}$

$$\text{Relative error } E_R = \frac{E_A}{\text{True value}} = \frac{0.21356 \times 10^{-3}}{\sqrt{2}} = 0.151 \times 10^{-3}$$

Example 14: If it is required to obtain the roots of $x^2 - 2x + \log_{10}2 = 0$ to four decimal places. To what accuracy sould $\log_{10}2$ be given?

Solution: Roots of quadratic equation are given by

$$x = \frac{2 \pm \sqrt{4 - 4\log_{10}2}}{2} = 1 \pm \sqrt{1 - \log_{10}2}$$

$\therefore$
$$|\Delta x| = \frac{1}{2} \frac{\Delta(\log 2)}{\sqrt{1 - \log 2}} < 0.5 \times 10^{-4}$$

or
$$\Delta(\log 2) < 2 \times 0.5 \times 10^{-4}(1 - \log 2)^{1/2} < 0.83604 \times 10^{-4}$$
$$\approx 8.3604 \times 10^{-5}.$$

Example 15: Find the relative error, absolute error and percentage error if 7/9 is approximated to 0.778.

Solution: True value = 7/9 = 0.777777

Approximate value = 0.778

Absolute error E_A = |True value – Approximate value|

$\qquad$ = |0.777777 – 0.778| = 0.000223

$$\text{Relative error } E_R = \frac{E_A}{\text{True value}} = \frac{0.000223}{0.777777}$$

$\qquad$ = 0.00029

Percentage error $E_p = E_R \times 100$

$\qquad$ = 0.00029 × 100 = 0.029%.

Example 16: Round off the number 85471 to four significant digits and then calculate the absolute error and percentage error.

Solution: Number round off to four significant digits = 85470

Absolute error E_A = |85471 – 85470| = 1

$$\text{Relative error } E_R = \frac{E_A}{85471} = \frac{1}{85471} = 0.000011$$

Percentage error $E_p = E_R \times 100 = 0.000011 \times 100 = 0.0011$

Example 17: Find the relative error of the number 7.8 if both of its digits are correct.

Solution: Here $\qquad E_A = 1/2 \times 10^{-1} = 0.05$

$\therefore$
$$E_R = \frac{E_A}{\text{True value}} = \frac{0.05}{7.8} = 0.0064$$

Example 18: Evaluate the sum $S = \sqrt{5} + \sqrt{7} + \sqrt{11}$ to 4 significant digits and find its absolute and relative errors.

Solution: $\sqrt{5} = 2.236, \sqrt{7} = 2.646, \sqrt{11} = 3.317$

Hence $\qquad S = 2.236 + 2.646 + 3.317 = 8.199$

and $\qquad E_A = 0.0005 + 0.0005 + 0.0005 = 0.0015$

The total absolute error shows that the sum is correct to 3 significant figure only.

$\therefore$ We take, $S = 8.19$

then $E_r = \dfrac{0.0015}{8.19} = 0.0001.$

Example 19: Find the absolute, relative and percentage errors if x is rounded off to three decimal digits. Given $x = 0.004997$.

Solution: Number rounded off to three decimal digits = 0.005

$$\text{Error} = 0.004997 - 0.005 = -0.000003$$

$$\text{Absolute error } E_A = |\,\text{error}\,| = 0.000003$$

$$\text{Relative error } E_R = \frac{E_A}{0.004997} = \frac{0.000003}{0.004997} = 0.0006$$

$$\text{Percentage error} = E_R \times 100 = 0.06.$$

Example 20: Given the solution of a problem as $x_a = 35.25$ with the relative error in the solution at most 2%. Find to four decimal digits, the range of values within which the exact value of the solution must lie.

Solution: Given that:

  i. Maximum relative error in the solution = 2% = 0.02.

  ii. Approximate value of the solution is $x_a = 35.25$.

If x is the exact value of the solution, then according to the problem, we have

$$\left|\frac{x - x_a}{x}\right| < 0.02$$

i.e.

$$\left|1 - \frac{x_a}{x}\right| < 0.02$$

which can be written as

$$-0.02 < \left(1 - \frac{x_a}{x}\right) < 0.02$$

Now, if $\left(1 - \dfrac{x_a}{x}\right) > -0.02$, then

$$\frac{-x_a}{x} > -1 - 0.02$$

or $\dfrac{-x_a}{x} > -1.02$ or $\dfrac{x_a}{x} < 1.02$

or $x_a < 1.02x$

or $x > \dfrac{x_a}{1.02} = \dfrac{35.25}{1.02} = 34.558823594$

However, if $\left(1 - \dfrac{x_a}{x}\right) < 0.02$, then

$$\frac{-x_a}{x} < -1 + 0.02 \text{ or } \frac{-x_a}{x} < -0.98$$

or $\qquad\qquad\qquad \dfrac{x_a}{x} > 0.98$ or $x_a > 0.98x$

or $\qquad\qquad\qquad x < \dfrac{x_a}{0.98} = \dfrac{35.25}{0.98} = 35.9693877551$

Therefore, we have

$$34.5588235294 < x < 35.9693877551$$

Hence, correct to four decimal digits, the range of values within which the exact value of the solution lies, is

$$34.5588 < x < 35.9694.$$

EXERCISE 1.2

1. If 0.333 is the approximate value of $1/3$, find absolute, relative and percentage errors.   [Ans. 0.000333, 0.000999, 0.099%]
2. If true value = $10/3$, approximate value = 3.33, then find the absolute and relative errors.   [Ans. $E_A = 0.003333$, $E_R = 0.000999$]
3. Round off the numbers correct to four significant digits: 2.26325, 35.46735, 4585561, 0.60035, 0.000023317.   [Ans. 2.263, 35.47, 458600, 0.6004, 0.00002332]
4. If $x = 2.536$, find the absolute error and relative error when (i) x is rounded off, ii x is truncated to two decimal digits.   [Ans. 0.004, 0.0015772]
5. Define absolute, relative and percentage error.
6. What do you mean by truncation error? Explain with example.
7. Find the relative error of the number 8.6 if both of its digits are correct.
   [Ans. $E_A = 0.05$, $E_R = 0.0058$]
8. Find the percentage error if 625.483 is approximated to three significant digits.
   [Ans. $E_A = 0.483$, $E_R = 0.000772$, $E_p = 0.077\%$]
9. Rounded off the number 75462 to four significant digits and then calculate the absolute error and percentage error.   [Ans. $E_A = 2$, $E_R = 0.0000265$, $E_p = 0.00265$]
10. The height of an observation tower was estimated to be 47 metre whereas its actual height was 45 metre. Calculate the percentage relative error in the measurement.   [Ans. 4.44%]
11. If the number x is correct to four decimal places, what will be the error.
    [Ans. 0.00005]

1.11 MACHINE EPSILON

A floating point number system within a computer is always limited by the finite word length of computers. This means that only a finite number of digits can be represented. As a result, numbers that are too large or too small can not be represented. Hence even in string an exact decimal number in its converted form in the computer memory, an error is occurred. This error is machine dependent and is called machine epsilon.

$$\text{Error} = \text{True value} - \text{Approximate value}$$

1.12 GENERAL FORMULA FOR TERMS

Let $u = g(x_1, x_2, x_3, ..., x_n)$ be a function of n variables $x_1, x_2, ..., x_n$ and let the error in any x_i be Δx_i, where $i = 1, 2, ..., n$. Then, the error Δu in u is given by

$$u + \Delta u = g(x_1 + \Delta x_1, x_2 + \Delta x_2, ..., x_n + \Delta x_n)$$

Expanding RHS by Taylor's series, we get

$$u + \Delta u = g(x_1, x_2, ..., x_n) + \sum_{i=1}^{n} \frac{\partial g}{\partial x_i} \Delta x_i + \text{terms involving } (\Delta x_i)^2 \text{ etc.}$$

Assuming that the error in Δx_2 are small so that $\dfrac{\Delta x_i}{x_i} \ll 1$. We can neglect the square and higher powers of Δx_i in the above expression and get

$$\Delta u \approx \sum_{i=1}^{n} \frac{\partial g}{\partial x_i} \Delta x_i$$

or

$$\Delta u \approx \frac{\partial g}{\partial x_1} \Delta x_1 + \frac{\partial g}{\partial x_2} \Delta x_2 + \frac{\partial g}{\partial x_3} \Delta x_3 + ... + \frac{\partial g}{\partial x_n} \Delta x_n$$

and relative error in u is

$$E_R = \frac{\Delta u}{u} = \frac{\partial u}{\partial x_1} \frac{\Delta x_1}{u} + \frac{\partial u}{\partial x_2} \frac{\Delta x_2}{u} + ... + \frac{\partial u}{\partial x_n} \cdot \frac{\Delta x_n}{u}.$$

1.13 ERRORS IN ARITHMETIC OPERATION

1. **Error in addition:** Let $\quad x = x_1 + x_2$

$\therefore \qquad\qquad x + \Delta x = (x_1 + \Delta x_1) + (x_1 + \Delta x_1)$

Hence the absolute error in x is given by

$$\Delta x = \Delta x_1 + \Delta x_2$$

Further more, $\qquad \dfrac{\Delta x}{x} = \dfrac{\Delta x_1}{x} + \dfrac{\Delta x_2}{x}$

This gives relative error in addition.
Maximum relative error is

$$\left| \frac{\Delta x}{x} \right| \le \left| \frac{\Delta x_1}{x} \right| + \left| \frac{\Delta x_2}{x} \right|.$$

and maximum absolute error is

$$|\Delta x| \le |\Delta x_1| + |\Delta x_2|$$

2. **Error in subtraction:** Let $x = x_1 - x_2$

It is assumed that $x_1 > x_2 > 0$.

$\therefore \qquad\qquad x + \Delta x = (x_1 + \Delta x_1) - (x_2 + \Delta x_2)$

$$= (x_1 - x_2) + (\Delta x_1 - \Delta x_2) = x + \Delta x_1 - \Delta x_2$$

$\Rightarrow \qquad\qquad \Delta x = \Delta x_1 - \Delta x_2$ is the absolute error.

Furthermore, $\qquad \dfrac{\Delta x}{x} = \dfrac{\Delta x_1}{x} - \dfrac{\Delta x_2}{x}$

This gives relative error in subtraction.
Maximum relative error is

$$\frac{\Delta x}{x} \le \left| \frac{\Delta x_1}{x} \right| + \left| \frac{\Delta x_2}{x} \right|$$

and maximum absolute error is

$$|\Delta x| = |\Delta x_1| + |\Delta x_2|$$

3. **Error in multiplication:** Let $x = x_1 x_2$

   We know that if x is a function of x_1, x_2, then

   $$\Delta x = \frac{\partial x}{\partial x_1} \Delta x_1 + \frac{\partial x}{\partial x_2} \Delta x_2$$

   Now $\qquad \dfrac{\Delta x}{x} = \dfrac{1}{x} \dfrac{\partial x}{\partial x_1} \Delta x_1 + \dfrac{1}{x} \dfrac{\partial x}{\partial x_2} \delta x_2$

   But $\qquad \dfrac{1}{x} \dfrac{\partial x}{\partial x_1} = \dfrac{x_2}{x_1 x_2} = \dfrac{1}{x_1}$

   $$\dfrac{1}{x} \dfrac{\partial x}{\partial x_2} = \dfrac{x_1}{x_1 x_2} = \dfrac{1}{x_2}$$

   Hence $\qquad \dfrac{\Delta x}{x} = \dfrac{\Delta x_1}{x_1} + \dfrac{\Delta x_2}{x_2}$

   $\therefore$ Absolute error and relative error is given by

   $$\text{Max. absolute error} = \left| \frac{\Delta x}{x} \right| x = \left| \frac{\Delta x_1}{x} \right| \cdot (x_1 x_2)$$

   $$\text{Max. relative error} = \left| \frac{\Delta x}{x} \right| \le \left| \frac{\Delta x_1}{x_1} \right| + \left| \frac{\Delta x_2}{x_2} \right|.$$

4. **Error in division:** Let $x = \dfrac{x_1}{x_2}$

   $\therefore \qquad \dfrac{\Delta x}{x} = \dfrac{1}{x} \dfrac{\partial x}{\partial x_1} \Delta x_1 + \dfrac{1}{x} \dfrac{\partial x}{\partial x_2} \Delta x_2$

   $$= \frac{\dfrac{1}{x_2} \Delta x_1}{\left(\dfrac{x_1}{x_2} \right)} + \frac{\Delta x_2}{\left(\dfrac{x_1}{x_2} \right)} \cdot \left(\frac{-x_1}{x_2^2} \right) = \frac{\Delta x_1}{x_1} - \frac{\Delta x_2}{x_2}$$

   $\Rightarrow \qquad \left| \dfrac{\Delta x}{x} \right| \le \left| \dfrac{\Delta x_1}{x_1} \right| + \left| \dfrac{\Delta x_2}{x_2} \right|$ which is maximum relative error

   $$\text{Max. absolute error} = |\Delta x| = \left| \frac{\Delta x}{x} \right| \cdot x = \left| \frac{\Delta x}{x} \right| \cdot \left(\frac{x_1}{x_2} \right).$$

5. **Error in a series approximation:** If a function $g(x)$ is expanded in the form of a convergent series in ascending powers of $(x - a)$ and values of $g(x)$ are evaluated from only the first n terms of the series, then error is said to be committed due to a series approximation.

   Taylor's expansion of $g(x)$ at $x = a$ be

   $$g(x) = g(a) + (x - a) g'(a) + \frac{(x - a)^2}{2!} g'(a) + \dots + \frac{(x - a)^{n - 1}}{(n - 1)!} g^{(n - 1)}(a) + R_n(x)$$

   where $R_n(x) = \dfrac{(x - a)^n}{n!} g^n(\eta), \ a < \eta < x.$

For a convergent series, $R_n(x) \to 0$ as $n \to \infty$. If we approximate $f(x)$ by first n terms of series, then maximum error committed is given by $R_n(x)$. As a result, if the accuracy is specified in advance, then it is possible to find n, i.e. the number of terms so that the finite series gives the desired accuracy. If the result is needed to correct to k decimal places then maximum error is $(1/2) \times 10^{-k}$.

Hence
$$|R_n(x)| = \left| \frac{(x-a)^n}{n!} g^{(n)}(\eta) \right| < \frac{1}{2} \times 10^{-k}$$

or
$$|R_n(x)| < \frac{1}{2} \times 10^{-k}$$

Example 21: If $f = \dfrac{5x^2y^3}{z^4}$ and errors in x, y, z be 0.001. Find maximum absolute error and relative error in f when $x = y = z = 1$.

Solution: Here
$$\frac{\partial f}{\partial x} = \frac{10xy^3}{z^4}, \frac{\partial f}{\partial y} = \frac{15x^2y^2}{z^4}, \frac{\partial f}{\partial z} = \frac{-20x^2y^3}{z^5}$$

$\therefore$
$$\Delta f = \frac{\partial f}{\partial x} \Delta x + \frac{\partial f}{\partial y} \Delta y + \frac{\partial f}{\partial z} \Delta z$$

$$= \left(\frac{10xy^3}{z^4} \right) \Delta x + \left(\frac{15x^2y^2}{z^4} \right) \Delta y - \left(\frac{20x^2y^3}{z^5} \right) \Delta z$$

Since the errors $\Delta x, \Delta y, \Delta z$ may be positive or negative, we take absolute value of the terms on the RHS. This gives

$$(\Delta f)_{max.} = \left| \frac{10xy^3}{z^4} \Delta x \right| + \left| \frac{15x^2y^2}{z^4} \Delta y \right| + \left| \frac{20x^2y^3}{z^5} \Delta z \right|$$

Since
$$\Delta x = \Delta y = \Delta z = 0.001 \text{ at } x = y = z = 1$$

$\therefore$
$$(\Delta f)_{max.} = |10 \times 0.001 + 15 \times 0.001 + 20 \times 0.001| = 0.045$$

Hence the maximum relative error

$$= \frac{(\Delta f)_{max.}}{f} = \frac{0.045}{5} = 0.009.$$

Example 22: Find smaller root of the equation $x^2 - 30x + 1 = 0$ correct to three decimal places. State different algorithms.

Solution: Roots of equation $x^2 - 30x + 1 = 0$ are

$$\frac{30 \pm \sqrt{(30)^2 - 4}}{2}$$

i.e.
$$\frac{30 + \sqrt{896}}{2}, \frac{30 - \sqrt{896}}{2}$$

The smallest root is
$$\frac{30 - \sqrt{896}}{2} = 15 - \sqrt{224}$$

I algorithm: Smaller root $= 15 - \sqrt{224} = 15 - 14.97 = 0.03$

II algorithm: Smaller root $= (15 - \sqrt{224})\left(\dfrac{15 + \sqrt{224}}{15 + \sqrt{224}}\right) = \dfrac{225 - 224}{15 + 14.97} = \dfrac{1}{29.97} = 0.033$

Second algorithm is clearly a better one as this gives the result correct to three figures.

Example 23: If $v = 3u^5 - 5u$, find the percentage error in v at $u = 1$, if the error in u is 0.05.

Solution:
$$v = 3u^5 - 5u$$

$$\Delta v = \frac{\partial v}{\partial u}\,\Delta u = (15u^4 - 5)\,\Delta u$$

$$\frac{\Delta v}{v} \times 100 = \left(\frac{15u^4 - 5}{3 \times 1^5 - 5 \times 1}\right) \times 0.05 \times 100$$

$$= \left(\frac{10}{-2}\right) \times 5 = -25\%$$

Maximum percentage error $= 25\%$.

Example 24: The error in the measurement of the area of a circle is not allowed to exceed 0.2%. How accurately should the diameter be measured?

Solution:
$$A = \frac{\pi r^2}{4}$$

$$\log A = \log \pi + 2 \log r - \log 4$$

$$\frac{\Delta A}{A} \times 100 = \frac{2}{r}(\Delta r \times 100)$$

$$\frac{\Delta r}{r} \times 100 = \frac{1}{2}\left(\frac{\Delta A}{A} \times 100\right) = \frac{1}{2} \times 0.2 = 0.1.$$

Example 25: In $\triangle ABC$, $a = 6$ cm, $c = 15$ cm, $\angle b = 90°$. Find the possible error in computed value of A if errors in measurements of a and c are 2 mm and 4 mm respectively.

Solution: Here
$$\tan A = \frac{a}{c}$$

$$A = \tan^{-1}\frac{a}{c}$$

$$\Delta A = \left(\frac{\partial A}{\partial a}\right)\Delta a + \left(\frac{\partial A}{\partial c}\right)\Delta c$$

$$= \frac{c}{a^2 + c^2}\,\Delta a - \frac{a}{a^2 + c^2}\,\Delta c$$

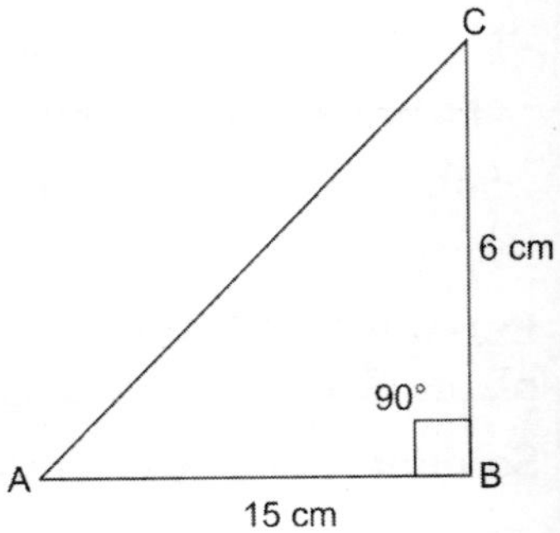

or
$$|\Delta A| \le \left|\frac{c}{a^2 + c^2}\,\Delta a\right| + \left|\frac{a}{a^2 + c^2}\,\Delta c\right|$$

$$= \frac{15}{261} \times 0.2 + \frac{6}{261} \times 0.4 = 0.0207 \text{ radians}$$

$\therefore \Delta A \le 0.0207$ radians.

Fig. 1.4

Example 26: Define the term 'absolute error' given that $w = 10.00 \pm 0.05$, $x = 18200 \pm 200$, $y = 0.0356 \pm 0.0002$, $z = 52000 \pm 500$.

Find the maximum value of absolute error in (i) $w + x + y + z$, (ii) $w + 4y - z$, (iii) z^4.

Solution: i. Max. absolute error $= \Delta w + \Delta x + \Delta y + \Delta z$

$$= 0.05 + 200 + 0.0002 + 500$$

$$= 700.0502$$

ii. Max. absolute error $= \Delta w + 4\Delta y + \Delta z$

$$= 0.05 + 4 \times 0.0002 + 500$$

$$= 500.0508$$

iii. Max. absolute error $= 4z^3 \Delta z$

$$= 4 \times (52000)^3 \times 500$$

$$= 2.81216 \times 10^{17}.$$

Example 27: $\sqrt{27} = 5.196$ and $\sqrt{11} = 3.317$ correct to four significant figure. Find the relative error in their sum and differences?

Solution: Number 5.196 and 3.317 are correct to four significant figures. The value of $x = x_1 + x_2 = 5.196 + 3.317 = 8.513$.

$\therefore$ Maximum error in each case is $(1/2) \times 10^{-3} = 0.0005$.

$\therefore \qquad\qquad \Delta x_1 = \Delta x_2 = 0.0005$

Relative error in their sum is

$$\left| \frac{\Delta x}{x} \right| \leq \left| \frac{\Delta x_1}{x} \right| + \left| \frac{\Delta x_2}{x} \right|$$

$$\leq \left| \frac{0.0005}{8.513} \right| + \left| \frac{0.0005}{8.513} \right| < 1.175 \times 10^{-4}$$

Relative error in their difference is

$$\left| \frac{\Delta x}{x} \right| \leq \left| \frac{\Delta x_1}{x} \right| + \left| \frac{\Delta x_2}{x} \right|, \text{ where } x = x_1 - x_2 = 5.196 - 3.317 = 1.879$$

$$\leq \left| \frac{0.0005}{1.879} \right| + \left| \frac{0.0005}{1.879} \right| < 5.322 \times 10^{-4}$$

Example 28: Find the product of 246.1 and 765.2. State how many figures of the result are trustworthy? Given that the numbers are correct to four significant figures.

Solution: Maximum error in each case $= 1/2 \times 10^{-1} = 0.05$

$\therefore \qquad\qquad \Delta x_1 = \Delta x_2 = 0.05$

$$x = 246.1 \times 765.2 = 188316 \text{ (correct to six digit)}$$

Max. relative error $(E_R) \leq \left| \frac{\Delta x_1}{x_1} \right| + \left| \frac{\Delta x_2}{x_2} \right|$

$$\leq \left| \frac{0.05}{246.1} \right| + \left| \frac{0.05}{765.2} \right|$$

$$= 0.000203 + 0.000065 = 0.000268$$

$\therefore \qquad\qquad$ Absolute error $= E_R X = 0.000268 \times 188316 \simeq 50$

$\therefore$ True value of the product of the numbers given lies between $188316 - 50 = 188266$

and $188316 + 50 = 188366$. Mean of these values is $\dfrac{188266 + 188366}{2} = 188316$ which is

188.3×10^3 correct to four significant digits. There is some uncertainty about last digit.

Example 29: Find the relative error in calculation of 6.431/0.232. Number are correct to three decimal places. Determine the smallest interval in which true result lies.

Solution: Maximum error in each case $= (1/2) \times 10^{-3} = 0.0005$

$$\therefore \qquad \Delta x_1 = \Delta x_2 = 0.0005$$

$$\text{Relative error} \leq \left|\frac{0.0005}{6.431}\right| + \left|\frac{0.0005}{0.232}\right| \leq 0.0005\left(\frac{1}{6.431} + \frac{1}{0.232}\right)$$

$$= 0.0022$$

$$\text{Absolute error} = 0.0022 \times \frac{x_1}{x_2} = 0.0022 \times \frac{6.431}{0.232} = 0.0609$$

Now, $$\frac{x_1}{x_2} = 27.7198$$

$\therefore$ True value of $\dfrac{x_1}{x_2}$ lies between 27.7198 ± 0.0609 i.e. 27.6589 and 27.7807.

Example 30: Use the series $\log_e\left(\dfrac{1+x}{1-x}\right) = 2\left(x + \dfrac{x^3}{3} + \dfrac{x^5}{5} + ...\right)$ to compute the value of log (1.2) correct to seven decimal places and find the number of terms retained.

Solution: $$\log_e\left(\frac{1+x}{1-x}\right) = 2\left(x + \frac{x^3}{3} + \frac{x^5}{5} + ... + \frac{x^{2n-1}}{2n-1}\right) + R_n(x)$$

If we retain n terms then,

$$(n+1)^{\text{th}} \text{ term} = \frac{2}{2n+1} x^{2n+1}$$

Let $$\frac{1+x}{1-x} = 1.2 \Rightarrow x = \frac{1}{11}$$

Hence $$(E_R)_{\text{max.}} = \frac{1}{11} = \frac{2}{2n+1}\left(\frac{4}{11}\right)^{2n+1}$$

For seven decimal accuracy,

$$\frac{2}{2n+1}\cdot\left(\frac{1}{11}\right)^{2n+1} < \frac{1}{2}\times 10^{-7} \Rightarrow (2n+1)(11)^{2n+1} > 4 \times 10^7$$

which gives $n \geq 3$.

Hence retaining the first three terms of the given series, we get

$$\log_e(1.2) = 2\left(x + \frac{x^3}{3} + \frac{x^5}{5}\right)_{\text{at } x = \frac{1}{11}} = 0.1823215.$$

Example 31: The function $f(x) = \tan^{-1}x$ can be expanded as

$$\tan^{-1}x = x - \frac{x^3}{3} + \frac{x^5}{5} - ... + (-1)^{n-1}\frac{x^{2n-1}}{2n-1} + ...$$

Find n such that series determines $\tan^{-1}(1)$ correct to eight significant digits.

Solution: If we retain n terms then $(n+1)^{\text{th}}$ term $= (-1)^n \dfrac{x^{2n+1}}{2n+1}$

For $\qquad\qquad\qquad x = 1, (n+1)^{\text{th}}$ term $= \dfrac{(-1)^n}{2n+1}$

For the determination of $\tan^{-1}(1)$ correct to eight significant digit accuracy

$$\left| \frac{(-1)^n}{2n+1} \right| < \frac{1}{2} \times 10^{-8}$$

$\Rightarrow \qquad\qquad\qquad (2n+1) > 2 \times 10^8$

which is satisfied by $n = 10^8 + 1$.

Example 32: Find the number of terms of the exponential series such that their sum gives the value of e^x correct to six decimal places at $x = 1$.

Solution: $\qquad e^x = 1 + x + \dfrac{x^2}{2!} + \dfrac{x^3}{3!} + ... + \dfrac{x^{n-1}}{(n-1)!} + R_n(x)$

where $\qquad\qquad R_n(x) = \dfrac{x^n}{n!} e^w, 0 < w < x$

Maximum absolute error at $(w = x) = \dfrac{x^n}{n!} e^x$ and maximum relative error $= \dfrac{x^n}{n!}$

Hence, $\qquad (E_R)_{\text{max.}}$ at $x = 1$ is $\dfrac{1}{n!}$

For a six decimal accuracy at $x = 1$, we have

$$\frac{1}{n!} < \frac{1}{2} \times 10^{-6} \text{ i.e. } n! > 2 \times 10^6$$

which gives $n = 10$.

Hence we need 10 terms of series (1) in order that its sum is correct to 6 decimal places.

EXERCISE 1.3

1. If $u = \dfrac{4x^2 y^3}{z^4}$ and errors in x, y, z be 0.001, compute the relative maximum error in u when $x = y = z = 1$. [Ans. 0.009]

2. The discharge Q over a notch for head H is calculated by the formula $Q = kH^{5/2}$, where k is a given constant. If the head is 75 cm and an error of 0.15 cm is possible in its measurement, estimate the percentage error in computing the discharge. [Ans. 0.5]

3. Compute the percentage error in the time period $T = 2\pi\sqrt{l/g}$ for $l = 1$ m if the error in the measurement of l is 0.01. [Ans. 0.5%]

4. If $y = 3x^7 - 6x$, find the percentage error in y at $x = 1$ if the error in x is 0.05. [Ans. −0.25%]

5. Find the relative error in the computation of $x - y$ for $x = 12.05$ and $y = 8.02$ having the absolute error $\Delta x = 0.005$ and $\Delta y = 0.001$. [Ans. 0.00029]

6. Sum the following numbers: 0.1532, 15.45, 0.000354, 305.1, 8.12, 143.3, 0.0212, 0.643 and 0.1734, where digits are correct. [Ans. $S = 472.6 \pm 0.15$]

7. In a $\triangle ABC$, $a = 30$ cm, $b = 80$ cm, $\angle B = 90°$, find the maximum error in computed value of A if possible errors in a and b are 1/3% and 1/4% respectively.

[Ans. $\Delta A < 0.00235$]

8. Find the product of numbers 56.54 and 12.4 which are both correct to the significant digits given.

9. The function $f(x) = \cos x$ can be expanded as

$$\cos x = 1 - \frac{x^2}{2!} + \frac{x^4}{4!} - \frac{x^6}{6!} + \dots$$

Compute the number of terms required to estimate $\cos(\pi/4)$ so that the result is correct to at least two significant digits. [Ans. $n = 3$]

10. Find the relative error in calculation of $7.342/0.241$. Numbers are correct to three decimal places. Determine the smallest interval in which true result lies.

[Ans. 30.4008, 30.5286]

1.14 NUMERICAL INSTABILITY

While adding or multiplying two or more appropximate numbers, errors are always added. Thus, if an input $x = 0.3216$ is an error of 0.01 cm, then $100x$ would be in an error of 1 cm. 10^4x would contain an error of 1 m. The output would, therefore be very unrealistic. This is a type of numerical instability or we can say that behaviour by reason of which a small error during computation gets so large as to make computed result totally redundant. We often encounter problems with two types of numerical instability:

i. **Inherent instability:** It may arise due to ill conditioning of the problem. Inherent instability can be avoided by suitable reformation of the problem.

ii. **Induced instability:** It may arise due to the wrong selection of the method of the solution of the problem. This type of instability can be avoided by a suitable change of the method of solution.

Roots of Equations

2.1 INTRODUCTION

The equation of the form $f(x) = 0$ occurs very frequently in many branches of science and engineering.

The equations of the form $f(x) = 0$ are called algebraic or transcendental according as $f(x)$ is purely a polynomial in x or contains some other functions such as exponential, trigonometric and logarithmic functions, etc. for example the equations $x^3 + x^2 - 1 = 0$ and $x - e^{-x} = 0$ are called algebraic and transcendental respectively.

2.2 SOLUTION OF AN EQUATION

The root of an equation $f(x) = 0$ is the value of x for which $f(x) = 0$. Geometrically, $x = \alpha$ is a root of the equation $f(x) = 0$ if the graph of $f(x)$ cuts the x-axis at the point $x = \alpha$.

2.3 MEAN VALUE THEOREM

The mean value theorem is one of the most important theoretical tool in calculus. According to this theorem if $f(x)$ is defined and continuous on the interval $[a, b]$ and differentiable on (a, b), then there is at least one number c in the interval (a, b) such that

$$f'(c) = \frac{f(b) - f(a)}{b - a}.$$

2.4 TAYLOR'S THEOREM

If the function $f(x)$ is continuous and has continuous derivatives up to the nth order in the interval $[x, x + h]$, then

$$f(x + h) = f(x) + hf'(x) + \frac{h^2}{2!} f''(x) + \dots + \frac{h^{n-1}}{(n-1)!} f^{(n-1)}(x) + \frac{h^n}{n!} f^{(n)}(x + \theta h)$$

where $\theta \in [0, 1]$. Assume that the function $f(x)$ has continuous derivatives of any order in the interval $[x, x + h]$. Therefore, we can choose n in the above equation arbitrarily large. Then, if the limit $n \to \infty$, we have $f^n(x + \theta h) \to 0$, the function can be expanded in a power series

$$f(x + h) = f(x) + hf'(x) + \frac{h^2}{2!} f''(x) + \dots + \frac{h^n}{n!} f^n(x) + \dots$$

This expansion is called Taylor's series. Taylor's series also can be written in different form if we denote the fixed point x by a and then $x = a + h$, thus

$$f(x) = f(a) + (x - a) f'(a) + \frac{(x - a)^2}{2!} f''(a) + ... + \frac{(x - a)^n}{n!} f^n(a) + ...$$

Theorem: If $f(x)$ is continuous in the interval (a, b) and if $f(a)$ and $f(b)$ are of opposite sign, then the equation $f(x) = 0$ will have at least one real root between a and b.

The above theorem enable us to locate the approximate position of a real root of equation $f(x) = 0$.

2.5 METHODS FOR FINDING THE ROOT (ZEROS) OF AN EQUATION

Methods for finding root(s) of an equation can be classified in two ways: (1) Direct methods, (2) Iterative methods.

If $f(x)$ is a polynomial of degree two, three or four, exact formulae (or direct methods) of finding roots are available. But for higher order polynomial equations and transcendental equation, it is very difficult and in many cases impossible to get the solution.

To solve polynomial of degree higher than four or transcendental equation, we apply numerical methods to find the value of root of $f(x) = 0$ approximately. These numerical methods are termed as iterative methods. These methods are based on the idea of successive approximations.

By using these methods to find the real root of an equation, we first find an approximate value of the root of the given equation and then successively improve it to some desired degree of accuracy. The general technique is that we start with an initial approximate value, say x_0, and then find the better approximations successively $x_1, x_2, ..., x_n$ by repeating the same method.

2.6 ORDER OF CONVERGENCE OF ITERATIVE METHODS

A sequence $<x_n>$ of successive approximation of a root $x = \alpha$ of $f(x) = 0$ is said to have convergence of order k, if $|x_{n+1} - \alpha| \le c |x_n - \alpha|^k$ where c being a finite constant, called the rate of convergence. If $k = 1$, the convergence is linear and if $k = 2$, it is quadratic.

The convergence is said to be fast or slow accordingly as c is nearer to 0 or 1.

It may be noted that for convergence, $<x_n>$ must be monotonic and bounded. Then

$$\alpha = \lim_{n \to \infty} x_n$$

2.7 BISECTION METHOD (OR BOLZANO'S METHOD)

Let $f(x)$ be continuous in $[a, b]$. Two points a and b enclose a root if the values of $f(a)$ and $f(b)$ are of opposite signs, i.e. $f(a) \cdot f(b) < 0$. For convenience, let $f(a) > 0$ and $f(b) < 0$.

Then the first approximation to the root is

$$x_1 = \frac{a + b}{2} \text{ (mid point of the ends of the range)}$$

If $f(x_1) = 0$, then x_1 is a root of $f(x) = 0$, otherwise find the sign of $f(x_1)$. If $f(x_1)$ is negative, the root lies between x_1 and a. If $f(x_1)$ is positive, the root lies between b and x_1. Out of these only one must be true. Let $f(x_1)$ is positive as shown in Fig. 2.1, then the root lies between x_0 and b. The second approximation to the root is $x_2 = \dfrac{x_1 + b}{2}$. Let

$f(x_2)$ is negative as shown in Fig. 2.1. Hence root lies between x_1 and x_2. So the next approximate root will be $x_3 = \dfrac{x_1 + x_2}{2}$ and so on. In this way, taking the mid point of the range as the approximate root, we construct a sequence of approximate roots $x_1, x_2, \ldots$ whose limit of convergence is the exact root. However, depending on the precision required, we stop the process after some steps. Though, the convergence of this method is slow but sure.

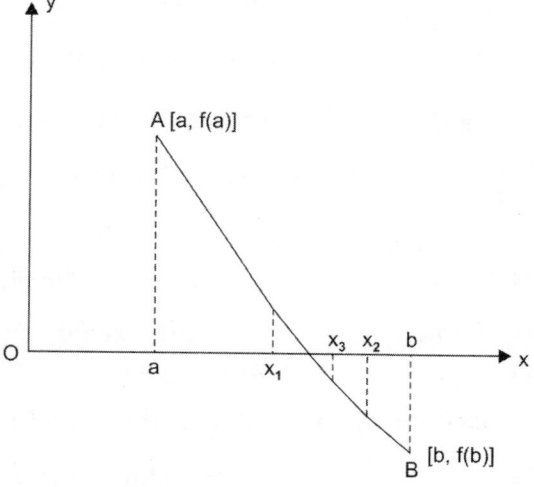

Fig. 2.1

Remarks: The number of iterations required may be determined from the following relation

$$\frac{|b - a|}{2^n} \leq \varepsilon$$

where ε is the permissible error

or

$$n \geq \frac{\log\left(\dfrac{|b - a|}{\varepsilon}\right)}{\log 2}$$

2.7.1 Convergence of Bisection Method

Initially consider that $p_1 = a$ and $q_1 = b$ and x_1 be the mid point of $[a, b]$.

First approximation is

$$x_1 = \frac{1}{2}(p_1 + q_1)$$

$\Rightarrow$ $$p_1 < x_1 < q_1$$

Now either the root lies in $[a, x_1]$ or in $[x_1, b]$

$\Rightarrow$ either $[p_2, q_2] = [p_1, x_1]$ or $[p_2, q_2] = [x_1, q_1]$

$\Rightarrow$ either $p_2 = p_1, q_2 = x_1$ or $p_2 = x_1, q_2 = q_1$

$\Rightarrow$ $p_1 \leq p_2, q_2 \leq q_1$

Second approximation is

$$x_2 = \frac{p_2 + q_2}{2}$$

So that $p_2 < x_2 < q_2$

Repeating the process, we obtain that at nth step

$$x_n = \frac{p_n + q_n}{2}, p_n < x_n < q_n$$

and $$p_1 \leq p_2 \leq \ldots \leq p_n \text{ and } q_1 \leq q_2 \leq \ldots \leq q_n$$

$\therefore$ $<p_n>$ is bounded nondecreasing sequence of numbers bounded by b and $<q_n>$ is a bounded nondecreasing sequence of numbers bounded by a.

Hence, both these sequences $<p_n>$ and $<q_n>$ converge.

Let
$$\text{Lim}_{n \to \infty} p_n = p \text{ and } \text{Lim}_{n \to \infty} q_n = q$$

The length of the interval at each iteration is halved, so that
$$\text{Lim}_{n \to \infty} (q_n - p_n) = 0 \Rightarrow q = p$$

Also
$$p_n < x_n < q_n$$

$\Rightarrow$
$$\text{Lim } p_n \leq \lim x_n \leq \lim q_n$$

$\Rightarrow$
$$p \leq \lim x_n \leq q$$

$\Rightarrow$
$$\text{Lim } x_n = p = q$$

Next, the root lie in $[p_n, q_n]$, therefore $f(p_n) \cdot f(q_n) < 0$

$\Rightarrow$
$$0 \geq \text{Lim}_{n \to \infty} [f(p_n) \cdot f(q_n)]$$

$\Rightarrow$
$$0 \geq f(p) \cdot f(q) \Rightarrow 0 \geq [f(p)]^2$$

As the square of any number is positive

$\Rightarrow$
$$f(p) = 0$$

$\Rightarrow p$ is a root of $f(x) = 0$

Hence the sequence $<x_n>$ converges necessarily to *a* root of equation $f(x) = 0$.

2.7.2 Algorithm for Bisection Method

1. Read x_0, x_1, e

   Remarks: x_0 and x_1 are two initial values that enclose the desired root, e is the error permitted in the answer. Steps 2, 3 and 4 are called initialisation steps which prepare the ground for the succeeding steps in the algorithm.

2. $y_0 \leftarrow f(x_0)$
3. $y_1 \leftarrow f(x_1)$
4. $i \leftarrow 0$
5. If $\sin(y_0) = \sin(y_1)$ then

   begin write 'starting values unsuitable'

   Write x_0, x_1, y_0, y_1

   stop end

6. While $|x_1 - x_0)/x_1| > e$ do

   begin

7. $x_2 \leftarrow (x_0 + x_1)/2$
8. $y_2 \leftarrow f(x_2)$
9. $i \leftarrow i + 1$
10. If $(\text{sign}(y_0) = \text{sign}(y_2))$ then $x_0 \leftarrow x_2$ else $x_1 \leftarrow x_2$

    end

11. Write 'solution converges to a root'
12. Write 'Number of iterations = ', i
13. Write x_2, y_2
14. Stop

Example 1: Using bisection method, find a real root of the equation $3x - \sqrt{1 + \sin x} = 0$.

Solution: Let
$$f(x) = 3x - \sqrt{1 + \sin x}$$

Since
$$f(0) = -1 \text{ i.e. } -ve$$

and
$$f(1) = 1.6429919, \text{ i.e. } +ve$$

Hence, root lies between 0 and 1.

First approximation to the root is

$$x_1 = \frac{0 + 1}{2} = 0.5$$

Now
$$f(x_1) = f(0.5) = 0.2836836, \text{ i.e. } +ve$$

Hence, root lies between 0 and 0.5.

Second approximation to the root is

$$x_2 = \frac{0 + 0.5}{2} = 0.25$$

Now
$$f(x_2) = f(0.25) = -0.3668724, \text{ i.e. } -ve$$

Hence, root lies between 0.25 and 0.5.

Third approximation to the root is

$$x_3 = \frac{0.25 + 0.5}{2} = 0.375$$

Now
$$f(x_3) = f(0.375) = -0.0438766, \text{ i.e. } -ve$$

Hence, root lies between 0.375 and 0.5.

Fourth approximation to the root is

$$x_4 = \frac{0.375 + 0.5}{2} = 0.4375$$

Now
$$f(x_4) = f(0.4375) = 0.1193209, \text{ i.e. } +ve$$

Hence, root lies between 0.375 and 0.4375.

Fifth approximation to the root is

$$x_5 = \frac{0.375 + 0.4375}{2} = 0.40625$$

Now
$$f(x_5) = f(0.40625) = 0.0375779, \text{ i.e. } +ve$$

Hence, root lies between 0.375 and 0.40625.

Sixth approximation to the root is

$$x_6 = \frac{0.375 + 0.40625}{2} = 0.390625$$

Now
$$f(x_6) = f(0.390625) = -0.031851, \text{ i.e. } -ve$$

Hence, root lies between 0.390625 and 0.40625.

Seventh approximation to the root is

$$x_7 = \frac{0.390625 + 0.40625}{2} = 0.3984375$$

Since x_6 and x_7 are same upto two decimal places, hence the approximate real root is 0.39

Example 2: Find the real root of the equation $x \log_{10} x = 1.2$ by using bisection method correct to three decimal places.

Solution: Let $\qquad f(x) = x \log_{10} x - 1.2 = 0$

Since $\qquad\qquad\qquad f(2.74) = -0.000563$, i.e. $-$ve

and $\qquad\qquad\qquad f(2.75) = 0.0081649$, i.e. $+$ve

Hence, root lies between 2.74 and 2.75.

First approximation to the root is

$$x_1 = \frac{2.74 + 2.75}{2} = 2.745$$

Now $\qquad\qquad f(x_1) = f(2.745) = 0.003798$, i.e. $+$ve

Hence, root lies between 2.74 and 2.745.

Second approximation to the root is

$$x_2 = \frac{2.74 + 2.745}{2} = 2.7425$$

Now $\qquad\qquad f(x_2) = f(2.7425) = 0.001617$, i.e. $+$ve

Hence, root lies between 2.74 and 2.7425.

Third approximation to the root is

$$x_3 = \frac{2.74 + 2.7425}{2} = 2.74125$$

Now $\qquad\qquad f(x_3) = f(2.74125) = 0.005267$, i.e. $+$ve

Hence, root lies between 2.74 and 2.74125.

Fourth approximation to the root is

$$x_4 = \frac{2.74 + 2.74125}{2} = 2.740625$$

Now $\qquad\qquad f(x_4) = f(2.740625) = -0.00001839$, i.e. $-$ve

Hence, root lies between 2.740625 and 2.74125.

Fifth approximation to the root is

$$x_5 = \frac{2.740625 + 2.74125}{2} = 2.7409375$$

Since x_4 and x_5 are same up to three decimal places, hence the approximate real root is 2.740.

EXERCISE 2.1

1. Apply bisection method to find a root of the equation $x^4 + 2x^3 - x - 1 = 0$ in the interval [0, 1]. [Ans. 0.8633]
2. Find the real root of the equation $x^3 - 9x + 1 = 0$, for $2 < x < 4$. [Ans. 2.943]
3. Find the real root of theequation using bisection method.
   $$\tan x + \tan hx = 0 \qquad\qquad\qquad\qquad \text{[Ans. 2.3650]}$$
4. Find the positive root of $x - \cos x = 0$ by using bisection method. [Ans. 0.7388]
5. Find a real root of $\cos x = xe^x$ up to three decimal places using bisection method. [Ans. 0.517]

2.8 REGULA-FALSI METHOD OR METHOD OF FALSE POSITION

We have already seen in the previous method that convergence of bisection method is slow but sure. Thus attempts have been made to speed up bisection method retaining its guaranteed convergence. A method of doing this is called the Regula-Falsi or method of false position.

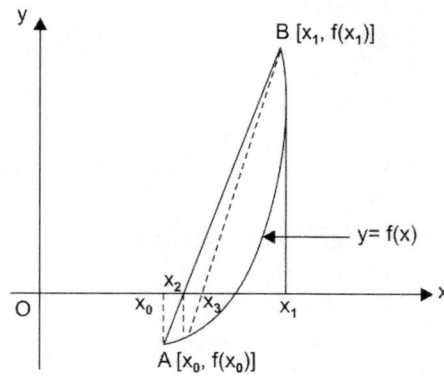

Fig. 2.2

In this method, two points x_0 and x_1 are chosen such that $f(x_0)$ and $f(x_1)$ are of opposite signs, i.e., the graph $y = f(x)$ crosses the x axis between these points. Hence, a root lies between x_0 and x_1 and $f(x_0) \cdot f(x_1) < 0$.

The equation of the chord joining the points $A[x_0, f(x_0)]$ and $B[x_1, f(x_1)]$ is

$$y - f(x_0) = \frac{f(x_1) - f(x_0)}{x_1 - x_0}(x - x_0)$$

The method consists in replacing the curve AB by the chord AB and taking the point of intersection of the chord with the x-axis as an approximation to the root which is shown in Fig. 2.2. The abscissa of the point where the chord cuts the x-axis ($y = 0$) is given by

$$x_2 = x_0 - \frac{x_1 - x_0}{f(x_1) - f(x_0)} f(x_0) = \frac{x_0 f(x_1) - x_1 f(x_0)}{f(x_1) - f(x_0)}$$

which is an approximation to the root of $f(x) = 0$.

If $f(x_0)$ and $f(x_2)$ are of opposite sign, the root lies between x_0 and x_2, and the next approximation x_3 is obtained as

$$x_3 = \frac{x_0 f(x_2) - x_2 f(x_0)}{f(x_2) - f(x_0)}$$

If $f(x_1)$ and $f(x_2)$ are of opposite sign, the root lies between x_1 and x_2, and the next approximation x_3 is given as

$$x_3 = \frac{x_2 f(x_1) - x_1 f(x_2)}{f(x_1) - f(x_2)}$$

In the same way, we get $x_4, x_5, \ldots$. This procedure is repeated till the root is found to the desired accuracy.

2.8.1 Convergence of Regula-Falsi Method

The convergence of Regula-Falsi method is exactly similar to Secant method. So, we have solve it for Secant method that is given in next section.

2.8.2 Algorithm for Regula-Falsi Method

1. Read x_0, x_1, e, n

   Remarks: x_0 and x_1 are two initial guesses to the root such that sign ($f(x_0)$) ≠ sign ($f(x_1)$). The prescribed precision is e and n is the maximum number of iterations. Step 2 and 3 are initialization steps.

2. $f_0 \leftarrow f(x_0)$
3. $f_1 \leftarrow f(x_1)$
4. for $i = 1$ to n in steps of 1 do
5. $x_2 \leftarrow (x_0 f_1 - x_1 f_0)/(f_1 - f_0)$
6. $f_2 \leftarrow f(x_2)$
7. If $|f_2| \leq e$ then
8. begin write 'convergent solution', x_2, f_2
9. Stop end
10. If sign $(f_2) \neq$ sign (f_0)
11. then begin $x_1 \leftarrow x_2$
12. $f_1 \leftarrow f_2$ end
13. else begin $x_0 \leftarrow x_2$
14. $f_0 \leftarrow f_2$ end
   end for
15. Write 'Does not converge in n iterations'
16. Write x_2, f_2
17. Stop

Example 3: Find the root of the equation $\tan x + \tan h x = 0$ which lies in the interval (1.6, 3.0) correct to four significant digits using method of false position.

Solution: Let $\qquad f(x) = \tan x + \tan h x = 0$

Since $\qquad f(2.35) = -0.03$ and $f(2.37) = 0.009$

Hence, the root lies between 2.35 and 2.37. Take $x_0 = 2.35$ and $x_1 = 2.37$. Using Regula-Falsi method

$$x_2 = x_0 - \left(\frac{x_1 - x_0}{f(x_1) - f(x_0)} \right) f(x_0)$$

$$= 2.35 - \left(\frac{2.37 - 2.35}{0.009 + 0.03} \right)(-0.03)$$

$$= 2.35 + \frac{0.02}{0.039}(0.03) = 2.365$$

Now $\qquad f(x_2) = f(2.365) = -0.00004$, i.e. $-$ve

Hence, the root lies between 2.365 and 2.37.

$$x_3 = 2.365 - \left(\frac{2.37 - 2.365}{0.009 + 0.00004} \right)(-0.00004)$$

$$= 2.365 + \frac{0.005}{0.00904} \times 0.00004 = 2.365$$

Hence the required root is 2.365 correct to four significant digits.

EXERCISE 2.2

Find a real root of the following equations correct to three decimal places by using Regula-Falsi method.

1. $xe^x = 2$ [Ans. 0.852]
2. $2x - \log_{10} x = 7$ [Ans. 3.789]
3. $x \tan x = -1$ [Ans. 2.798]

4. $e^{-x} - \sin x = 0$ [Ans. 0.5885]

5. $x^2 - \log_e x = 12$ [Ans. 3.646]

2.9 SECANT METHOD

This method is almost same as that of Regula–Falsi method. But it does not require the condition $f(x_0) \cdot f(x_1) < 0$. Here the graph of the function $y = f(x)$ in the neighbourhood of the root is approximated by a secant line (chord) but it is not necessary that the interval at each iteration must contain the root.

Secant method is illustrated in Fig. 2.3.

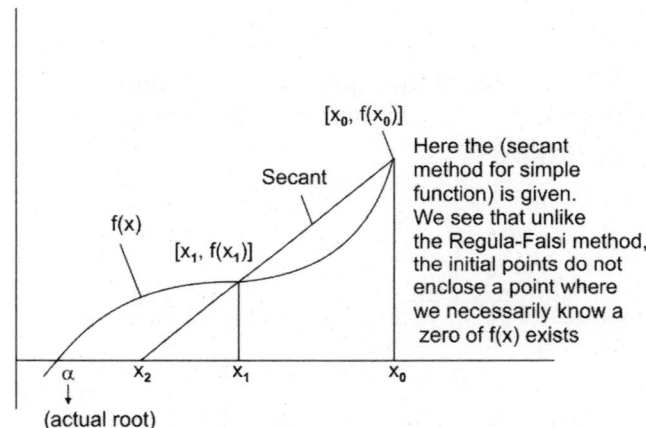

Fig. 2.3: Secant method for simple function

We can see that the function $f(x)$ is being approximated by a straight line which is an extrapolation based on the two points, i.e. x_0 and x_1.

Then the equation of line passing through the two points $[x_0, f(x_0)]$ and $[x_1, f(x_1)]$ is given by

$$y - f(x_1) = \frac{f(x_1) - f(x_0)}{x_1 - x_0} (x - x_1)$$

At the point where the line crosses the axis of x, the abscissa is given by

$$x_2 = x_1 - \frac{(x_1 - x_0)}{f(x_1) - f(x_0)} f(x_1)$$

which is an approximation to the root.

Thus, the general formula for successive approximation is given by

$$x_{n+1} = x_n - \frac{(x_n - x_{n-1}) f(x_n)}{f(x_n) - f(x_{n-1})}$$

But the above formula is not suitable for computation because it involves division by small quantity which would be loss of significant digits. So, we use the following formula for computation

$$x_{n+1} = \frac{x_{n-1} f(x_n) - x_n f(x_{n-1})}{f(x_n) - f(x_{n-1})}$$ (2.1)

In Eq. (2.1), we observe that if $f(x_n) = f(x_{n-1})$. This method will fail.

Thus, this method does not converge necessarily while Regula–Falsi method always converges. The only advantage lies in fact that if Secant method converges, then it will converge more faster than the Regula–Falsi method.

2.9.1 Rate of Convergence of Secant Method

We have the following formula for Regula–Falsi method

$$x_{n+1} = x_n - \frac{(x_n - x_{n-1})}{f(x_n) - f(x_{n-1})} f(x_n) \tag{2.2}$$

where x_{n-1}, x_n and x_{n+1} are three approximations to the root of $f(x) = 0$.

Consider α be the actual root of the equation $f(x) = 0$ and e_{n-1}, e_n and e_{n+1} be the errors corresponding to the $(n-1)^{th}$, n^{th} and $(n+1)^{th}$ iterations, then

$$f(\alpha) = 0$$

and

$$x_{n-1} = \alpha + e_{n-1}$$
$$x_n = \alpha + e_n$$
$$x_{n+1} = \alpha + e_{n+1}$$

Substituting the values of x_{n-1}, x_n and x_{n+1} in Eq. (2.2), we get

$$\alpha + e_{n+1} = (\alpha + e_n) - \frac{[(\alpha + e_n) - (\alpha + e_{n-1})]}{f(\alpha + e_n) - f(\alpha + e_{n-1})} f(\alpha + e_n)$$

or

$$e_{n+1} = e_n - \frac{(e_n - e_{n-1}) f(\alpha + e_n)}{f(\alpha + e_n) - f(\alpha + e_{n-1})}$$
$$= \frac{e_{n-1} f(\alpha + e_n) - e_n f(\alpha + e_{n-1})}{f(\alpha + e_n) - f(\alpha + e_{n-1})}$$

By using Taylor's series expansion, we have

$$e_{n+1} = \frac{e_{n-1}\left[f(\alpha) + e_n f'(\alpha) + \frac{e_n^2}{2!} f''(\alpha) + ... \right] - e_n\left[f(\alpha) + e_{n-1} f'(\alpha) + \frac{e_{n-1}^2}{2!} f''(\alpha) + ... \right]}{\left[f(\alpha) + e_n f'(\alpha) + \frac{e_n^2}{2!} f''(\alpha) + ... \right] - \left[f(\alpha) + e_{n-1} f'(\alpha) + \frac{e_{n-1}^2}{2!} f''(\alpha) + ... \right]}$$

$$= \frac{e_{n-1}\left[e_n f'(\alpha) + \frac{e_n^2}{2} f''(\alpha) + ... \right] - e_n\left[e_{n-1} f'(\alpha) + \frac{e_{n-1}^2}{2} f''(\alpha) + ... \right]}{\left[e_n f'(\alpha) + \frac{e_n^2}{2} f''(\alpha) + ... \right] - \left[e_{n-1} f'(\alpha) + \frac{e_{n-1}^2}{2} f''(\alpha) + ... \right]} \quad \text{as } f(\alpha) = 0$$

$$= \frac{e_{n-1}\left[e_n f'(\alpha) + \frac{e_n^2}{2} f''(\alpha) \right] - e_n\left[e_{n-1} f'(\alpha) + \frac{e_{n-1}^2}{2} f''(\alpha) \right]}{\left[e_n f'(\alpha) + \frac{e_n^2}{2} f''(\alpha) \right] - \left[e_{n-1} f'(\alpha) + \frac{e_{n-1}^2}{2} f''(\alpha) \right]}$$

On neglecting 3rd and higher order terms, we have

$$e_{n+1} = \frac{\dfrac{e_{n-1}\, e_n^2}{2} f''(\alpha) - \dfrac{e_n\, e_{n-1}^2}{2} f''(\alpha)}{(e_n - e_{n-1})\, f'(\alpha) + \dfrac{1}{2}(e_n^2 - e_{n-1}^2)\, f''(\alpha)}$$

$$= \frac{\dfrac{e_{n-1}\, e_n}{2}(e_n - e_{n-1})\, f''(\alpha)}{(e_n - e_{n-1})\, f'(\alpha)}$$

Again, neglecting e_n^2 and e_{n-1}^2 if e_n and e_{n-1} are very small, we obtain

$$e_{n+1} = \frac{e_{n-1}\, e_n}{2} \frac{f''(\alpha)}{f'(\alpha)}$$

which can be written as

$$e_{n+1} = e_{n-1} e_n k \tag{2.3}$$

where $k = \dfrac{1}{2} \dfrac{f''(\alpha)}{f'(\alpha)}$

If we want to find the order of convergence then it is necessary to find a formula of the form

$$e_{n+1} = k_1 e_n^m \tag{2.4}$$

with an appropriate value of m.

The Eq. (2.4) can be written as

$$e_n = k_1 e_{n-1}^m \tag{2.5}$$

where m is an order of convergence.

Now, putting the values of e_{n+1} and e_{n-1} from Eq. (2.4) and Eq. (2.5) in Eq. (2.3)

$$k_1 e_n^m = \left(\frac{e_n}{k_1}\right)^{\frac{1}{m}} e_n k$$

or

$$e_n^m = \frac{(e_n)^{\frac{1}{m}+1}}{(k_1)^{\frac{1}{m}+1}} \cdot k$$

Now choose k and k_1 such that $k = k_1^{1+\frac{1}{m}}$

then

$$e_n^m = (e_n)^{\frac{1}{m}+1} \Rightarrow m = \frac{1}{m} + 1$$

$$\Rightarrow \qquad m^2 - m - 1 = 0$$

$$\Rightarrow \qquad m = \frac{1}{2}(1 \pm \sqrt{5})$$

By taking only the positive root, we have

$$m = \frac{1}{2}(1 + \sqrt{5}) = 1.62 \text{ approx.}$$

Hence the order of convergence of Secant method is 1.62.

2.9.2 Algorithm for Secant Method

1. Read x_0, x_1, e, delta, n

   Remarks: x_0, x_1 are two initial guesses to the root. e is the prescribed precision, delta the minimum allowed value of slope and n the maximum number of itertions to be permitted.

2. $g_0 \leftarrow g(x_0)$

3. $g_1 \leftarrow g(x_1)$

   Remarks: Step 5 to 12 are repeated until the procedure converges to a root or iteration exceed n.

4. For $i = 1$ to n in step of 1 do

5. If $|g_1 - g_0| <$ delta then Go To 15

6. $x_2 \leftarrow (x_0 g_1 - x_1 g_0)/(g_1 - g_0)$

7. $g_2 \leftarrow g(x_2)$

8. If $|g_2| < e$ then Go To 17

9. $g_0 \leftarrow g_1$

10. $g_1 \leftarrow g_2$

11. $x_0 \leftarrow x_1$

12. $x_1 \leftarrow x_2$

    end for

13. Write 'Does not converge', x_0, x_1, g_0, g_1

14. Stop

15. Write 'Slope too small', i, g_0, g_1, x_0, x_1

16. Stop

17. Write 'Convergent solution', i, x_2, g_2

18. Stop.

Example 4: A real root of the equation $f(x) = x^3 - 5x + 1 = 0$ lies in the interval $(0, 1)$. Perform four iteration of the secant method.

Solution: Given that $x_0 = 0, x_1 = 1$

then $f(x_0) = 1, f(x_1) = -3$

Now by secant method, we have

$$x_2 = x_1 - \frac{(x_1 - x_0)}{f(x_1) - f(x_0)} f(x_1)$$

$$= 1 - \frac{(1-0)}{-3-1}(-3) = 0.25$$

$$f(x_2) = f(0.25) = -0.234375$$

The second approximation to the root is given by

$$x_3 = x_2 - \frac{x_2 - x_1}{f(x_2) - f(x_1)} f(x_2)$$

$$= 0.25 - \frac{(0.25 - 1)}{(-0.234375) - (-3)} \cdot (-0.234375)$$

$$= 0.186441$$

$$f(x_3) = f(0.186441) = 0.074276$$

The third approximation is

$$x_4 = x_3 - \frac{x_3 - x_2}{f(x_3) - f(x_2)} f(x_3)$$

$$= 0.186441 - \frac{(0.186441 - 0.25)}{(0.074276) - (-0.234375)} \times 0.074276$$

$$= 0.201736$$

$$f(x_4) = f(0.201736) = -0.000470$$

The fourth approximation is

$$x_5 = x_4 - \frac{x_4 - x_3}{f(x_4) - f(x_3)} f(x_4)$$

$$= 0.201736 - \frac{(0.201736 - 0.186441)}{(-0.000470) - (0.074276)} \times (-0.000470)$$

$$= 0.201640$$

Example 5: Compute root of the equation $f(x) = \cos x - xe^x = 0$ using Secant method up to four decimal places.

Solution: Let $x_0 = 0$ and $x_1 = 1$ as initial approximation.

Then $f(x_0) = 1, f(x_1) = -2.1779795$

By Secant method, first approximation is

$$x_2 = x_1 - \frac{(x_1 - x_0)}{f(x_1) - f(x_0)} f(x_1)$$

$$= 1 - \frac{(1 - 0)}{(-2.1779795) - 1} (-2.1779795)$$

$$= 0.3146653$$

$$f(x_2) = f(0.3146653) = 0.5198711$$

Second approximation is

$$x_3 = x_2 - \frac{(x_2 - x_1)}{f(x_2) - f(x_1)} f(x_2)$$

$$= 0.3146653 - \frac{(0.3146653 - 1)}{0.5198711 - (-2.1779795)} \times 0.5198711$$

$$= 0.4467281$$

$$f(x_3 = f(0.4467281) = 0.2035448$$

Third approximation is

$$x_4 = x_3 - \frac{(x_3 - x_2)}{f(x_3) - f(x_2)} f(x_3)$$

$$= 0.4467281 - \frac{(0.4467281 - 0.3146653)}{(0.2035448 - 0.5198711)} (0.2035448)$$

$$= 0.5317058$$

$$f(x_4) = f(0.5317058) = -0.042931$$

Fourth approximation is

$$x_5 = x_4 - \frac{(x_4 - x_3)}{f(x_4) - f(x_3)} f(x_4)$$

$$= 0.5317058 - \frac{(0.5317058 - 0.4467281)}{(-0.042931 - 0.2035448)} (-0.042931)$$

$$= 0.5169044$$

$$f(x_5) = f(0.5169044) = 0.0025929$$

Fifth approximation is

$$x_6 = x_5 - \frac{(x_5 - x_4)}{f(x_5) - f(x_4)} f(x_5)$$

$$= 0.5169044 - \frac{(0.5169044 - 0.5317058)}{(0.0025929) - (-0.042931)} (0.0025929)$$

$$= 0.5177474$$

$$f(x_6) = f(0.5177474) = 0.0000302$$

Sixth approximation is

$$x_7 = x_6 - \frac{(x_6 - x_5)}{f(x_6) - f(x_5)} f(x_6)$$

$$= 0.5177474 - \frac{(0.5177474 - 0.5169044)}{(0.0000302 - 0.0025929)} (0.0000302)$$

$$= 0.5177376$$

Hence x_6 and x_7 are same up to four decimal places. So, the required root is 0.5177.

Example 6: Determine the root of the equation $f(x) = \cos x + 2\sin x + x^2$, using secant method, correct to three decimal places.

Solution: Let $x_0 = 0$ and $x_1 = -0.1$ as initial approximation.

Then
$$f(x_0) = f(0) = 1$$
$$f(x_1) = f(-0.1) = 0.8053$$

By secant method, first approximation is

$$x_2 = x_1 - \frac{(x_1 - x_0)}{f(x_1) - f(x_0)} f(x_1)$$

$$= -0.1 - \frac{(-0.1 - 0)}{(0.8053 - 0)} (0.8053)$$

$$= -0.5136$$

$$f(x_2) = f(-0.5136) = 0.1522$$

The second approximation is

$$x_3 = x_2 - \frac{(x_2 - x_1)}{f(x_2) - f(x_1)} f(x_2)$$

$$= -0.5136 - \frac{[-0.5136 - (-0.1)]}{(0.1522 - 0.8053)} (0.1522)$$

$$= -0.6100$$

$$f(x_3) = f(-0.6100) = 0.0457$$

The third approximation is

$$x_4 = x_3 - \frac{(x_3 - x)}{f(x_3) - f(x_2)} f(x_3)$$

$$= -0.6100 - \frac{[-0.6100 - (-0.5136)]}{(0.0457 - 0.1522)} (0.0457)$$

$$= -0.6514$$

$$f(x_4) = f(-0.6514) = 0.0065$$

The fourth approximation is

$$x_5 = x_4 - \frac{(x_4 - x_3)}{f(x_4) - f(x_3)} f(x_4)$$

$$= -0.6514 - \frac{[-0.6514 - (-0.6100)]}{(0.0065 - 0.0457)} (0.0065)$$

$$= -0.6582$$

$$f(x_5) = f(-0.6582) = 0.0013$$

The fifth approximation is

$$x_6 = x_5 - \frac{(x_5 - x_4)}{f(x_5) - f(x_4)} f(x_5)$$

$$= -0.6582 - \frac{[-0.6582 - (-0.6514)]}{(0.0013 - 0.0065]} (0.0013)$$

$$= -0.6598$$

$$f(x_6) = f(-0.6598) = 0.0006$$

The sixth approximation is

$$x_7 = x_6 - \frac{(x_6 - x_5)}{f(x_6) - f(x_5)} f(x_6)$$

$$= -0.6598 - \frac{[-0.6598 - (-0.6582)]}{(0.0006 - 0.0013)} (0.0006)$$

$$= -0.6595$$

Hence x_6 and x_7 are same upto three decimal places. So, the required root is -0.659.

EXERCISE 2.3

1. Find the root of the equation lying between 2 and 4 by secant method.
$$x^2 - 4x - 10 = 0$$ [Ans. 5.6182]

2. Find the root of the equation by secant method correct to three decimal places.
$$2x + \log x = 6$$ [Ans. 2.7065]

3. Find a root of the $x^2 e^{-x_{12}} = 1$ in the interval [0, 2] by using secant method.
[Ans. 1.429]

4. Find the root of the following equations correct to three decimal places by secant method.

i. $x^3 + x^2 + x + 7 = 0$ [Ans. −2.0625]

ii. $x + e^{-x} = 0$ [Ans. 0.567]

iii. $\cos x - 3x - 1 = 0$ [Ans. 0.6071]

2.10 FIXED POINT ITERATION METHOD (OR METHOD OF SUCCESSIVE APPROXIMATION)

Suppose we have to approximate the root of the equation

$$f(x) = 0 \tag{2.6}$$

Now, rewrite the Eq. (2.6) in the form

$$x = g(x) \tag{2.7}$$

It can be done by many ways, e.g. let us take the equation

$$x^3 = 2x + 5$$

or $$x^3 - 2x - 5 = 0$$

which can be rewrite in the following ways

$$x = \frac{x^3 - 5}{2} \text{ and } x = (2x + 5)^{1/3}$$

Assume x_0 to be the starting approximate value to the actual root α of $x = g(x)$. Putting it for x on the right hand side of $x = g(x)$, we get the first approximation

$$x_1 = g(x_0)$$

Now, putting $x = x_1$ in right hand side of Eq. (2.7), we get

$$x_2 = g(x_1)$$

Next successive approximations are given by

$$x_3 = g(x_2)$$
$$x_4 = g(x_3)$$
$$\dots \dots \dots$$
$$x_n = g(x_{n-1})$$

which is known as the method of fixed point iteration method (Fig. 2.4).

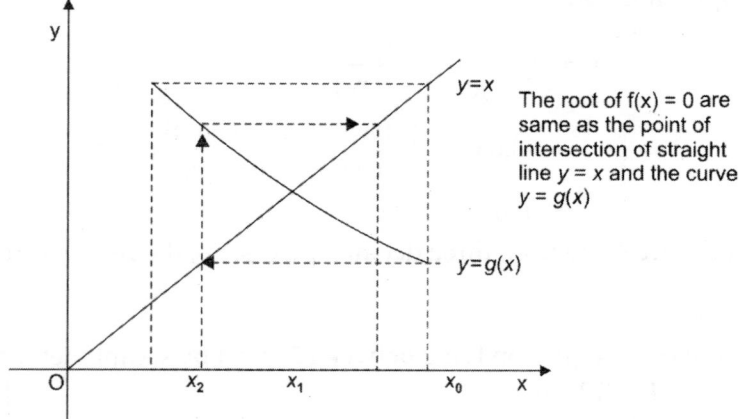

The root of f(x) = 0 are same as the point of intersection of straight line $y = x$ and the curve $y = g(x)$

Fig. 2.4: Root approximation by fixed point iteration method

The sequence of approximate roots $x_1, x_2, \dots, x_n$, if it converge to α is taken as the root of the equation $f(x) = 0$.

Here it is noted that it is not sure whether the sequence of successive approximations $x_1, x_2, \dots, x_n$ always converge to the same number which is a root of $f(x) = 0$ or not.

The successive approximation $x_1, x_2, \dots, x_n$ will converge to the root α if we will select the initial approximation x_0 suitably.

2.10.1 Condition for Convergence of Fixed Point Iteration Method

Let $$f(x) = 0 \tag{2.8}$$

We can rewrite the Eq. (2.8) in the following form
$$x = g(x) \tag{2.9}$$

Let α is the actual root of Eq. (2.9), and $\alpha \in (a, b)$, thus
$$\alpha = g(\alpha) \tag{2.10}$$

Further x_0 be the initial approximatioon of actual root α, then by fixed point iteration method, we get the first approximation x_1 as
$$x_1 = g(x_0) \tag{2.11}$$

Subtracting Eq. (2.11) from Eq. (2.10), we get
$$\alpha - x_1 = g(\alpha) - g(x_0) \tag{2.12}$$

By mean value theorem of differential calculus,

$$g'(\alpha_0) = \frac{g(\alpha) - g(x_0)}{\alpha - x_0}, \quad x_0 < \alpha_0 < \alpha$$

or $\qquad g(\alpha) - g(x_0) = (\alpha - x_0)\, g'(\alpha_0), \quad x_0 < \alpha_0 < \alpha \tag{2.13}$

From Eq. (2.12) and Eq. (2.13)
$$\alpha - x_1 = (\alpha - x_0)\, g'(\alpha_0), \quad x_0 < \alpha_0 < \alpha$$
$$\alpha - x_2 = (\alpha - x_1)\, g'(\alpha_1), \quad x_1 < \alpha_1 < \alpha$$
$$\alpha - x_3 = (\alpha - x_2)\, g'(\alpha_2), \quad x_2 < \alpha_2 < \alpha$$
$$\cdots \cdots \cdots$$
$$\cdots \cdots \cdots$$
$$\alpha - x_{n+1} = (\alpha - x_n)\, g'(\alpha_n), \quad x_n < \alpha_n < \alpha$$

Multiplying all these relations, we have
$$(\alpha - x_1)(\alpha - x_2) \cdots (\alpha - x_{n+1}) = (\alpha - x_0)(\alpha - x_1) \cdots (\alpha - x_n)\, g'(\alpha_0)\, g'(\alpha_1) \cdots g'(\alpha_n)$$

or $\qquad (\alpha - x_{n+1}) = (\alpha - x_0)\, g'(\alpha_0)\, g'(\alpha_1) \cdots g'(\alpha_n) \tag{2.14}$

Let $|g'(x)| \leq k$ for all x in the interval (a, b), then from Eq. (2.14), we get
$$(\alpha - x_{n+1}) = (\alpha - x_0)\, k^n \tag{2.15}$$

If $\qquad k < 1,\ k^n \to 0$ as $n \to \infty$

Hence $\qquad |\alpha - x_{n+1}| \to 0$ as $n \to \infty$

i.e. $\qquad \underset{n \to \infty}{\text{Lim}}\, x_{n+1} = \alpha$

which shows that the sequence of approximations converge to the root α. Hence the required condition for convergence of fixed point iteration method is
$$|g'(x)| < k < 1, \quad \forall\, x \in I = (a, b)$$

Again, the relation
$$\alpha - x_{n+1} = (\alpha - x_n)\, g'(\alpha_n), x_n < \alpha_n < \alpha$$

gives $\qquad \alpha - x_{n+1} < (\alpha - x_n)\, k,\ k < 1 \tag{2.16}$

We see from relation (2.16) that the convergence of fixed point iteration method is linear and order is one.

2.10.2 Order of Convergence of Fixed Point Iteration Method

Consider $x_0, x_1, x_2, \ldots, x_n, \ldots$ be the successive approximations of the root α of $f(x) = 0$. Consider e_n and e_{n+1} be the error in the corresponding root x_n and x_{n+1}.

If α is the exact root, then

$$e_n = x_n - \alpha \text{ and } e_{n+1} = x_{n+1} - \alpha$$

If $k \geq 1$ can be find out such that

$$|e_{n+1}| \leq |e_n|^k \cdot c,$$

where c is a constant then k is called the order of convergence.

If $k = 1$, the convergence is linear and if $k = 2$, it is quadratic.

2.10.3 Algorithm for Fixed Point Iteration Method

1. Read x_0, e, n

   Remarks: x_0 is the initial guess, e is the allowed error in the root and n the total iterations to be allowed for convergence.

2. $x_1 \leftarrow g(x_0)$

   Remarks: Steps 4 to 6 are repeated until the procedure convergence to a root or iterations reach n.

3. For $i = 1$ to n in the steps of 1 do.

4. $x_0 \leftarrow x_1$

5. $x_1 \leftarrow g(x_0)$

6. If $\left| \dfrac{(x_1 - x_0)}{x_1} \right| \leq e$ then Go To 9

   end for

7. Write 'Does not converge to a root', x_0, x_1

8. Stop

9. Write 'Converges to a root', i, x_1

10. Stop

Example 7: Find a real root of $2x - \log_{10} x = 7$ correct to four decimal places using the iteration method.

Solution: We have

$$f(x) = 2x - \log_{10} x - 7$$
$$f(3) = 6 - \log 3 - 7 = -1.4471$$
$$f(4) = 6 - \log 4 - 7 = 0.398$$

So, root lies between 3 and 4.

Now, the given equation can be rewritten as

$$x = \frac{1}{2}(\log_{10} x + 7) = g(x)$$

We have

$$g'(x) = \frac{1}{2x} \log_{10} e$$

$\therefore$

$$|g'(x)| < 1 \text{ when } 3 < x < 4.$$

Hence for this choice of $g(x)$, fixed point iteration method will converge.

Consider the initial approximation $x_0 = 3.5$. Now by iteration method, we have

$$x_{n+1} = g(x_n), \text{ for } n = 0, 1, 2, 3, \ldots$$

$\Rightarrow$

$$x_1 = g(x_0) = \frac{1}{2}(\log_{10} x_0 + 7) = \frac{1}{2}(\log_{10} 3.5 + 7)$$
$$= 3.7720340$$

$$x_2 = g(x_1) = \frac{1}{2}(\log_{10} x_1 + 7) = \frac{1}{2}(\log_{10} 3.7720340 + 7)$$

$$= 3.7882878$$

$$x_3 = g(x_2) = \frac{1}{2}(\log_{10} x_2 + 7) = \frac{1}{2}(\log_{10} 3.7882878 + 7)$$

$$= 3.7892214$$

$$x_4 = g(x_3) = \frac{1}{2}(\log_{10} x_3 + 7) = \frac{1}{2}(\log_{10} 3.7892214 + 7)$$

$$= 3.7892749$$

Here $x_4 \cong x_5$ upto four decimal places.

Hence the required real root is 3.7892.

Example 8: Find a positive root of $3x - \sqrt{1 + \sin x} = 0$ by iteration method.

Solution: We have $\quad f(x) = 3x - \sqrt{1 + \sin x}$

$$f(0) = 3 \times 0 - \sqrt{1 + \sin 0} = -1 = - \text{ve}$$

$$f(1) = 3 \times 1 - \sqrt{1 + \sin 1} = + \text{ve}$$

So, the root lies between 0 and 1.

Now the given equation can be rewritten as

$$x = \frac{1}{3}\sqrt{1 + \sin x} = g(x)$$

We have $\quad\quad g'(x) = \frac{1}{6} \frac{\cos x}{\sqrt{1 + \sin x}}$

$\therefore\quad\quad\quad |g'(x) < 1 \quad$ when $\quad 0 < x < 1.$

Hence for this choice of $g(x)$, fixed point iteration method will converge.

Consider the initial approximation $x_0 = 0.4$. Now by iteration method, we have

$$x_{n+1} = g(x_n)$$

$\Rightarrow\quad\quad x_1 = g(x_0) = \frac{1}{3}\sqrt{1 + \sin(0.4)} = 0.39291$

$$x_2 = g(x_1) = \frac{1}{3}\sqrt{1 + \sin(0.39291)} = 0.39199$$

$$x_3 = g(x_2) = \frac{1}{3}\sqrt{1 + \sin(0.39199)} = 0.39187$$

$$x_4 = g(x_3) = \frac{1}{3}\sqrt{1 + \sin(0.39187)} = 0.39185$$

$$x_5 = g(x_4) = \frac{1}{3}\sqrt{1 + \sin(0.39185)} = 0.39185$$

Here $x_4 \cong x_5$ up to five decimal places. Hence the required root is 0.39185.

Example 9: Find the root of the equation $\sin x = 10(x-1)$ by iteration method.

Solution: Choose $\quad f(x) = \sin x - 10x + 10$

$$f(1) = \sin 1 - 10 + 10 = 0.84147$$

$$f(2) = \sin 2 - 10 \times 2 + 10 = -9.0907$$

So, the root lies between 1 and 2.

Now, the given equation can be rewritten as

$$x = \frac{\sin x + 10}{10} = g(x)$$

We have $\quad g'(x) = \dfrac{\cos x}{10}$

$\therefore \qquad |g'(x)| < 1 \quad$ when $1 < x < 2$

Hence for this choice of $g(x)$, fixed point iteration method will converge.

Now by iteration method, we have

$$x_{n+1} = g(x_n)$$

$$\Rightarrow \qquad x_1 = g(x_0) = \frac{1}{10}(\sin 1 + 10) = 1.08414$$

$$x_2 = g(x_1) = \frac{1}{10}(\sin 1.08414 + 10) = 1.08839$$

$$x_3 = g(x_2) = \frac{1}{10}(\sin 1.08839 + 10) = 1.08858$$

$$x_4 = g(x_3) = \frac{1}{10}(\sin 1.08858 + 10) = 1.08859$$

$$x_5 = g(x_4) = \frac{1}{10}(\sin 1.08859 + 10) = 1.08859$$

Here $x_4 \cong x_5$ upto five decimal places. Hence the required root is 1.08859.

Example 10: Use the iteration method to find the real root correct to five significant figure of the equation $e^{-x} = 10x$.

Solution: We have $\quad f(x) = -e^{-x} + 10x$

$$f(0) = 10 \times 0 - e^{-0} = -1$$

$$f(1) = 10 \times 1 - e^{-1} = 9.63212$$

So, the root lies between 0 and 1.

Now, the given equation can be rewritten as

$$x = \frac{e^{-x}}{10} = g(x)$$

We have $\quad g'(x) = -\dfrac{e^{-x}}{10}$

$\therefore \qquad |g'(x)| < 1 \quad$ when $0 < x < 1$

Hence for this choice of $g(x)$, fixed point iteration method will converge.

Consider the initial approximation $x_0 = 0$. Now by iteration method, we have

$$x_{n+1} = g(x_n)$$

$\Rightarrow$

$$x_1 = g(x_0) = \frac{e^{-0}}{10} = 0.1$$

$$x_2 = g(x_1) = \frac{e^{-1}}{10} = 0.0367879$$

$$x_3 = g(x_2) = \frac{e^{-0.0367879}}{10} = 0.0963880$$

$$x_4 = g(x_3) = \frac{e^{-0.0963880}}{10} = 0.0908111$$

$$x_5 = g(x_4) = \frac{e^{-0.0908111}}{10} = 0.0913190$$

$$x_6 = g(x_5) = \frac{e^{-0.0913190}}{10} = 0.0912726$$

$$x_7 = g(x_6) = \frac{e^{-0.0912726}}{10} = 0.0912768$$

Here $x_6 \cong x_7$ up to five decimal places. Hence the required root is 0.09127.

EXERCISE 2.4

1. The equation $f(x) = 3x^3 + 4x^2 + 4x + 1 = 0$ has a root in interval $[-1, 0]$. Find this root with an accuracy of 10^{-4} using iteration method.   [Ans. -0.333316]
2. Starting with $x = 0.12$, solve $x = 0.21 \sin(0.5 + x)$ by using iteration method.
     [Ans. 0.1224327]
3. Find a real root of the equation $x^3 + x^2 - 1 = 0$ on the interval $[0, 1]$ with an accuracy of 10^{-4} by the iteration method.   [Ans. 0.7548]
4. Show that the rearrangement $x = \dfrac{(10 - 3x^2 - x^3)}{5}$ of the equation $x^3 + 3x^2 + 5x - 10 = 0$
   does not yield a convergent sequence of successive approximation by iteration method near $x = 1$.
5. Use the iteration method to find real root of the following equations correct to four decimal places:

   i. $x = \operatorname{cosec} x$ ii. $\sin x = \dfrac{x+1}{x-1}$ iii. $3x + \sin x = e^x$

   iv. $\sin^2 x = x^2 - 1$ v. $x^3 + 2x^2 + 10x = 20$

   [Ans. i. 1.1142 ii. -5.5174 iii. 0.3604 iv. 1.4044 v. 1.3688]

2.11 AITKEN's Δ^2 METHOD

We know that the iteration method is linearly convergent. This slow rate of convergence can be improved with the help of the Aitken's Δ^2 method.

Consider x_{i-1}, x_i and x_{i+1} are three successive approximations to the actual root α of the equation $x = g(x)$. Then we have

$$\alpha - x_i = (\alpha - x_{i-1}) k$$

and

$$\alpha - x_{i+1} = (\alpha - x_i) k$$

on dividing, we get

$$\frac{\alpha - x_i}{\alpha - x_{i+1}} = \frac{\alpha - x_{i-1}}{\alpha - x_i}$$

$$(\alpha - x_i)^2 = (\alpha - x_{i-1})(\alpha - x_{i+1})$$

Solving for α, we get

$$\alpha = \frac{x_{i-1} x_{i+1} - x_i^2}{x_{i+1} - 2x_i + x_{i-1}}$$

$$= x_{i+1} - \frac{(x_{i+1} - x_i)^2}{x_{i+1} - 2x_i + x_{i-1}} \qquad \begin{array}{l} \Delta x_i = x_{i+1} - x_i \\ \Delta^2 x_{i-1} = x_{i+1} - 2x_i + x_{i-1} \end{array}$$

$$= x_{i+1} - \frac{(\Delta x_i)^2}{\Delta^2 x_{i-1}}$$

where α gives an improved value of the approximation x_{i+1}.

Example 11: Find a real root of the equation $2x - \log_{10}x = 7$, correct to three decimal places using Aitken's method and iteration method. Also show how the rate of convergence of Aitken's method is rapid then iteration method.

Solution: Here $\quad f(x) = 2x - \log_{10}x - 7$

$$f(3) = 6 - \log_{10}3 - 7 = -1.4771212$$

$$f(4) = 8 - \log_{10}4 - 7 = +0.39794$$

Hence, the root lies in between 3 and 4.
Rewriting the given equation as

$$x = \frac{1}{2}[\log_{10} x + 7] = g(x)$$

$$g'(x) = \frac{1}{2}\left(\frac{1}{x} \log_{10} e\right)$$

and $\quad |g'(x)| < 1$ when $3 < x < 4$.

Hence for this choice of $g(x)$, iteration method will converge.

Let $\quad x_0 = 3.5$

$$x_1 = g(x_0) = \frac{1}{2}(\log_{10} 3.5 + 7) = 3.7720340$$

$$x_2 = g(x_1) = \frac{1}{2}(\log_{10} 3.7720340 + 7) = 3.7882878$$

Now we will use Aitken's Δ^2 method

$$x_3 = x_2 - \frac{(\Delta x_1)^2}{\Delta^2 x_0} = x_2 - \frac{(x_2 - x_1)^2}{(x_2 - 2x_1 + x_0)}$$

$$= 3.7882878 - \frac{(3.7882878 - 3.7720340)^2}{(3.7882878 - 2 \times 3.7720340 + 3.5)}$$

$$= 3.789320$$

So the root is 3.789. We can see that the above value can be find by iterative method after many iteration. So, the rate of convergence of Aitken's method is rapid than iterative method.

Example 12: Perform two iterations of the linear iteration method followed by one iteration of the Aitken's Δ^2 method to find the root of the equation

$$f(x) = x - e^{-x} = 0, x_0 = 1$$

Solution: Here

$$f(x) = x - e^{-x}$$
$$f(0) = -1, f(1) = 1 - e^{-1} = 0.6321205$$

Therefore, the root lies in between 0 and 1.

Rewriting the given equation as

$$x = e^{-x} = g(x)$$

$\therefore$
$$g'(x) = -e^{-x}$$
$$|g'(x)| < 1, \quad \text{when } 0 < x < 1$$

Hence for this choice of $g(x)$, iteration method will converge.

Let
$$x_0 = 1$$

Then
$$x_1 = g(x_0) = e^{-1} = 0.367879$$
$$x_2 = g(x_1) = e^{-0.367879} = 0.692200$$

Now we will use Aitken's Δ^2 method

$$x_3 = x_2 - \frac{(\Delta x_1)^2}{\Delta^2 x_0} = x_2 - \frac{(x_2 - x_1)^2}{x_2 - 2x_1 + x_0}$$

$$= 0.692200 - \frac{(0.692200 - 0.367879)^2}{(0.692200 - 2 \times 0.367879 + 1)}$$

$$= 0.692200 - \frac{0.105184}{0.956442}$$

$$= 0.582225$$

Example 13: Find a real root of the equation $2x = \cos x + 3$ by using Aitken's Δ^2 method.

Solution: Here
$$f(x) = 2x - \cos x - 3$$
$$f(0) = -4$$
$$f(1) = -1.5403023$$
$$f(2) = 1.4161468$$

Therefore, the root lies in between 1 and 2.

Let
$$x_0 = 1.5$$

The given equation can be rewrite as

$$x = \frac{1}{2}(\cos x + 3) = g(x)$$

$\therefore$
$$g'(x) = -\frac{1}{2} \sin x$$
$$|g'(x)| < 1, \quad \text{when } 1 < x < 2$$

Thus for this choice of $g(x)$, iterative method will converge.

$$x_1 = g(x_0) = g(1.5) = \frac{1}{2}[\cos(1.5) + 3] = 1.535$$

$$x_2 = g(x_1) = g(1.535) = \frac{1}{2}[\cos(1.535) + 3] = 1.518$$

Now by Aitken's Δ^2 method

$$x_3 = x_2 - \frac{(\Delta x_1)^2}{\Delta^2 x_0} = x_2 - \frac{(x_2 - x_1)^2}{x_2 - 2x_1 + x_0}$$

$$= 1.518 - \frac{(1.518 - 1.535)^2}{(1.518 - 2 \times 1.535 + 1.5)}$$

$$= 1.518 - \frac{(-0.017)^2}{-0.052} = 1.518 + 0.00555$$

$$= 1.524$$

EXERCISE 2.5

1. Perform two iteration of the linear iteration method followed by one iteration of the Aitken's Δ^2 method to find the root of the equation
   $$f(x) = x^3 - 5x + 1 = 0; \quad x_0 = 0.5$$ [Ans. 0.200232]
2. Perform two iteration of the linear iteration method followed by one iteration of the Aitken's Δ^2 method to find the root of the following equation:
   i. $(15)^{1/3}$ ii. $\sin x = 5x - 2$ [Ans. i. 2.46661 ii. 0.4014]

2.12 NEWTON–RAPHSON METHOD

Let α be the exact root of the equation $f(x) = 0$ which is either algebraic or transcendental equation. Consider x_0 be the approximate root of the given equation $f(x) = 0$.

Then $\alpha = x_0 + h$

where h is very small, positive or negative.

Since α is exact root, $\therefore$ $f(\alpha) = 0$

i.e. $f(x_0 + h) = 0$

By Taylor's expansion for function of one variable, we get

$$f(\alpha) = f(x_0 + h)$$

$$= f(x_0) + hf'(x_0) + \frac{h^2}{2!} f''(x_0) + \dots = 0$$

Neglecting the second and higher order derivative of $f(x)$ at $x = x_0$, we get

$$f(x_0) + hf'(x_0) = 0$$

$\Rightarrow$ $h = -\dfrac{f(x_0)}{f'(x_0)}$, approximately.

Hence an approximation x_1 better than x_0 is given by

$$x_1 = x_0 + h$$

$$= x_0 - \frac{f(x_0)}{f'(x_0)}$$

Starting with this x_1, a better approximation x_2 is given by

$$x_2 = x_1 - \frac{f(x_1)}{f'(x_1)}$$

In general, we get the relation

$$x_{n+1} = x_n - \frac{f(x_n)}{f'(x_n)}, \quad n = 0, 1, 2, \ldots$$

From this formula, we can find successive better values of the root. This formula is known as Newton–Raphson formula.

2.12.1 Geometrical Explanation of Netwon–Raphson's Method

In Fig. 2.5, the curve $y = f(x)$ meets the x-axis at A. Let x_0 be the approximate root of α. The tangent line to $y = f(x)$ at the point $P[x_0, f(x_0)]$ has equation

$$y - f(x_0) = f'(x_0)(x - x_0)$$

This tangent line cuts the x-axis at $x = x_1$. To get the point, solve the equation of the tangent at P with $y = 0$

$\therefore$

$$0 - f(x_0) = f'(x_0)(x_1 - x_0)$$

or

$$x_1 = x_0 - \frac{f(x_0)}{f'(x_0)}$$

which is Newton–Raphson formula for first approximation. It means the point A given by exact value $x = \alpha$ is approximated to $x = x_1$ corresponding to the point where tangent meet the x axis.

Therefore, the curve $y = f(x)$ is approximated successively to the tangents at points on the curve corresponding to successive approximation $x_0, x_1, x_2, \ldots, x_n, x_{n+1}, \ldots$

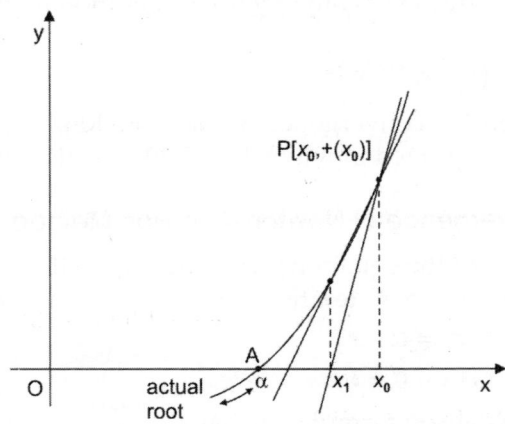

Fig. 2.5

2.12.2 Criterion for the Convergence in Newton–Raphson's Method

In Newton–Raphson method, we have

$$x_{n+1} = x_n - \frac{f(x_n)}{f'(x_n)} \tag{2.17}$$

and in iteration method, we have

$$x_{n+1} = g(x_n) \tag{2.18}$$

On comparing Eq. (2.17) and Eq. (2.18), we get

$$g(x_n) = x_n - \frac{f(x_n)}{f'(x_n)}$$

Hence the equation is

$$x = g(x) \text{ where } g(x) = x - \frac{f(x)}{f'(x)} \tag{2.19}$$

We know that the iteration method converges if $|g'(x)| < 1$.
Now differentiating Eq. (2.19), we get

$$g'(x) = 1 - \left[\frac{f'(x) \cdot \dfrac{d}{dx} f(x) - f(x) \dfrac{d}{dx} f'(x)}{\{f'(x)\}^2} \right]$$

$$g'(x) = 1 - \left[\frac{\{f'(x)\}^2 - f(x) f''(x)}{\{f'(x)\}^2} \right]$$

or

$$g'(x) = \left[\frac{\{f'(x)\}^2 - \{f'(x)\}^2 + f(x) f''(x)}{\{f'(x)\}^2} \right]$$

or

$$g'(x) = \left[\frac{f(x) f''(x)}{\{f'(x)\}^2} \right]$$

According to the condition of convergence of iteration method, i.e. $|g'(x)| < 1$, we get

$$|f(x) f''(x)| < [f'(x)]^2 \tag{2.20}$$

This is the criterion for convergence of Newton–Raphson method. The interval containing α should be selected in which Eq. (2.20) is satisfied.

2.12.3 Order of Convergence of Newton–Raphson Method

Let α be the actual root of the equation $f(x) = 0$. Let x_n be the nth approximate root and e_n be the error at the nth stage of iteration. Then

$$x_n = \alpha + e_n$$

and

$$x_{n+1} = \alpha + e_{n+1}$$

Now by Newton–Raphson formula, we have

$$x_{n+1} = x_n - \frac{f(x_n)}{f'(x_n)}$$

or

$$(\alpha + e_{n+1}) = (\alpha + e_n) - \frac{f(\alpha + e_n)}{f'(\alpha + e_n)}$$

$$e_{n+1} = e_n - \frac{f(\alpha + e_n)}{f'(\alpha + e_n)}$$

By using Taylor's equation, we get

$$e_{n+1} = e_n - \frac{f(\alpha) + e_n f'(\alpha) + \dfrac{e_n^2}{2!} f''(\alpha) + \dots}{f'(\alpha) + e_n f''(\alpha) + \dfrac{e_n^2}{2!} f'''(\alpha) + \dots}$$

α is the actual root of $f(x) = 0$. So, $f(\alpha) = 0$, then

$$e_{n+1} = e_n - \frac{e_n f'(\alpha) + \dfrac{e_n^2}{2!} f''(\alpha) + \dots}{f'(\alpha) + e_n f''(\alpha) + \dfrac{e_n^2}{2!} f'''(\alpha) + \dots}$$

or

$$e_{n+1} = e_n - \frac{e_n f'(\alpha) + \dfrac{e_n^2}{2!} f''(\alpha)}{f'(\alpha) + e_n f''(\alpha)} \quad \text{(Neglecting higher power of } e_n)$$

or

$$e_{n+1} = \frac{e_n f'(\alpha) + e_n^2 f''(\alpha) - e_n f'(\alpha) - \dfrac{e_n^2}{2} f''(\alpha)}{f'(\alpha)\left[1 + e_n \dfrac{f''(\alpha)}{f'(\alpha)}\right]}$$

$$= \frac{\dfrac{e_n^2}{2} f''(\alpha)}{f'(\alpha)} \left[1 + e_n \frac{f''(\alpha)}{f'(\alpha)}\right]^{-1}$$

$$= \frac{e_n^2}{2} \frac{f''(\alpha)}{f'(\alpha)} \left[1 - e_n \frac{f''(a)}{f'(a)} + \dots\right]$$

$$= \frac{e_n^2}{2} \frac{f''(\alpha)}{f'(\alpha)} - \frac{e_n^3}{2} \left[\frac{f''(\alpha)}{f'(\alpha)}\right]^2 + \dots$$

$$= \frac{e_n^2}{2} \frac{f''(\alpha)}{f'(\alpha)} \quad \text{(Neglecting higher power of } e_n)$$

$\therefore$ $$e_{n+1} \propto e_n^2$$

$\therefore$ Hence by definition, the convergence is quadratic and is of order 2.

Note:

1. Newton–Raphson method is applicable only when h is very small, i.e. when $f'(x)$ is large.
2. When $f'(x)$ is zero or nearly zero, then method fails.
3. This method is also used to find the complex root.
4. The method is also called method of tangents.

2.12.4 Algorithm for Newton-Raphson Method

1. Read x_0, epsilon, delta, n

   Remarks: x_0 is the initial guess, epsilon is the prescribed relative error, delta is the prescribed lower bound for f' and n the maximum number of iterations to be allowed. Statement 3 to 8 are repeated until the procedure converges to a root or iterations equal n.

2. For $i = 1$ to n in steps of 1 do
3. $f_0 \leftarrow f(x_0)$
4. $f_0' \leftarrow f'(x_0)$
5. If $| f_0' | \leq$ delta then Go To 11
6. $x_1 \leftarrow x_0 - \left(\dfrac{f_0}{f_0'} \right)$
7. If $|(x_1 - x_0)/x_1| <$ epsilon then Go To 13
8. $x_0 \leftarrow x_1$
   end for
9. Write 'Does not converge in n iterations', f_0, f_0', x_0, x_1
10. Stop
11. Write 'Slope too small' x_0, f_0, f_0', i
12. Stop
13. Write 'Convergent solution', $x_1, f(x_1), i$
14. Stop

Example 14: Find a real root of the equation $x = e^{-x}$ using the Newton–Raphson method, correct to four decimal places.

Solution: We have $\quad f(x) = x - e^{-x}$

then $\qquad\qquad\quad f'(x) = 1 + e^{-x}$

Since $\qquad\qquad f(0.5) = -0.1065$

$\qquad\qquad\qquad f(0.6) = 0.05118$

Hence the root lies between 0.5 and 0.6.

Since $\qquad\qquad |f(0.6)| < |f(0.5)|$

$\therefore$ Root is nearer to 0.6.

Let $\qquad\qquad\qquad x_0 = 0.55$

By Newton–Raphson method

$$x_{n+1} = x_n - \frac{f(x_n)}{f'(x_n)}$$

$$= x_n - \frac{(x_n - e^{-x_n})}{(1 + e^{-x_n})} = \frac{x_n + x_n e^{-x_n} - x_n + e^{-x_n}}{1 + e^{-x_n}}$$

$$= \frac{x_n e^{-x_n} + e^{-x_n}}{1 + e^{-x_n}}$$

Putting $n = 0$, we get

$$x_1 = \frac{x_0 e^{-x_0} + e^{-x_0}}{1 + e^{-x_0}} = \frac{0.55 e^{-0.55} + e^{-0.55}}{1 + e^{-0.55}}$$

$$= \frac{e^{-0.55}(0.55 + 1)}{1 + e^{-0.55}} = \frac{0.576949 \times 1.55}{1 + 0.576949}$$

$$= \frac{0.894270}{1.576949} = 0.567088$$

Putting $n = 1$

$$x_2 = \frac{x_1 e^{-x_1} + e^{-x_1}}{1 + e^{-x_1}} = \frac{(1 + x_1)e^{-x_1}}{1 + e^{-x_1}}$$

$$= \frac{0.567174 \times 1.567088}{1.567174} = 0.567142$$

Putting $n = 2$, we get

$$x_3 = \frac{(1 + x_2)e^{-x_2}}{1 + e^{-x_2}} = \frac{1.567142 \times 0.567144}{1.567144} = 0.567143$$

Since x_2 and x_3 are same up to four decimal places, hence the required root is 0.5671.

Example 15: The equation $2e^{-x} = \dfrac{1}{x+2} + \dfrac{1}{x+1}$ has two roots greater than -1. Calculate these roots correct to five decimal places.

Solution: We have $\quad f(x) = 2e^{-x} - \dfrac{1}{x+2} - \dfrac{1}{x+1}$

Also $\quad f'(x) = -2e^{-x} + \dfrac{1}{(x+2)^2} - \dfrac{1}{(x+1)^2}$

Since $\quad f(0) = 2 - \dfrac{1}{2} - 1 = 0.5$

$$f(1) = 2e^{-1} - \frac{1}{3} - \frac{1}{2} = -0.0976$$

$$f(-0.8) = 2e^{-(-0.8)} - \frac{1}{1.2} - \frac{1}{0.2} = -1.38$$

Hence two roots of $f(x) = 0$ which are greater than -1 lie in the intervals $(-0.7, 0)$ and $(0, 1)$. By using Newton–Raphson method

$$x_{n+1} = x_n - \frac{f(x_n)}{f'(x_n)}$$

$$= x_n - \frac{\left(2e^{-x_n} - \dfrac{1}{x_n + 2} - \dfrac{1}{x_n + 1}\right)}{-2e^{-x_n} + \dfrac{1}{(x_n + 2)^2} + \dfrac{1}{(x_n + 1)^2}}$$

Put $n = 0$
$$x_1 = x_0 - \frac{2e^{-x_0} - \dfrac{1}{x_0 + 2} - \dfrac{1}{x_0 + 1}}{-2e^{-x_0} + \dfrac{1}{(x_0 + 2)^2} + \dfrac{1}{(x_0 + 1)^2}}$$

First Root

Let $x_0 = -0.6$, we get

$$x_1 = -0.737984$$

Putting $n = 1$, the second approximation of root is

$$x_2 = -0.699338$$

Putting $n = 2$, the third approximation of root is

$$x_3 = -0.690163$$

Putting $n = 3$, the fourth approximation of root is

$$x_4 = -0.689753$$

Putting $n = 4$, the fifth approximation of root is

$$x_5 = -0.689752$$

Since x_4 and x_5 are same up to five decimal places. So the required root is -0.68975.

Second Root

Let $x_0 = 0.8$

Then
$$x_1 = 0.8 - \frac{2e^{-0.8} - \dfrac{1}{0.8 + 2} - \dfrac{1}{0.8 + 1}}{-2e^{-0.8} + \dfrac{1}{(2.8)^2} + \dfrac{1}{(1.8)^2}} = 0.769640$$

for $n = 1$, the second approximation of root is $x_2 = 0.770091$
for $n = 2$, the third approximation of root is $x_3 = 0.770091$.

Since x_2 and x_3 are same upto five decimal places. Hence the *second* required root is 0.77009.

Example 16: Show that the equation $f(x) = \cos\left[\dfrac{\pi(x + 1)}{8}\right] + 0.148\,x - 0.9062 = 0$ has one root in the interval $(-1, 0)$ and one in $(0, 1)$. Calculate the negative root correct to 4 decimals.

Solution: Given that $f(x) = \cos\left[\dfrac{\pi(x + 1)}{8}\right] + 0.148\,x - 0.9062$

Also $f'(x) = -\sin\left[\dfrac{\pi(x + 1)}{8}\right] + 0.148$

Since
$$f(0) = 0.0177$$
$$f(1) = -0.0511$$
$$f(-1) = -0.0542.$$

Hence, we observe that one root lies in the interval $(-1, 0)$ and one root in the interval $(1, 0)$. To find negative root, we assume that $x_0 = -0.5$.

By Newton–Raphson method

$$x_{n+1} = x_n - \frac{f(x_n)}{f'(x_n)}$$

$$= x_n - \frac{\cos\left(\dfrac{\pi(x_n+1)}{8}\right) + 0.148\,x_n - 0.9062}{-\sin\left(\dfrac{\pi(x_n+1)}{8}\right) + 0.148}$$

Put $n = 0$, $\quad x_1 = x_0 - \dfrac{\cos\left(\dfrac{\pi(x_0+1)}{8}\right) + 0.148\,x_0 - 0.9062}{-\sin\left(\dfrac{\pi(x_0+1)}{8}\right) + 0.148}$

$$= -0.5 - \frac{\cos\left[\dfrac{\pi(-0.5+1)}{8}\right] + (0.148 \times -0.5) - 0.9062}{-\sin\left[\dfrac{\pi(-0.5+1)}{8}\right] + 0.148}$$

$$= -0.508199$$

Put $n = 1$, $\quad x_2 = x_1 - \dfrac{\cos\left(\dfrac{\pi(x_1+1)}{8}\right) + 0.148\,x_1 - 0.9062}{-\sin\left(\dfrac{\pi(x_1+1)}{8}\right) + 0.148}$

$$= -0.508199 - \frac{\cos\left[\dfrac{\pi(-0.508199+1)}{8}\right] + (0.148 \times -0.508199) - 0.9062}{-\sin\left[\dfrac{\pi\,(-0.508199+1)}{8}\right] + 0.148}$$

$$= -0.508129$$

Since x_1 and x_2 are same upto four decimal places. Hence the required root is -0.5081.

Example 17: Find all the positive roots to the equation $10\int_0^x e^{-x^2}\,dt = 1$ with six correct decimals by using Newton–Raphson method.

Solution: We have $\qquad f(x) = 10xe^{-x^2} - 1$

Also $\qquad\qquad\quad f'(x) = 10(1 - 2x^2)e^{-x^2}$

Since $\qquad\qquad\quad f(0) = -1,\ f(1) = 2.6788,\ f(2) = -0.6337$

Hence the given equation $f(x) = 0$ has two positive roots, one in the interval $(0, 1)$ and the other in the interval $(1, 2)$.

By Newton–Raphson method

$$x_{n+1} = x_n - \frac{f(x_n)}{f'(x_n)}$$

For $n = 0$ $\qquad\qquad x_1 = x_0 - \dfrac{f(x_0)}{f'(x_0)} = x_0 - \dfrac{(10x_0\,e^{-x_0^2} - 1)}{10(1 - 2x_0^2)\,e^{-x_0^2}}$

Let $x_0 = 0.1$, we have

$$x_1 = 0.1 - \frac{(10 \times 0.1e^{-0.01} - 1)}{10(1 - 2 \times 0.01)e^{-0.01}} = 0.10102553$$

For $n = 1$ $x_2 = x_1 - \dfrac{f(x_1)}{f'(x_1)} = x_1 - \dfrac{(10x_1 e^{-x_1^2} - 1)}{10(1 - 2x_1^2) e^{-x_1^2}} = 0.10102585$

For $n = 2$ $x_3 = x_2 - \dfrac{f(x_2)}{f'(x_2)} = x_2 - \dfrac{(10x_2 e^{-x_2^2} - 1)}{10(1 - 2x_2^2) e^{-x_2^2}} = 0.10102585$

Hence, the root correct to 6 decimal places is 0.101026.

Second Root

Let $x_0 = 1.6$

For $n = 0$ $x_1 = 1.6 - \dfrac{(10 \times 1.6 e^{-2.56} - 1)}{10(1 - 2 \times 2.56) e^{-2.56}} = 1.67437337$

For $n = 1$ $x_2 = 1.67960443$
For $n = 2$ $x_3 = 1.67963061$
For $n = 3$ $x_4 = 1.67963061$

Hence, the another root correct to 6 decimal places is 1.679631.

Example 18: Find a real root of the equation $2x \tan x - 1 = 0$, $0.6 < x < 0.7$ using the Newton–Raphson method.

Solution: We have $f(x) = 2x \tan x - 1$ (i)
Also $f'(x) = 2[x \sec^2 x + \tan x]$ (ii)
Since $f(0.6) = 2 \times 0.6 \tan(0.6) - 1 = -0.1790$
   $f(0.7) = 2 \times 0.7 \tan(0.7) - 1 = 0.1792$

Hence the root lies between 0.6 and 0.7.

By Newton–Raphson method

$$x_{n+1} = x_n - \frac{f(x_n)}{f'(x_n)} \qquad \text{(iii)}$$

Let $x_0 = 0.65$, so from Eq. (iii)

$$x_1 = x_0 - \frac{f(x_0)}{f'(x_0)} = 0.65 - \frac{2 \times 0.65 \tan(0.65) - 1}{2[0.65 \sec^2(0.65) + \tan 0.65]}$$

$$= 0.65 - \frac{(-0.011734)}{3.571693} = 0.653285$$

$$x_2 = x_1 - \frac{f(x_1)}{f'(x_2)}$$

$$= 0.653285 - \frac{2 \times 0.653285 \tan(0.653285) - 1}{2[0.653285 \sec^2(0.653285) + \tan 0.653285]}$$

$$= 0.653298$$

Since x_1 and x_2 are same upto four decimal places. Hence the root is 0.6532.

Example 19: Find $\sqrt{14}$ to five places of decimal point by Newton–Raphson method.

Solution: Let $x = \sqrt{14}$
or $x^2 - 14 = 0$
i.e. the root of equation $x^2 - 14 = 0$ will give the value of $\sqrt{14}$.

Let $\qquad f(x) = x^2 - 14$

then $\qquad f'(x) = 2x$

Now $\qquad f(3) = 3^2 - 14 = -5$

$\qquad\qquad f(4) = 4^2 - 14 = 2$

So, the root lies between 3 and 4.

Let $\qquad\qquad x_0 = 3$

By Newton–Raphson method

$$x_{n+1} = x_n - \frac{f(x_n)}{f'(x_n)} = x_n - \frac{x_n^2 - 14}{2x_n}$$

$$= \frac{2x_n^2 - x_n^2 + 14}{2x_n} = \frac{x_n^2 + 14}{2x_n}$$

$\Rightarrow \qquad\qquad x_{n+1} = \dfrac{x_n^2 + 14}{2x_n}$ $\qquad\qquad\qquad$ (i)

by putting $n = 0, 1, 2, 3, \ldots$ in Eq. (i), we get

$$x_1 = \frac{x_0^2 + 14}{2x_0} = \frac{(3)^2 + 14}{2 \times 3} = \frac{23}{6} = 3.833333$$

$$x_2 = \frac{x_1^2 + 14}{2x_1} = \frac{(3.833333)^2 + 14}{2 \times 3.833333} = 3.742754$$

$$x_3 = \frac{x_2^2 + 14}{2x_2} = \frac{(3.742754)^2 + 14}{2 \times 3.742754} = 3.741658$$

$$x_4 = \frac{x_3^2 + 14}{2x_3} = \frac{(3.741658)^2 + 14}{2 \times 3.742754} = 3.741657$$

Since x_3 and x_4 are same up to five decimal places. Hence the root is 3.74165.

EXERCISE 2.6

1. Using Newton–Raphson method, find the real root of the following equations correct to four decimal places.
   - i. $x \sin x + \cos x = 0$ $\qquad\qquad\qquad\qquad\qquad$ [Ans. 2.7065]
   - ii. $x^3 - 5x + 3 = 0$ $\qquad\qquad\qquad\qquad\qquad\qquad$ [Ans. 0.6566]
   - iii. $x \log x = 12$ $\qquad\qquad\qquad\qquad\qquad\qquad\quad$ [Ans. 2.7406]
   - iv. $\cos x - xe^x = 0$ $\qquad\qquad\qquad\qquad\qquad\quad$ [Ans. 0.5178]
2. Show that the following two sequences, both have convergence of the second order with the same limit $\sqrt{a}$.

3. Show that the square root of $N = AB$ is given by $\sqrt{N} \approx \dfrac{S}{4} + \dfrac{N}{S}$ where $S = A + B$.

4. Use Newton–Raphson method to obtain a root. Correct to three decimal places of the following equations.
   - i. $\tan x = x$ $\qquad\qquad\qquad\qquad\qquad\qquad\qquad\quad$ [Ans. 4.493]
   - ii. $4(x - \sin x) = 1$ $\qquad\qquad\qquad\qquad\qquad\qquad$ [Ans. 1.171]
   - iii. $x + \log x = 2$ $\qquad\qquad\qquad\qquad\qquad\qquad\quad$ [Ans. 1.756]

5. Find all roots of $\cos x - x^2 - x = 0$ to five decimal places.   [Ans. 0.55001, 1.25115]

6. The equation $f(x) = 0$, where $f(x) = 0.1 - x + \dfrac{x^2}{(2!)^2} - \dfrac{x^3}{(3!)^2} + \dfrac{x^4}{(4!)^2} - \dots$ has one root in interval $(0, 1)$. Calculate this root correct to five decimal places. [Ans. 0.10260]

2.13 MODIFIED NEWTON–RAPHSON METHOD FOR MULTIPLE ROOTS

Let $\qquad\qquad f(x) = 0 \qquad\qquad\qquad\qquad\qquad\qquad$ (2.21)

which is either algebraic or transcendental equation. If α is the repeated root of Eq. (2.21), then the Eq. (2.21) can be written as

$$f(x) = (x - \alpha)^m \, g(x) = 0 \qquad\qquad\qquad (2.22)$$

where $g(x)$ is bounded and $g(\alpha) \neq 0$. The root α is called a multiple root of multiplicity m. We get from Eq. (2.22)

$$f(\alpha) = f'(\alpha) = \dots = f^{m-1}(\alpha) = 0, f^m(\alpha) \neq 0$$

For $m = 1$, the number α is said to be a simple root.

If α is a root of Eq. (2.21) with multiplicity m, then the new formula can be written as

$$x_{n+1} = x_n - m \frac{f(x_n)}{f'(x_n)} \qquad\qquad\qquad (2.23)$$

This means that $\dfrac{1}{m} f'(x_n)$ is the slope of the line through (x_n, y_n) and intersecting the x-axis at $(x_{n+1}, 0)$.

Since α is a root of $f(x) = 0$ with multiplicity m, it implies that α is also a root of $f'(x) = 0$ with multiplicity $(m - 1)$ and it is a root of $f''(x) = 0$ with multiplicity $(m - 2)$ and so on. Therefore $x_0 - \dfrac{mf(x_0)}{f'(x_0)}, x_0 - \dfrac{(m-1)f'(x_0)}{f''(x_0)}, x_0 - \dfrac{(m-2)f''(x_0)}{f'''(x_0)}$ will all have the same value if the initial approximation x_0 is sufficiently close to the actual root.

Example 20: Find the double root of the equation $x^3 - 7x^2 + 16x - 12 = 0$.

Solution: Let $\qquad\qquad f(x) = x^3 - 7x^2 + 16x - 12$

$\therefore \qquad\qquad\qquad f'(x) = 3x^2 - 14x + 16$

Starting with $x_0 = 1.8$, we have

$$x_0 - 2\frac{f(x_0)}{f'(x_0)} = 1.8 - 2\left[\frac{(1.8)^3 - 7 \times (1.8)^2 + 16 \times 1.8 - 12}{3 \times (1.8)^2 - 14 \times 1.8 + 16}\right]$$

$$= 1.8 - \frac{2 \times -0.048}{0.52} = 1.9846$$

and $\qquad x_0 - \dfrac{(2-1)f'(x_0)}{f''(x_0)} = 1.8 - \dfrac{[3 \times (1.8)^2 - 14 \times 1.8 + 16]}{6 \times 1.8 - 14} = 1.9625$

The closedness of these values implies that there is double root near to $x = 2$. Choosing $x_1 = 1.99$

$$x_1 - \frac{2f(x_1)}{f'(x_1)} = 1.99 - \frac{2[(1.99)^3 - 7 \times (1.99)^2 + 16 \times 1.99 - 12]}{3 \times (1.99)^2 - 14 \times 1.99 + 16}$$

$$= 1.97 - \frac{(-0.000202)}{0.0203} = 1.9999$$

and $\quad x_1 - \frac{(2-1) f'(x_1)}{f''(x_1)} = 1.99 - \frac{[3 \times (1.99)^2 - 14 \times 1.99 + 16]}{6 \times 1.99 - 14} = 1.9998$

This shows that there is a double root at $x = 1.9999$ which is quite near the actual root $x = 2$.

Example 21: Show that the equation $f(x) \equiv 1 - xe^{1-x} = 0$ has a double root at $x = 1$. The root is obtained by using the modified Newton–Raphson method with $m = 2$ starting with $x_0 = 0$.

Solution: Given that $\quad f(x) = 1 - xe^{1-x}$

$\therefore \qquad\qquad f'(x) = -[-xe^{1-x} + e^{1-x}] = -e^{1-x}(1-x)$
$$= (x-1) e^{1-x}$$

and $\qquad\qquad f''(x) = -(x-1) e^{1-x} + e^{1-x} = e^{1-x}(2-x)$

Starting with $x_0 = 0$, we have

$$x_0 - 2\frac{f(x_0)}{f'(x_0)} = 0 - 2\left[\frac{1 - x_0 e^{1-x_0}}{(x_0 - 1)e^{1-x_0}}\right]$$

$$= -2\left[\frac{1 - 0 \times e^{1-0}}{(0-1) e^{1-0}}\right] = \frac{-2}{-e^1} = \frac{2}{e} = 0.7357$$

and $\qquad x_0 - (2-1)\frac{f'(x_0)}{f''(x_0)} = 0 - \frac{1 \times (-e)}{2e} = 0.5$

Again taking $x_1 = 0.8$

$$x_1 - 2\frac{f(x_1)}{f'(x_1)} = 0.8 - \frac{2[1 - 0.8 e^{1-0.8}]}{(0.8-1) e^{1-0.8}} = 0.98728$$

and $\qquad x_1 - (2-1)\frac{f'(x_1)}{f''(x_1)} = 0.8 - \frac{(0.8-1)e^{1-0.8}}{(2-0.8) e^{1-0.8}} = 0.96667$

Again choosing $x_2 = 0.99$

$$x_2 - 2\frac{f(x_2)}{f'(x_2)} = 0.99 - \frac{2[1 - 0.99 e^{1-0.99}]}{(0.99-1) e^{1-0.99}}$$

$$= 0.99 - \frac{0.0001007}{(-0.0101005)} = 0.99997$$

$$x_2 - (2-1)\frac{f'(x_2)}{f''(x_2)} = 0.99 - \frac{(0.99-1)e^{1-0.99}}{(2-0.99)e^{1-0.99}} = 0.99990$$

This shows that there is a double root at $x = 0.9999$ which is quite nearer to the actual root $x = 1$.

EXERCISE 2.7

1. Find the double root of $x^3 - 5.4x^2 + 9.24x - 5.096 = 0$ given that it is nearer to 1.5.
[Ans. 1.4]

2. Find the double root of the equation $x^3 - x^2 - x + 1 = 0$.　　[Ans. 1.001]

3. Find the double root of $x^3 - 5x^2 + 7x - 3 = 0$.　　[Ans. 1.000]

2.14 POLYNOMIAL EQUATIONS

Consider a real polynomial equation of degree n

$$f(x) = a_0 x^n + a_1 x^{n-1} + a_2 x^{n-2} + \ldots + a_n x^n = 0; \quad a_0 \neq 0$$

where $a_0, a_1, \ldots, a_n$ are real numbers.

If all the roots (real and complex) of a polynomial equation are to be determined, it is always useful to aware of some facts about the polynomial equations, which are given below.

2.14.1 Some Facts About Polynomial Equations $f(x) = 0$

- The maximum number of roots of an equation always be equal to the degree of that equation.
- If α is a root of $f(x) = 0$ then $f(x)$ is divisible by $(x - \alpha)$.
- The number of positive roots of $f(x) = 0$ can not exceeds the number of changes of sign from positive to negative and from negative to positive in $f(x)$. For example consider the polynomial

$f(x)$	$4x^7$	$+3x^6$	$-4x^4$	$-2x^3$	$-6x^2$	$+x$	-1
sign	+	+	−	−	−	+	−

There are three signs changes, so the maximum number of positive roots will be three.

- To determine the maximum number of negative roots of a polynomial $f(x)$, replace x by $(-x)$ and then count the changes in sign as before. For example, the polynomial $f(x) = 4x^7 + 3x^6 - 4x^4 - 2x^3 - 6x^2 + x - 1$ has four maximum negative root, because there are four sign changes.

$f(-x)$	$-4x^7$	$+3x^6$	$-4x^4$	$+2x^3$	$-6x^2$	$-x$	-1
sign	−	+	−	+	−	−	−

- If the degree of $f(x) = 0$ is n and it has m positive roots and k negative roots, then at least $n - (m + k)$ roots will be imaginary.
- If $a + ib$ is a root of $f(x) = 0$, then $a - ib$ will always be the root of $f(x) = 0$.
- If $a + \sqrt{b}$ is a root of $f(x) = 0$, then $a - \sqrt{b}$ will always be the root of $f(x) = 0$.

2.15 BIRGE–VIETA METHOD FOR POLYNOMIALS

In this method, we consider that the polynomial has real coefficients and all roots are real.

So, we consider a real polynomial equation of degree n

$$f(x) = a_0 + a_1 x + a_2 x^2 + a_3 x^3 + \ldots + a_n x^n = 0 \qquad (2.24)$$

where $a_0, a_1, a_2, \ldots, a_n$ are real numbers.

If p is a real root it means $(x - p)$ is a factor of $f(x)$. After dividing the Eq. (2.24) by $(x - p)$, we get

$$f(x) = (x - p)\, g(x) + R \qquad (2.25)$$

where $g(x)$ is the quotient and it will be written as

$$g(x) = b_0 x^{n-1} + b_1 x^{n-2} + b_2 x^{n-3} + \dots + b_{n-2}x + b_{n-1} \tag{2.26}$$

and R is a residue in Eq. (2.25) which depends on p.

Now starting with the initial approximation p_0 to p, we can use same iterative method to improve the value of p such that

$$f(p) = R(p) = 0 \tag{2.27}$$

If we apply the Newton–Raphson method with a starting value p_0, the Newton–Raphson formula becomes

$$p_{i+1} = p_i - \frac{f(p_i)}{f'(p_i)}, i = 0, 1, 2, \dots \tag{2.28}$$

Now by comparing the coefficient of $f(x)$ and $(x - p)\, g(x) + R$, we get

$$a_0 x^n + a_1 x^{n-1} + a_2 x^{n-2} + \dots + a_{n-1}x + a_n$$
$$= (x - p)(b_0 x^{n-1} + b_1 x^{n-2} + b_2 x^{n-3} + \dots + b_{n-2}x + b_{n-1}) + R$$

Equating various power of x in both the sides, we get

$$\begin{array}{ll}
a_0 = b_0 & b_0 = a_0 \\
a_1 = b_1 - pb_0 & b_1 = a_1 + pb_0 \\
a_2 = b_2 - pb_1 & b_2 = a_2 + pb_1 \\
\dots \dots \dots & \dots \dots \dots \\
\dots \dots \dots & \dots \dots \dots \\
a_n = R - pb_{n-1} & b_n = R = a_n + pb_{n-1}
\end{array}$$

In general $\qquad b_i = a_i + pb_{i-1}$ where $i = 1, 2, \dots$ and $b_0 = a_0$. $\tag{2.29}$

Now differentiating Eq. (2.29), w.r.t. p, we get

$$\frac{d\,b_i}{dp} = 0 + b_{i-1} + p \cdot \frac{d\,b_{i-1}}{dp} = b_{i-1} + p\frac{db_{i-1}}{dp} \tag{2.30}$$

Let $\dfrac{d\,b_i}{dp} = c_{i-1}$ $\tag{2.31}$

then from Eq. (2.30) $\quad c_{i-1} = b_{i-1} + p\, c_{i-2}$

or $\qquad\qquad\qquad c_i = b_i + p\, c_{i-1}$ where $i = 1, 2, \dots, n - 1$. $\tag{2.32}$

Put $i = 1$ in Eq. (2.31), we get

$$\frac{db_1}{dp} = c_0$$

By using Eq. (2.29)

$$c_0 = \frac{db_1}{dp} = \frac{d(a_1 + pb_0)}{dp} = 0 + b_0$$

$\Rightarrow \qquad\qquad c_0 = b_0$

We put $x = p$ in Eq. (2.2) and then differentiate w.r.t. p, we get

$$\frac{d}{dp}f(p) = \frac{dR}{dp} = \frac{db_n}{dp}$$

$\Rightarrow \qquad\qquad f'(p) = c_{n-1} \qquad \left| \qquad \therefore \frac{db_i}{dp} = c_{i-1} \right.$

Now the Newton–Raphson method in Eq. (2.28) becomes

$$p_{i+1} = p_i - \frac{b_n}{c_{n-1}}, \quad i = 0, 1, 2, \ldots \quad \Big| \quad \because f(p) = R = b_n$$

On converging, this iterative process will give one root p of the polynomial equation $f(x) = 0$. Now, the deflated polynomial equation $g(x) = 0$ can be used to find the other real roots.

This method is often called Birge–Vieta method.

We can also obtain b_n and c_{n-1} by using synthetic division method as given below:

p	a_0	a_1	a_2	...	a_{n-1}	a_n
		pb_0	pb_1	...	pb_{n-2}	pb_{n-1}
p	b_0	b_1	b_2	...	b_{n-1}	$b_n = R = f(p)$
		pc_0	pc_1	...	pc_{n-2}	
	c_0	c_1	c_2	...	$c_{n-1} = \dfrac{dR}{dp} = f'(p)$	

2.15.1 Algorithm for Birge-Vieta Method

1. Read x_0, e, n, N

   Remarks: x_0 is the initial consideration of the root, e the allowed error, n the order of polynomial and N the total number of iterations. x_0 is guessed using the points mentioned earlier.

2. For $i = 0$ to n in steps of 1 do Read a_i
   end for

3. For $i = 0$ to $n - 1$ in steps of 1 do Read b_i
   end for

4. $P \leftarrow a_n$

5. $b_{n-1} \leftarrow a_n$

6. $S \leftarrow b_{n-1}$

7. For $k = 1$ to N in steps of 1 do

8. For $i = 1$ to $n - 1$ in steps of 1 do

9. $b_{n-(i+1)} \leftarrow a_{n-i} + x_0 b_{n-i}$

10. $S \leftarrow b_{n-(i+1)} + x_0 S$
    end for

11. $P \leftarrow a_0 + b_0 x_0$

12. $x_1 \leftarrow x_0 - (P/S)$

13. If $\left| x_1 - \dfrac{x_0}{x_1} \right| \leq$ Go To 18

    else

14. $x_0 \leftarrow x_1$
    end if
    end for

15. Write "Root not found in N iterations"

16. Write S, P, x_1, x_0

17. Stop

18. Write "Root found in k iterations"

19. $x_0 \leftarrow x_1$
20. Write x_0, S, P
21. Stop.

Example 22: Use Birge-Vieta method to find a real root correct to three decimal places of the equation $x^3 - 11x^2 + 32x - 22 = 0$, $p = 0.5$. Also find the deflated polynomials.

Solution:

First iteration:

0.5	1	-11	32	-22
		0.5	-5.25	13.375
0.5	1	-10.5	26.75	$-8.625 = b_3$
		0.5	-5.00	
	1	-10.00	$21.75 = c_2$	

$$p_1 = p_0 - \frac{b_3}{c_2} = 0.5 - \frac{(-8.625)}{21.75} = 0.8966$$

Second iteration:

0.8966	1	-11	32	-22
		0.8966	-9.0587	20.5692
0.8966	1	-10.1034	22.9413	$-1.4308 = b_3$
		0.8966	-8.2548	
	1	-9.2068	$14.6865 = c_2$	

$$p_2 = p_1 - \frac{b_3}{c_2} = 0.8966 - \frac{(-1.4308)}{14.6865} = 0.9940$$

Third iteration:

0.9940	1	-11	32	-22
		0.9940	-9.9460	21.9217
0.9940	1	-10.0060	22.0540	$-0.0783 = b_3$
		0.9940	-8.9579	
	1	-9.0120	$13.0961 = c_2$	

$$p_3 = p_2 - \frac{b_3}{c_2} = 0.9940 - \frac{(-0.0783)}{13.0961} = 0.99998$$

The root correct to three decimal places is 1.00.

Deflated Polynomial

1	1	-11	32	-22
		1	-10	
	1	-10	22	

The deflated polynomial is $x^2 - 10x + 22$.

Example 23: Use the Birge–Vieta method to find smallest positive roots of the following equation: $x^4 - 3x^3 + 3x^2 - 3x + 2 = 0$, $p_0 = 0.5$. Also tell the deflated polynomial for this equation. [APJAKTU 2014 (carry over)]

Solution

First iteration:

0.5	1	−3	3	−3	2
		0.5	−1.25	0.875	−1.0625
0.5	1	− 2.5	1.75	−2.125	**0.9375** = b_4
		+0.5	− 1	0.375	
	1	−2	0.75	**−1.750** = c_3	

$$p_1 = p_0 - \frac{b_4}{c_3} = 0.5 - \frac{0.9375}{(-1.750)} = 1.0357$$

Second iteration:

1.0357	1	−3	3	−3	2
		1.0357	−2.0344	1	−2.0714
1.0357	1	− 1.9643	0.9656	−2	**− 0.0714** = b_4
		1.0357	− 0.9618	0.0039	
	1	−0.9286	0.0038	**−1.9961** = c_3	

$$p_2 = p_1 - \frac{b_4}{c_3} = 1.0357 - \frac{(-0.0714)}{(-1.9961)} = 0.9999$$

The root correct to three decimal place is 1.000.

Deflated Polynomial

1	1	−3	3	−3	2
		1	−2	1	
	1	−2	1	−2	

The deflated polynomial is $x^3 - 2x^2 + x - 2$.

Example 24: Use synthetic division and perform 2 iterations of the Birge–Vieta method to find smallest positive root of the polynomial $P_3(x) = 2x^3 - 5x + 1 = 0$. Use $p_0 = 0.5$.

Solution:

First iteration:

0.5	2	0	−5	1
		1	0.5	−2.25
0.5	2	1	−4.5	− 1.25 = b_3
		1	1	
	2	2	−3.5 = c_2	

$$p_1 = p_0 - \frac{b_3}{c_2} = 0.5 - \frac{(-1.25)}{(-3.5)} = 0.142857$$

Second iteration:

0.142857	2	0	−5	1
		0.285714	0.040816	−0.708454
0.142857	2	0.285714	−4.959184	$0.291546 = b_3$
		0.285714	0.081632	
	2	0.571428	$-4.877552 = c_2$	

$$p_2 = p_1 - \frac{b_3}{c_2} = 0.142857 - \frac{(0.291546)}{(-4.877552)} = 0.202630$$

Hence the root is 0.202630.

Example 25: Find the real root of $x^3 - x^2 - x + 1 = 0$.

Solution: Let the initial approximation $p_0 = 0.5$. By using synthetic division method.

0.5	1	−1	−1	1
		0.5	−0.25	−0.625
0.5	1	−0.5	−1.25	$0.375 = b_3$
		+0.5	0	
	1	0	$-1.25 = c_2$	

$$p_1 = p_0 - \frac{b_3}{c_2} = 0.5 - \frac{(0.375)}{(-1.25)} = 0.8$$

Again

0.8	1	−1	−1	1
		0.8	−0.16	−0.928
0.8	1	−0.2	−1.16	$0.072 = b_3$
		+0.8	0.48	
	1	0.6	$-0.68 = c_2$	

$$p_2 = p_1 - \frac{b_3}{c_2} = 0.8 - \frac{0.072}{(-0.68)} = 0.9059$$

Again

0.9059	1	−1	−1	1
		0.9059	−0.0852	−0.9831
0.9059	1	−0.0941	−1.0852	$0.0169 = b_3$
		+0.9059	0.7354	
	1	0.8118	$-0.3498 = c_2$	

$$p_3 = p_2 - \frac{b_3}{c_2} = 0.9059 - \frac{0.0169}{(-0.3498)} = 0.9533$$

Again

0.9533	1	−1	−1	1
		0.9533	−0.0445	−0.9957
0.9533	1	−0.0467	−1.0445	**0.0043 = b_3**
		+0.9533	0.8643	
	1	0.9066	**−0.1802 = c_2**	

$$p_4 = p_3 - \frac{b_3}{c_2} = 0.9533 - \frac{0.0043}{(-0.1802)} = 0.9772$$

So the root is 0.9772 which is near to actual root 1.

EXERCISE 2.8

1. Use Birge–Vieta method to find a real root correct to three decimal places of the equation $x^4 - x - 10 = 0$, $p_0 = 1.5$. [Ans. 1.856]
2. Use Birge–Vieta method to find a real root correct to three decimal places of the equation $x^6 - x^4 - x^3 - 1 = 0$, $p = 1.5$. Perform three iterations. [Ans. 1.404]
3. Use Birge–Vieta method to find a real root correct to three decimal places of the following equations:
   i. $x^5 - x + 1 = 0$ [Ans. −1.167]
   ii. $x^3 - 6x^2 + 11x - 6 = 0$ [Ans. 1]
   iii. $x^3 - 4x^2 + 5x - 2 = 0$ [Ans. 2]
   iv. $x^4 - x^3 + 3x^2 + x - 4 = 0$ [Ans. 1]
   v. $x^3 - x - 4 = 0$ [Ans. 1.796]

2.16 LIN-BAIRSTOW'S METHOD FOR QUADRATIC FACTOR

This is an iterative method used to find both the real and complex roots of a polynomial. Since complex roots occur in pair of the form $a \pm ib$. Each such pair corresponds to a quadratic factor

$$[x - (a + ib)] [x - (a - ib)] = x^2 - 2ax + (a^2 + b^2)$$

which can be written in the form $x^2 + px + q$, where p and q are real numbers. If

$$f(x) = a_0 + a_1 x + a_2 x^2 + a_3 x^3 + \ldots + a_n x^n \tag{2.33}$$

and we divide it by the quadratic factor $x^2 + px + q$, we obtain a quotient

$$g(x) = b_0 + b_1 x + b_2 x^2 + \ldots + b_{n-2} x^{n-2} \tag{2.34}$$

of degree $n - 2$ and a remainder $Rx + S$ of degree 1 where all b's are real and R, S are functions of p and q.

Thus $$f(x) = (x^2 + px + q)(b_0 x^{n-2} + b_1 x^{n-3} + b_2 x^{n-4} + \ldots + b_{n-2}) + Rx + S \tag{2.35}$$

If $x^2 + px + q$ divides $f(x)$ completely, then the remainder $Rx + S$ should be zero i.e. $R = 0$, $S = 0$. Clearly R and S both depend on p and q.

So, our problem is to find p_c and q_c such that

$$R(p_c q_c) = 0, \ S(p_c, q_c) = 0 \tag{2.36}$$

In this case $(x^2 + p_c x + q_c)$ is a quadratic factor. We will now try to get expressions for $R(p, q)$ and $S(p, q)$. From Eq. (2.35)

$$a_0 + a_1 x + a_2 x^2 + \ldots + a_n x^n = (x^2 + px + q)(b_0 + b_1 x + b_2 x^2 + \ldots + b_{n-2} x^{n-2}) + Rx + S$$

Equating coefficient of various powers of x, we get

$$\left. \begin{aligned}
& a_n = b_{n-2}; \ a_{n-1} = b_{n-3} + p b_{n-2} \\
& a_{n-2} = b_{n-4} + p b_{n-3} + q b_{n-2}, \ldots \\
& a_{n-i} = b_{n-(i+2)} + p b_{n-(i+1)} + q b_{n-i}; \ldots \\
& a_2 = b_0 + p b_1 + q b_2, \\
& a_1 = R + p b_0 + q b_1, a_0 = S + q b_0
\end{aligned} \right\} \tag{2.37}$$

Equation (2.35) can be written as

$$b_{n-2} = a_n; \ b_{n-3} = a_{n-1} - p b_{n-2} \tag{2.38}$$

$$b_{n-4} = a_{n-2} - p b_{n-3} - q b_{n-2}; \ldots$$

$$b_{n-(i+2)} = a_{n-i} - p b_{n-(i+1)} - q b_{n-i}; \ldots$$

$$b_0 = a_2 - p b_1 - q b_2; \ R = a_1 - p b_0 - q b_1, \ S = a_0 - q b_0 \tag{2.39}$$

In order to find p and q which make $R(p, q)$ and $S(p, q)$ simultaneously zero we write

$$R(p_c, q_c) = R(p + \Delta p, q + \Delta q)$$

$$S(p_c, q_c) = S(p + \Delta p, q + \Delta q)$$

By using Taylor's series expansion for function of two variable, we get

$$R(p_c, q_c) = R(p, q) + \left(\frac{\partial R}{\partial p}\right) \Delta p + \left(\frac{\partial R}{\partial q}\right) \Delta q = 0 \tag{2.40}$$

$$S(p_c, q_c) = S(p, q) + \left(\frac{\partial S}{\partial p}\right) \Delta p + \left(\frac{\partial S}{\partial q}\right) \Delta q = 0 \tag{2.41}$$

where

$$p_c = p + \Delta p; \ q_c = q + \Delta q \tag{2.42}$$

As $R(p_c, q_c) = 0 = S(p_c, q_c) = 0$, we have

$$\left(\frac{\partial R}{\partial p}\right) \Delta p + \left(\frac{\partial R}{\partial q}\right) \Delta q = -R(p, q) \tag{2.43}$$

and

$$\left(\frac{\partial S}{\partial p}\right) \Delta p + \left(\frac{\partial S}{\partial q}\right) \Delta q = -S(p, q) \tag{2.44}$$

On solving Eqs (2.43) and (2.44), we get the value of Δp and Δq.

Use of these values in Eq. (2.42) will give the next approximation to p_c and q_c. The process can be repeated until successive value of p_c and q_c show no significant changes.

Note: Bairstow's method works well only if the initial values of p and q are close to the correct values. In this case the convergence is quite rapid. If the starting values are arbitrarily choosen, then the method does not converge but very often diverges.

Alternative Method

According to the previous method, let

$$f(x) = x^n + a_1 x^{n-1} + a_2 x^{n-2} + \dots + a_n \tag{2.45}$$

If we divide Eq. (2.45) by the quadratic factor $x^2 + px + q$, we get

$$g(x) = x^{n-2} + b_1 x^{n-3} + \dots + b_{n-2}$$

and the remainder $Rx + S$.

Thus,
$$f(x) = (x^2 + px + q)\ g(x) + Rx + S \tag{2.46}$$

or $x^n + a_1 x^{n-1} + a_2 x^{n-2} + \dots + a_n = (x^2 + px + q)(x^{n-2} + b_1 x^{n-3} + b_2 x^{n-4} + \dots + b_{n-2}) + Rx + S$

If the quadratic factor divides $f(x)$ completely, then $Rx + S = 0$, i.e. $R = 0$, $S = 0$. Clearly R and S both depend on p and q.

So, we want to find p and q such that

$$R(p, q) = 0, \quad S(p, q) = 0 \tag{2.47}$$

Let $p + \Delta p$ and $q + \Delta q$ be the actual values of p and q which satisfy Eq. (2.47) then

$$R(p + \Delta p, q + \Delta q) = 0, \quad S(p + \Delta p, q + \Delta q) = 0$$

To find the corrections Δp, Δq, we have the following equations

$$c_{n-2}\Delta p + c_{n-3}\Delta q = b_{n-1}$$

$$(c_{n-1} - b_{n-1})\,\Delta p + C_{n-2}\Delta q = b_n$$

After finding the values of $b_i'^s$ and $c_i'^s$ by synthetic division scheme, we obtain approximate values of Δp and Δq, say Δp_0 and Δq_0.

If p_0 and q_0 be the initial values then

$$p_1 = p_0 + \Delta p_0$$

$$q_1 = q_0 + \Delta q_0$$

Now taking p_1 and q_1 as the initlal values and repeat the process until we get better values of p and q.

2.16.1 Algorithm for Lin-Bairstow's Method

1. Read n, p, q, e, N

   Remarks: n is the order of polynomial, p and q are initial approximation.

   e is the allowed error in Δp and Δq.

   N is the maximum number of iteration to be allowed.

2. For $i = 0$ to n do

3. Read a_i

   end for

4. For $k = 1$ to N do

5. $b_{n-2} \leftarrow a_n$

6. $b_{n-3} \leftarrow a_{n-1} \cdot p \cdot b_{n-2}$

7. For $i = 2$ to $(n-2)$ do

8. $b_{n-(i+2)} \leftarrow a_{n-i} - p \cdot b_{n-(i+1)} - q \cdot b_{n-i}$

   end for

9. $R \leftarrow a_1 - p \cdot b_0 - q \cdot b_1$

10. $S \leftarrow a_0 - q \cdot b_0$

11. If $(|R| \le e)$ and; $|S| \le e)$ then Go To 33

Remarks: $\left(\dfrac{\partial b}{\partial p}\right), \left(\dfrac{\partial b}{\partial q}\right)$ calculated now

$\left(\dfrac{\partial b}{\partial p}\right)$ is named dpb and $\left(\dfrac{\partial b}{\partial q}\right)$ is named dqb

12. $dpb_{n-2} \leftarrow 0$

13. $dpb_{n-3} \leftarrow -b_{n-2}$

14. For $i = 2$ to $(n-2)$ do

15. $dp\,b_{n-(i+2)} \leftarrow -b_{n-(i+1)} - p \cdot dpb_{n-(i+1)} - q \cdot dpb_{n-i}$
    end for

16. $dqb_{n-2} \leftarrow 0$

17. $dqb_{n-3} \leftarrow 0$

18. $dqb_{n-4} \leftarrow -b_{n-2}$

19. For $i = 3$ to $(n-2)$ do

20. $dq\,b_{n-(i+2)} \leftarrow -p \cdot dq\,b_{n-(i+1)} - q \cdot dq\,b_{n-i} - b_{n-i}$
    end for

Remarks: $\left(\dfrac{\partial R}{\partial p}\right)$ is represented by dpR. $\left(\dfrac{\partial R}{\partial q}\right)$ is represented by dqR, $\left(\dfrac{\partial S}{\partial p}\right)$ is represented by dpS and $\left(\dfrac{\partial S}{\partial q}\right)$ is represented by dqS.

21. $dpR \leftarrow -b_0 - p \cdot dpb_0 - q \cdot dpb_1$

22. $dqR \leftarrow -pdq\,b_0 - b_1 - q \cdot dqb_1$

23. $dpS \leftarrow -qdp\,b_0$

24. $dqS \leftarrow b_0 - q\,dq\,b_0$

Remarks: We now find increments for p and q using Eqs (2.43) and (2.44)

25. $D = (dpR \cdot dqS - dpS \cdot dqR)$

26. delta $p = (-R \cdot dqS + S \cdot dqR)/D$

27. delta $q = (-S \cdot dpR + R \cdot dpS)/D$

28. $p \leftarrow p + $ delta p

28. $q \leftarrow q + $ delta q
    end for

30. Write 'Does not converge in N iterations'

31. Write $p, q,$ delta $p,$ delta q, R, S

32. Stop

33. Write 'Converges in k iterations'

34. Write R, S, p, q

35. Stop

Example 26: Find a quadratic factor of the polynomial $x^4 + 5x^3 + 3x^2 - 5x - 9 = 0$ starting with $p_0 = 3$, $q_0 = -5$ by using Bairstow's method.

Solution: We have $\quad f(x) = x^4 + 5x^3 + 3x^2 - 5x - 9 = 0$ $\hfill$ (i)

Consider $\quad\quad\quad f(x) = a_0 + a_1x + a_2x^2 + a_3x^3 + a_4x^4$ $\hfill$ (ii)

On comparison of Eqs (i and ii), we get

$$a_0 = -9, a_1 = -5, a_2 = 3, a_3 = 5 \text{ and } a_4 = 1$$

Let $\quad\quad\quad f(x) = (x^2 + px + q)(b_0 + b_1x + b_2x^2) + Rx + S$ $\hfill$ (iii)

Now $\quad\quad\quad b_{n-2} = a_n \Rightarrow b_2 = a_4$

$$b_1 = a_3 - pb_2; \ b_0 = a_2 - pb_1 - qb_2$$

$$R = a_1 - pb_0 - qb_1, \ S = a_0 - qb_0$$

Substituting the values of a_0, a_1, a_2, a_3 and a_4

$$b_2 = 1, b_1 = 5 - p, b_0 = 3 - pb_1 - q$$

$$R = -5 - pb_0 - qb_1, \ S = -9 - qb_0$$

Put the value of b_0 and b_1 in R and S, we get

$$R = -5 - p[3 - pb_1 - q] - q[5 - p]$$
$$= -5 - 3p + p^2b_1 + pq - 5q + pq$$
$$= -5 - 3p + p^2(5 - p) + 2pq - 5q$$
$$= -5 - 3p + 5p^2 - p^3 + 2pq - 5q$$

$$S = -9 - q[3 - p(5 - p) - q] = -9 - 3q + 5pq - p^2q + q^2$$

Now $\quad\quad\quad \dfrac{\partial R}{\partial p} = -3 + 10p - 3p^2 + 2q; \ \dfrac{\partial R}{\partial q} = 2p - 5$

$$\dfrac{\partial S}{\partial p} = 5q - 2pq; \ \dfrac{\partial S}{\partial q} = -3 + 5p - p^2 + 2q$$

$$D = \left(\dfrac{\partial R}{\partial p}\right)\left(\dfrac{\partial S}{\partial q}\right) - \left(\dfrac{\partial R}{\partial q}\right)\left(\dfrac{\partial S}{\partial p}\right)$$

$$\Delta p = \left(-R\dfrac{\partial S}{\partial q} + S\dfrac{\partial R}{\partial q}\right)\bigg/ D; \quad \Delta q = \left(-S\dfrac{\partial R}{\partial p} + R\dfrac{\partial S}{\partial p}\right)\bigg/ D$$

Given that $p_0 = 3$ and $q_0 = -5$.

So, $\quad\quad\quad R = -5 - 3 \times 3 + 5(3)^2 - (3)^3 + 2 \times 3 \times -5 - 5 \times -5 = -1$

$$S = -9 - 3 \times -5 + 5 \times 3 \times -5 - 3^2 \times (-5) + (-5)^2 = 1$$

$$\dfrac{\partial R}{\partial p} = -3 + 10 \times 3 - 3 \times 3^2 + 2 \times (-5) = -10$$

$$\dfrac{\partial R}{\partial q} = 2 \times 3 - 5 = 1$$

$$\dfrac{\partial S}{\partial p} = 5 \times (-5) - 2 \times 3 \times (-5) = 5$$

$$\dfrac{\partial S}{\partial q} = -3 + 5 \times 3 - 3^2 + 2 \times (-5) = -7$$

$$D = (-10) \times (-7) - 1 \times 5 = 65$$

$$\Delta p = \frac{-(-1) \times (-7) + 1 \times 1}{65} = -0.092$$

$$\Delta q = \frac{-(1) \times (-10) + (-1) \times 5}{65} = 0.077$$

$$p_1 = p_0 + \Delta p = 3 - 0.092 = 2.908$$

$$q_1 = q_0 + \Delta q = -5 + 0.077 = -4.923$$

Again
$$R = -5 - 3 \times 2.908 + 5 \times (2.908)^2 - (2.908)^3 + 2 \times 2.908 \times$$
$$(-4.923) - 5 \times (-4.923) = -0.0502$$

$$S = -9 - 3 \times (-4.923) + 5 \times 2.908 \times (-4.923) - (2.908)^2 \times$$
$$(-4.923) + (-4.923)^2 = 0.056$$

and
$$\frac{\partial R}{\partial p} = -3 + 10 \times 2.908 - 3 \times (2.908)^2 + 2 \times (-4.923) = -9.1354$$

$$\frac{\partial R}{\partial q} = 2 \times 2.908 - 5 = 0.816$$

$$\frac{\partial S}{\partial p} = 5 \times (-4.923) - 2 \times 2.908 \times (-4.923) = 4.0172$$

$$\frac{\partial S}{\partial q} = -3 + 5 \times 2.908 - (2.908)^2 + 2 \times (-4.923) = -6.7625$$

$$D = (-9.1354) \times (-6.7625) - 0.816 \times 4.0172 = 58.5001$$

$$\Delta p = \frac{-(-0.0502) \times (-6.7625) + 0.056 \times 0.816}{58.5001}$$

$$= \frac{-0.2938}{58.5001} = -0.00502$$

$$\Delta q = \frac{-(0.056) \times (-9.1354) + (-0.0502) \times 4.0172}{58.5001}$$

$$= \frac{0.3099}{58.5001} = 0.0053$$

$$p_2 = p_1 + \Delta p_0 = 2.908 + (-0.00502) = 2.903$$

$$q_2 = q_1 + \Delta q_0 = -4.923 + 0.0053 = -4.9177$$

The values of R and S is almost equal to zero. So, we now stop iteration for finding p and q.

Thus
$$f(x) = (x^2 + 2.903x - 4.918)(b_0 + b_1 x + b_2 x^2)$$

Now we will calculate b_0, b_1 and b_2.

$$b_2 = a_4 = 1$$

$$b_1 = a_3 - p_2 b_2 = 5 - 2.903 \times 1 = 2.097$$

$$b_0 = a_2 - p_2 b_1 - q_2 b_2 = 3 - 2.903 \times 2.097 - (-4.918) \times 1$$
$$= 1.8304$$

So,
$$f(x) = (x^2 + 2.903x - 4.918)(1.830 + 2.097x + x^2)$$

Hence, $(x^2 + 2.903x - 4.918)$ and $(x^2 + 2.097x + 1.830)$ are two quadratic factors of (i).

Example 27: Find the quadratic factor of the equation $x^4 + 5x^3 + 3x^2 - 5x - 9 = 0$ by using synthetic division method. Take $p_0 = 3$, $q_0 = -5$.

Solution: Firstly, we are giving the synthetic division scheme as below:

a_0	a_1	a_2	$\cdots$	a_{n-1}	a_n	
	$-pb_0$	$-pb_1$	$\cdots$	$-pb_{n-2}$	$-pb_{n-1}$	$-p$
		$-qb_0$	$\cdots$	$-qb_{n-3}$	$-qb_{n-2}$	$-q$
b_0	b_1	b_2	$\cdots$	b_{n-1}	b_n	
	$-pc_0$	$-pc_1$	$\cdots$	$-pc_{n-2}$		$-p$
		$-qc_0$	$\cdots$	$-qc_{n-3}$		$-q$
c_0	c_1	c_2	$\cdots$	c_{n-1}		

We have

1	5	3	-5	-9	
	-3	-6	-6	3	-3
		5	10	10	5
1	2	2	$-1 (= b_{n-1})$	$4 (= b_n)$	
	-3	3	-30		-3
		5	-5		5
1	$-1 (= c_{n-3})$	$10 (= c_{n-2})$	$-36 (= c_{n-1})$		

$\therefore \qquad c_{n-1} - b_{n-1} = -36 + 1 = -35$

Corrections Δp_0 and Δq_0 are given by

$$c_{n-2}\Delta p_0 + c_{n-3}\Delta q_0 = b_{n-1}$$

$$\Rightarrow \qquad 10\,\Delta p_0 - \Delta q_0 = -1 \qquad\qquad\qquad\qquad (i)$$

and $\quad (c_{n-1} - b_{n-1})\,\Delta p_0 + c_{n-2}\Delta q_0 = b_n$

$$\Rightarrow \qquad [-36 - (-1)]\,\Delta p_0 + 10\Delta q_0 = 4$$

$$-35\,\Delta p_0 + 10\,\Delta q_0 = 4 \qquad\qquad\qquad\qquad (ii)$$

on solving Eqs (i and ii), we get

$$\Delta p_0 = -0.09, \quad \Delta q_0 = 0.08$$

Thus, $\qquad\qquad\qquad p_1 = p_0 + \Delta p_0 = 3 - 0.09 = 2.91$

$$q_1 = q_0 + \Delta q_0 = -5 - 0.08 = -4.92$$

Δp and Δq provide new guesses. Process is repeated until approximate error falls below prespecified tolerance.

$\therefore$

$$\left|\epsilon_p\right| = \left|\frac{\Delta p_0}{p_1}\right| \times 100\% = \left|\frac{-0.09}{2.91}\right| \times 100\% = 3.0927\%$$

$$\left|\epsilon_q\right| = \left|\frac{\Delta q_0}{q_1}\right| \times 100\% = \left|\frac{0.08}{-4.92}\right| \times 100\% = 1.6260\%$$

Repeating same process, i.e. dividing $f(x)$ by $x^2 + 2.91x - 4.92$, we get

1	5	3	−5	−9	
	−2.91	−6.08	−5.35	0.2	−2.91
		4.92	10.28	9.05	4.92
1	2.09	1.85	−0.07	0.25	
	−2.91	2.37	−26.57		−2.91
		4.92	−4.03		4.92
1	−0.82	9.13	−30.67		

At this step, the corrections Δp_1 and Δq_1 are given by

$$9.13\,\Delta p_1 - 0.82\,\Delta q_1 = -0.07 \tag{iii}$$
$$-30.60\,\Delta p_1 + 9.13\,\Delta q_1 = 0.25 \tag{iv}$$

On solving Eqs (iii) and (iv), we get

$$\Delta p_1 = -0.00745$$
$$\Delta q_1 = 0.00241$$

Hence, second approximation of p and q are given by

$$p_2 = p_1 + \Delta p_1 = 2.91 - 0.00745 = 2.90255$$
$$q_2 = q_1 + \Delta q_1 = -4.92 + 0.00241 = -4.91759$$

$$\left| \epsilon_p \right| = \left| \frac{\Delta p_1}{p_2} \right| \times 100\% = \left| \frac{-0.00745}{2.90255} \right| \times 100\% = 0.2566\%$$

$$\left| \epsilon_q \right| = \left| \frac{\Delta q_1}{q_2} \right| \times 100\% = \left| \frac{0.0024}{-4.91759} \right| \times 100\% = 0.04901\%$$

Thus a quadratic factor is

$$x^2 + 2.90255x - 4.91759$$

Dividing the given polynomial by it, the other quadratic factor is

$$x^2 + 2.09745x + 1.82964$$

EXERCISE 2.9

1. Solve $x^4 - 5x^3 + 20x^2 - 40x + 60 = 0$ given that all the roots of $f(x) = 0$ are complex, by using Lin–Bairstow method. Take the values as $p_0 = -4$, $q_0 = 8$.
   [Ans. $1.9149 \pm 1.9077i$, $0.585 \pm 2.8054i$]
2. Find the complex root of the equaion $x^4 - 8x^3 + 39x^2 - 62x + 50 = 0$ by Lin–Bairstow method.
   [Ans. $1 \pm i$, $3 \pm 4i$]
3. Find a quadratic factor of the equation $5x^6 + 4x^5 + 2x^4 + 3x^3 + 20x^2 + 3x + 1 = 0$ by Lin–Bairstow method.
   [Ans. $x^2 - 0.9102x + 0.5585$]
4. Apply Lin–Bairstow method to the equation $x^4 - 3x^3 + 20x^2 + 44x + 54 = 0$ to find a quadratic factor close to $x^2 + 2x + 2 = 0$.
   [Ans. $4.000, 3.001, 2.000, 0.999$]

3

Calculus of Finite Differences

3.1 INTRODUCTION

In general, the solution of problem in applied mathematics consists of number that satisfy some types of equations. Theoretically, these numbers may be specified by the numbers, but in practice, it is not always possible to write down an exact decimal representative of the solution. Numerical methods are very important tools to give practical methods for calculating the solution of problems to applied mathematics to a desired degree of accuracy. The use of electronic computers for solving problem in various fields of science has further enhance the scope of numerical methods.

Let $y = f(x)$ be an explicit equation. We can calculate values of $f(x)$ for some given values of x, simply by substitution. If it is not the case, even then we can find approximate representative of the function with the help of calculus of finite differences. The calculus of finite differences deals with the changes in the value of function (dependent variable) due to changes in the independent variable and this independent variable changes by finite jumps whether equal or unequal. In this chapter, we shall study the case when independent variable changes by equal intervals.

3.2 DIFFERENCES

Let $y = f(x)$ be a given function of x and let $f(a), f(a + h), f(a + 2h), ..., f(a + nh)$ be the values of y corresponding to a, $a + h, a + 2h, ..., a + nh$, the values of x. The independent variable x is called argument and the corresponding dependent value y is called the entry. Here h is the difference between two consecutive arguments.

3.2.1 Forward Difference Operator

The differences $f(a + h) - f(a), f(a + 2h) - f(a + h), ..., f(a + nh) - f(a + \overline{n-1}h)$ are called first forward difference and are denoted by

$$\Delta f(a) = f(a + h) - f(a)$$

$$\Delta f(a + h) = f(a + 2h) - f(a + h)$$

$$... ... ...$$

$$... ... ...$$

$$f(a + \overline{n-1}h) = f(a + nh) - f(a + \overline{n-1}h)$$

In general, the first forward difference is denoted by
$$\Delta f(x) = f(x + h) - f(x)$$

The difference of the first forward difference is called second forward differences and are denoted by $\Delta^2 f(a)$, $\Delta^2 f(a + h)$ etc.
$$\Delta^2 f(a) = \Delta f(a + h) - \Delta f(a)$$
$$= f(a + 2h) - f(a + h) - f(a + h) + f(a)$$
$$= f(a + 2h) - 2f(a + h) + f(a)$$

Again difference of the second forward difference is called third forward difference and is denoted by $\Delta^3 f(a)$, $\Delta^3 f(a + h)$, etc.
$$\Delta^3 f(a) = \Delta f(a + 2h) - 2\Delta f(a + h) + \Delta f(a)$$
$$= f(a + 3h) - f(a + 2h) - 2\{f(a + 2h) - f(a + h)\} + f(a + h) - f(a)$$
$$= f(a + 3h) - 3f(a + 2h) + 3f(a + h) - f(a)$$

and so on.

3.2.2 Backward Difference Operator

The difference $f(a) - f(a - h)$, $f(a - h) - f(a - 2h)$ ... are called first backward difference and are denoted by $\nabla f(a)$, $\nabla f(a - h)$ etc. where ∇ is called the backward difference operator.
$$\nabla f(a) = f(a) - f(a - h)$$
$$\nabla f(a - h) = f(a - h) - f(a - 2h)$$
$$\text{... ... ...}$$
$$\text{... ... ...}$$
$$\nabla f(a - nh) = f(a - nh) - f(a + \overline{n + 1}h)$$

In general, the first backward difference is denoted by
$$\nabla f(x) = f(x) - f(x - h)$$

and second backward difference is given by
$$\nabla^2 f(x) = \nabla f(x) - \nabla f(x - h)$$
$$= f(x) - f(x - h) - f(x - h) + f(x - 2h)$$
$$= f(x) - 2f(x - h) + f(x - 2h)$$

and so on.

3.2.3 Shift Operator

The shift operator is denoted by E and is defined as
$$E[f(x)] = f(x + h)$$
$$E^2 f(x) = f(x + 2h)$$

In general, $E^n f(x) = f(x + nh)$

We can also write it as
$$E^n y_r = y_{r + n}$$

The inverse shift operator is defined as
$$E^{-1} f(x) = f(x - h)$$
$$E^{-2} f(x) = f(x - 2h)$$

In general $E^{-n} f(x) = f(x - nh)$

or $E^{-n} y_r = y_{r - n}$

3.2.4 Central Difference Operator δ

The central difference operator δ is defined as

$$\delta f(x) = f(x + h/2) - f(x - h/2)$$

or
$$\delta y_x = y_{x + h/2} - y_{x - h/2}$$

3.2.5 Averaging Operator μ

It is defined by

$$\mu f(x) = \frac{f(x + h/2) + f(x - h/2)}{2}$$

or
$$\mu y_x = \frac{y_{x+h/2} + y_{x-h/2}}{2}$$

3.2.6 Relation between Operators

1. $\Delta = E - 1$ or $E = 1 + \Delta$

   Proof: We have $Ef(x) = f(x + h)$

   and $\Delta f(x) = f(x + h) - f(x)$
   $$= Ef(x) - f(x)$$
   $$= (E - 1)f(x)$$
   $\Rightarrow$ $\Delta = E - 1$

   or $E = 1 + \Delta$

2. $\nabla = 1 - E^{-1}$

   Proof: We have $E^{-1}f(x) = f(x - h)$

   and $\nabla f(x) = f(x) - f(x - h)$
   $$= f(x) - E^{-1}f(x)$$
   $\Rightarrow$ $\nabla = 1 - E^{-1}$

   or $E^{-1} = 1 - \nabla$

3. $\delta = E^{1/2} - E^{-1/2}$

   Proof: We have $\delta f(x) = f(x + h/2) - f(x - h/2)$
   $$= E^{1/2}f(x) - E^{-1/2}f(x)$$
   $$= (E^{1/2} - E^{-1/2})f(x)$$
   $\Rightarrow$ $\delta = E^{1/2} - E^{-1/2}$

4. $\mu = 1/2\,(E^{1/2} + E^{-1/2})$

   We have $\mu f(x) = \frac{f(x + h/2) + f(x - h/2)}{2}$

   $$= \frac{1}{2}[E^{1/2}f(x) + E^{-1/2}f(x)]$$

   $$= \frac{1}{2}(E^{1/2} + E^{-1/2})f(x)$$

   $\Rightarrow$ $\mu = \frac{1}{2}(E^{1/2} + E^{-1/2})$

5. $\Delta = E\nabla = \nabla E = \delta E^{1/2}$

   Proof: $E\nabla f(x) = E[f(x) - f(x - h)]$

$$= Ef(x) - Ef(x - h) = f(x + h) - f(x)$$
$$= \Delta f(x)$$

$\therefore$ $\qquad E\nabla = \Delta$

and $\qquad \nabla Ef(x) = \nabla f(x + h) = f(x + h) - f(x)$
$$= \Delta f(x)$$

$\therefore$ $\qquad \nabla E = \Delta$

and $\qquad \delta E^{1/2} f(x) = \delta f(x + h/2)$
$$= f(x + h/2 + h/2) - f(x + h/2 - h/2)$$
$$= f(x + h) - f(x) = \Delta f(x)$$

$\therefore$ $\qquad \delta E^{1/2} = \Delta$

6. $E = e^{hD}$

   Proof: $\qquad Ef(x) = f(x + h)$

   By Taylor's series expansion, we get

$$Ef(x) = f(x) + hf'(x) + \frac{h^2 f''(x)}{2!} + \dots$$

$$= f(x) + hDf(x) + \frac{h^2 D^2 f(x)}{2!} + \dots$$

$$= \left(1 + hD + \frac{h^2 D^2}{2!} + \dots\right) f(x)$$

$\Rightarrow$ $\qquad E = 1 + hD + \frac{h^2 D^2}{2!} + \dots = e^{hD}$

$\therefore$ $\qquad E = e^{hD}$

3.3 FACTORIAL NOTATION

A factorial polynomial of degree n is denoted by $x^{(n)}$ and is defined as
$$x^{(n)} = x (x - 1) (x - 2) (x - 3) \dots (x - n + 1)$$
In particular $\qquad x^{(1)} = x,\ x^{(2)} = x (x - 1),\ x^{(3)} = x (x - 1) (x - 2),\ \dots$ etc.
In particular $\qquad x^{(0)} = 1.$
In advancing difference of the argument is h, then
$$x^{(n)} = x (x - h) (x - 2h) (x - 3h) \dots (x - nh + h)$$

3.3.1 Differences of Factorial Notation

1. Find the difference of $x^{(n)}$.

   By definition of Δ, we have
   $$\Delta x^{(n)} = (x + h)^{(n)} - x^{(n)}$$
   $$= (x + h) (x + h - h) \dots (x + h - nh + h) - x (x - h) (x - 2h) \dots (x - nh + h)$$
   $$= (x + h) (x) (x - h) \dots (x - nh + 2h) - x (x - h) (x - 2h) \dots (x - nh + h)$$
   $$= x (x - h) \dots (x - nh + 2h) [(x + h) - (x - nh + h)]$$
   $$= x^{(n-1)} (x + h - x + nh - h) = nh x^{(n-1)} \qquad (3.1)$$
   $$= nx^{(n-1)} \Delta x \text{ as } h = x + h - x = \Delta x$$

$\Rightarrow$ $\qquad \dfrac{\Delta x^{(n)}}{\Delta x} = nx^{(n-1)}$

Again $\qquad \Delta^2 x^{(n)} = \Delta \, \Delta x^{(n)}$

$$= \Delta \; nh \, x^{(n-1)}$$

$$= nh \, \Delta x^{(n-1)}$$

$$= nh \, (n-1) \, hx^{(n-2)}$$

$$= n \, (n-1) \, h^2 x^{(n-2)} \qquad (3.2)$$

Proceeding in the same manner, we get

$$\Delta^{n-1} x^{(n)} = n \, (n-1) \, (n-2) \dots 2 \cdot h^{n-1} x^{(1)} \qquad (3.3)$$

This give $\qquad \Delta^n x^{(n)} = n \, (n-1) \, (n-2) \dots 2 \cdot 1 h^{n-1} \Delta x$

$$= n(n-1) \, (n-2) \, 2 \cdot 1 h^n$$

$$= n! \, h^n \qquad (3.4)$$

This implies $\quad \Delta^{n+1} x^{(n)} = \Delta(n! \, h^n) = 0 \qquad (3.5)$

2. Find $\Delta^{-1} x^{(n)}$.

We have $\qquad \Delta x^{(2)} = 2hx^{(1)}$

$$\Rightarrow \qquad \Delta \left(\frac{x^{(2)}}{2h} \right) = x^{(1)}$$

$$\Rightarrow \qquad \Delta^{-1} x^{(1)} = \frac{x^{(2)}}{2h} + t(x)$$

where $t(x)$ is a periodic function of period h.

Similarly $\qquad \Delta^{-1} x^{(2)} = \frac{x^{(3)}}{3h} + t(x)$

In general $\qquad \Delta^{-1} x^{(n)} = \frac{x^{(n+1)}}{(n+1)h} + t(x) \qquad (3.6)$

Note: Δ^{-1} behaves on a polynomial like integration.

3. Relation between $x^{(-n)}$ and $(x+nh)^{(n)}$.

For the interval of differencing is h, we have

$$x^{(n)} = x(x-h)(x-2h) \dots (x - \overline{n-1}h)$$

$$= x(x-h)(x-2h) \dots (x - \overline{n-2}h)(x - \overline{n-1}h)$$

$$= (x - \overline{n-1}h) \, x^{(n-1)} \qquad (3.7)$$

If $n = 0$, we have $\quad x^{(0)} = (x+h) \, x^{(-1)}$

$$\Rightarrow \qquad x^{(-1)} = \frac{1}{x+h} \text{ as } x^{(0)} = 1.$$

Again from $n = -1$, Eq. (3.7) gives

$$x^{(-1)} = (x+2h) \, x^{(-2)}$$

$$\Rightarrow \qquad x^{(-2)} = \frac{x^{(-1)}}{x+2h} = \frac{1}{(x+h)(x+2h)}$$

In general, we have

$$x^{(-n)} = \frac{1}{(x+h)(x+2h) \dots (x+nh)} = \frac{1}{(x+nh)^{(n)}} \qquad (3.8)$$

Note: 1. If the interval of differencing is unity, from Eq. (3.8), we have

$$x^{(-n)} = \frac{1}{(x+1)(x+2)(x+3)\dots(x+n)} = \frac{1}{(x+n)^{(n)}}$$

$$\Rightarrow \qquad x^{(-n)} = \frac{1}{(x+n)^{(n)}} \qquad\qquad (3.9)$$

2. $\quad \Delta x^{(-n)} = (x+h)^{(-n)} - x^{(-n)}$

$$= \frac{1}{(x+2h)(x+3h)\dots(x+nh)(x+\overline{n+1}h)} - \frac{1}{(x+h)(x+2h)\dots(x+nh)}$$

$$= \frac{1}{(x+2h)(x+3h)\dots(x+nh)}\left[\frac{1}{(x+\overline{n+1}h)} - \frac{1}{x+h}\right]$$

$$= \frac{1(-nh)}{(x+h)(x+2h)\dots(x+nh)(x+\overline{n+1}h)}$$

$$= -nh\,x^{(-\overline{n+1})} = -nh\,x^{(-n-1)} \qquad\qquad (3.10)$$

$$\Rightarrow \qquad \frac{\Delta x^{(-n)}}{\Delta x} = -nx^{(-n-1)}$$

Similarly

$$\Delta^2 x^{(-n)} = (-1)^2 n\,(n+1)\,h^2 x^{(-n-2)} \text{ and so on.}$$

3.3.2 Generalized Factorial Notation

If $f(x)$ is any function, then generalized factorial function is given by

$$[f(x)]^{(n)} = f(x)\,f(x+h)\,f(x+2h)\dots f(x-\overline{n-1}h)$$

and $\qquad [f(x)]^{(-n)} = \dfrac{1}{f(x+h)\,f(x+2h)\,f(x+2h)\dots f(x+nh)}$

for $n = 1, 2, 3, \dots$

3.4 LEIBNITZ'S RULE FOR DIFFERENCES

If u and v are two functions then nth difference of the product (uv) is given by

$$\Delta^n(uv) = (\Delta^n u)\,v + {}^nC_1(\Delta^{n-1}Eu)(\Delta v) + {}^nC_1(\Delta^{n-2}E^2u)(\Delta^2 v) + \dots + (E^n u)(\Delta^n v)$$

This rule is known as Leibnitz's rule for differences.

3.5 DIFFERENCES OF ZERO

If $f(x) = x^m$, then we have

$$\Delta^n f(x) = \Delta^n x^m = (E-1)^n x^m \text{ as } \Delta = E - 1$$

$$= [E^n - {}^nC_1 E^{n-1} + {}^nC_2 E^{n-2} - {}^nC_3 E^{n-3} + \dots + (-1)^n]x^m$$

$$= E^n x^m - {}^nC_1 E^{n-1} x^m + {}^nC_2 E^{n-2} x^m - {}^nC_3 E^{n-3} x^m + \dots + (-1)^n x^m$$

$$= (x+nh)^m - {}^nC_1(x+(n-1)h)^m + {}^nC_2(x+(n-2)h)^m -$$

$$\quad {}^nC_3(x+(n-3)h)^m + \dots + (-1)^n x^m$$

$$= (x+n)^m - {}^nC_1(x+(n-1))^m + {}^nC_2(x+(n-2))^m -$$

$$\quad {}^nC_3(x+(n-3))^m + \dots + (-1)^n x^m \text{ if } h = 1$$

For $x = 0$, we have

$$[\Delta^n f(x)]_{x=0} = [\Delta^n x^m]_{x=0}$$
$$= n^m - {}^nC_1 (n-1)^m + {}^nC_2 (n-2)^m - {}^nC_3(n-3)^m + ... + (-1)^{n-1} {}^nC_{n-1}$$

This expression in RHS is called difference of zero and is written as

$$\Delta^n O^m = n^m - {}^nC_1 (n-1)^m + {}^nC_2 (n-2)^m - {}^nC_3(n-3)^m + ... + (-1)^{n-1} {}^nC_{n-1}$$

Note: (i) For $n = 1 = m$, we have $\Delta^1 O^1 = 1^1 = 1$

(ii) $n = 2 = m$, we have $\Delta^2 O^2 = 2^2 - {}^2C_1(2-1)^2 = 4 - 2 = 2!$

(iii) $n = 3 = m$, we have $\Delta^3 O^3 = 3^3 - {}^3C_1(3-1)^3 + {}^3C_2(3-2)^3 = 27 - 24 + 3 = 6 = 3!$

In general $\qquad \Delta^n O^n = n!$ and $\Delta^{n+1} O^n = \Delta\Delta^n O^n = \Delta n! = 0$

i.e. $\qquad\qquad \Delta^m O^n = 0$ if $m > n$.

3.5.1 Recurrence Relation for Differences of Zero

We know that

$$\Delta^n O^m = n^m - {}^nC_1 (n-1)^m + {}^nC_2 (n-2)^m - ... + (-1)^{n-1} {}^nC_{n-1}$$

$$= n^m - n(n-1)^m + \frac{n(n-1)}{2!} (n-2)^m - \frac{n(n-1)(n-2)}{3!} (n-3)^m + ... + (-1)^{n-1} n$$

$$= n\left[n^{m-1} - (n-1)^m + \frac{(n-1)}{2!} (n-2)^m - \frac{(n-1)(n-2)}{3!} (n-3)^m + ... + (-1)^{n-1} \right]$$

$$= n\left[n^{m-1} - (n-1)(n-1)^{m-1} + \frac{(n-1)(n-2)}{2!} (n-2)^{m-1} - \right.$$
$$\left. \frac{(n-1)(n-2)(n-3)}{3!} (n-3)^{m-1} - ... + (-1)^{n-1} \right]$$

$$= n\left[(1+n-1)^{m-1} - {}^{n-1}C_1 (1+n-2)^{m-1} + {}^{n-1}C_2 (1+n-3)^{m-1} - \right.$$
$$\left. {}^{n-1}C_3 (1+n-4)^{m-1} + ... + (-1)^{n-1} \right]$$

$$= n\left[E^{n-1} (1)^{m-1} - {}^{n-1}C_1 E^{n-2} (1)^{m-1} + {}^{n-1}C_2 E^{n-3}(1)^{m-1} - ... + (-1)^{n-1} \right]$$

$$= n\left[E^{n-1} - {}^{n-1}C_1 E^{n-2} + {}^{n-1}C_2 E^{n-3} + ... (-1)^{n-1} \right](1)^{m-1}$$

$$= n(E-1)^{n-1} (1)^{m-1} = n\Delta^{n-1} EO^{m-1}$$

$$= n\Delta^{n-1} (1+\Delta) O^{m-1} = n[\Delta^{n-1} + \Delta^4] O^{m-1}$$

$\Rightarrow \qquad = n(\Delta^{n-1}O^{m-1} + \Delta^4 O^m)$

This is the required recurrence formula for differences of zero.

Example 1: Prove the following identities:

i. $(E^{1/2} + E^{-1/2}) (1 + \Delta)^{1/2} = 2 + \Delta$

ii. $\mu = \sqrt{1 + \delta^2/4}$

iii. $\Delta = \delta^2/2 + \delta\sqrt{1 + \delta^2/4}$

iv. $\Delta + \nabla = \dfrac{\Delta}{\nabla} - \dfrac{\nabla}{\Delta}$

Solution: (i)

$$\begin{aligned}
\text{LHS} &= (E^{1/2} + E^{-1/2})(1 + \Delta)^{1/2} \\
&= (E^{1/2} + E^{-1/2}) E^{1/2} = E + 1 \\
&= 1 + \Delta + 1 = 2 + \Delta = \text{RHS}
\end{aligned}$$

ii.

$$\text{RHS} = \sqrt{1 + \delta^2/4} = \sqrt{1 + \frac{(E^{1/2} - E^{-1/2})^2}{4}}$$

$$= \sqrt{1 + \frac{(E + E^{-1} - 2)}{4}} = \sqrt{\frac{4 + E + E^{-1} - 2}{4}} = \frac{1}{2}\sqrt{E + E^{-1} + 2}$$

$$= \frac{1}{2}\sqrt{(E^{1/2} + E^{-1/2})^2} = \frac{1}{2}(E^{1/2} + E^{-1/2}) = \mu = \text{LHS}$$

iii.

$$\text{RHS} = \frac{\delta^2}{2} + \delta\sqrt{1 + \frac{\delta^2}{4}}$$

$$= \frac{(E^{1/2} - E^{-1/2})}{2} + (E^{1/2} - E^{-1/2})\sqrt{1 + \frac{(E^{1/2} - E^{-1/2})^2}{4}}$$

$$= \frac{(E^{1/2} - E^{-1/2})^2}{2} + (E^{1/2} - E^{-1/2})\sqrt{\frac{4 + E + E^{-1} - 2}{4}}$$

$$= \left(\frac{E + E^{-1} - 2}{2}\right) + (E^{1/2} - E^{-1/2})\sqrt{\frac{E + E^{-1} + 2}{4}}$$

$$= \frac{E + E^{-1} - 2}{2} + \frac{(E^{1/2} - E^{-1/2})(E^{1/2} + E^{-1/2})}{2}$$

$$= \frac{E + E^{-1} - 2}{2} + \frac{E - E^{-1}}{2} = \frac{E + E^{-1} - 2 + E - E^{-1}}{2}$$

$$= E - 1 = 1 + \Delta - 1 = \Delta = \text{LHS}$$

iv.

$$\text{RHS} = \frac{\Delta}{\nabla} - \frac{\nabla}{\Delta} = \frac{E - 1}{1 - E^{-1}} - \frac{1 - E^{-1}}{E - 1} = \frac{E - 1}{E - 1} - \frac{\dfrac{(E - 1)}{E}}{E - 1}$$

$$= E - \frac{1}{E} = E - E^{-1} = (1 + \Delta) - (1 - \nabla) = \Delta + \nabla = \text{LHS}$$

Example 2: Prove the following:

i. $hD = -\log(1 - \nabla) = \sin h^{-1}(\mu\delta)$

ii. $\mu\delta = \dfrac{1}{2}(\Delta + \nabla) = \dfrac{\Delta E^{-1}}{2} + \dfrac{\Delta}{2}$

iii. $\nabla - \Delta = -\nabla\Delta$

iv. $\delta^2 E = \Delta^2$

v. $\delta = \Delta(1 + \Delta)^{-1/2} = \nabla(1 - \nabla)^{-1/2}$

Solution: i. We have $\quad E = e^{hD}$

Taking log on both sides

$$hD = \log E = -\log E^{-1} = -\log(1 - \nabla)$$

Also
$$\mu = \frac{1}{2}(E^{1/2} + E^{-1/2}) \text{ and } \delta = E^{1/2} - E^{-1/2}$$

$$\mu\delta = \frac{1}{2}(E^{1/2} + E^{-1/2})(E^{1/2} - E^{-1/2})$$

$$= \frac{1}{2}(E - E^{-1}) = \frac{1}{2}(e^{hD} - e^{-hD})$$

$$= \sin h(hD)$$
$$hD = \sin h^{-1}(\mu\delta)$$

ii.
$$\mu\delta = \frac{1}{2}(E^{1/2} + E^{-1/2})(E^{1/2} - E^{-1/2})$$

$$= \frac{1}{2}(E - E^{-1}) = \frac{1}{2}[(1 + \Delta) - (1 - \nabla)]$$

$$= \frac{1}{2}(\Delta + \nabla)$$

Also
$$\mu\delta = \frac{1}{2}[\Delta + (E - 1)E^{-1}] = \frac{1}{2}(\Delta + \Delta E^{-1})$$

$$= \frac{\Delta E^{-1}}{2} + \frac{\Delta}{2}$$

iii.
$$\text{LHS} = \nabla - \Delta = (1 - E^{-1}) - (-1 + E) = 1 - E^{-1} + 1 - E$$
$$= -(1 - E^{-1})(E - 1)$$
$$\nabla - \Delta = -\nabla\Delta$$

iv.
$$\text{LHS} = \delta^2 E = (E^{1/2} - E^{-1/2})^2 E = (E + E^{-1} - 2)E$$
$$= E^2 + 1 - 2E = (E - 1)^2 = \Delta^2 = \text{RHS}$$

v.
$$\Delta(1 + \Delta)^{-1/2} = (E - 1)(1 + E - 1)^{-1/2} = (E - 1)E^{-1/2} = E^{1/2} - E^{-1/2} = \delta$$

Also
$$\nabla(1 - \nabla)^{-1/2} = (1 - E^{-1})[(1 - (1 - E^{-1})]^{-1/2} = (1 - E^{-1})E^{1/2}$$
$$= E^{1/2} - E^{-1/2} = \delta$$

Example 3: Prove that:

i. $\nabla^2 = h^2 D^2 - h^3 D^3 + \dfrac{7}{12} h^4 D^4 - \ldots$

ii. $E^{1/2} = \mu + \dfrac{1}{2}\delta$

Solution: i. We have $\quad E = e^{hD}$ and $\nabla = 1 - E^{-1}$

$$\nabla^2 = (1 - e^{-hD})^2 = \left[1 - \left(1 - hD + \frac{(hD)^2}{2!} - \frac{(hD)^3}{3!} + \frac{(hD)^4}{4!} \ldots\right)\right]^2$$

$$= \left[hD - \frac{(hD)^2}{2!} + \frac{(hD)^3}{3!} - \frac{(hD)^4}{4!} \ldots\right]^2$$

$$= h^2 D^2 \left[1 - \left(\frac{hD}{2!} - \frac{(hD)^2}{3!} + \dots \right) \right]^2$$

$$= h^2 D^2 \left[1 + \left(\frac{hD}{2} - \frac{(hD)^2}{6} + \dots \right)^2 - 2 \left(\frac{hD}{2} - \frac{(hD)^2}{6} + \dots \right) \right]$$

$$= h^2 D^2 \left[1 - hD + \left(\frac{1}{4} + \frac{1}{3} \right) h^2 D^2 - \dots \right]$$

$$= h^2 D^2 \left[1 - hD + \frac{7}{12} h^2 D^2 - \dots \right]$$

$$\nabla^2 = h^2 D^2 - h^3 D^3 + \frac{7}{12} h^4 D^4 - \dots$$

ii.

$$\text{RHS} = \mu + \frac{1}{2} \delta = \left(\frac{E^{1/2} + E^{-1/2}}{2} \right) + \frac{1}{2} (E^{1/2} - E^{-1/2})$$

$$= \frac{E^{1/2} + E^{-1/2} + E^{1/2} - E^{-1/2}}{2} = \frac{2(E^{1/2})}{2} = E^{1/2} = \text{LHS}$$

EXERCISE 3.1

1. If $hD = U$ and D, E, δ, μ be the operators with usual meaning and h is the interval of difference, then prove that
   (i) $E = e^u$
   (ii) $(E + 1) \delta = 2 (E - 1) \mu$
   (iii) $\delta = 2 \sinh \dfrac{U}{2}$
   (iv) $\mu = \cosh \dfrac{U}{2}$
   (v) $\mu \delta = \sinh U$

2. Prove that:
   (i) $\mu^2 = 1 + \dfrac{1}{4} \delta^2$
   (ii) $\sqrt{(1 + \delta^2 \mu^2)} = 1 + \dfrac{1}{2} \delta^2$
   (iii) $E^{-1/2} = \mu - \dfrac{1}{2} \delta$
   (v) $(E + 1) \delta = 2\mu (E - 1)$

3. If third differences are constant, prove that:

$$y_{x+\frac{1}{2}} = \frac{1}{2}[y_x + y_{x+1}] - \frac{1}{16}[\Delta^2 y_{x-1} + \Delta^2 y_x]$$

Interpolation

4.1 INTRODUCTION

In this chapter we shall discuss Gregory-Newton forward, backward interpolation formula, central difference interpolation etc. We shall also discuss Lagrange-Newton divided difference formulae. In last two interpolation formulae, the difference between two successive arguments is not constant, i.e. they are unequally spaced.

Interpolation according to Theile is "*the art of reading between the lines of the table*". It also means insertion or filling up intermediate terms of the series.

Interpolation is the technique of obtaining the value of the independent variable (i.e. argument) when the values of the function corresponding to number of values of the argument are given. The process of computing the value of the function outside the range of given values of the variable is called extrapolation. All interpolation formula are based on the hypothesis that the given function is polynomial or it can be represented by a polynomial with a good degree of approximation.

The following assumptions are to be made while doing interpolation:

1. There should be no sudden jumps or falls in the value of function during the period of interpolation, i.e. the value of function should be either in increasing or decreasing order.

2. The rise and fall in the values should be uniform, e.g. if we are given data regarding the details in various years in particular town and some of the observations are for the years in which epidemics or famine or war took place, the interpolation methods are not applicable for this type of data.

4.2 GREGORY-NEWTON FORWARD INTERPOLATION FORMULA

Let the function $y = f(x)$ take the values $f(a), f(a + h), f(a + 2h), ..., f(a + nh)$ corresponding to the values of x. Let the values of x viz. $a, a + h, ..., a + nh$ are equispaced. That is the difference between any two consecutive argument will remain same. Let $f(x)$ be a polynomial in x of degree n. This polynomial may be written as

$$f(x) = a_0 + a_1 (x - a) + a_2 (x - a) (x - a - h) + a_3 (x - a) (x - a - h)$$
$$(x - a - 2h) + ... + a_n (x - a) (x - a - h) ... (x - a - \overline{n-1}h) \tag{4.1}$$

where $a_0, a_1, a_2, a_3, ..., a_n$ are to be determined.

Put $x = a, a + h, a + 2h, ..., a + nh$ in Eq. (4.1) successively, we get

For $x = a$, $\qquad\qquad f(a) = a_0$ or $a_0 = f(a)$

For $x = a + h$, $\quad f(a + h) = a_0 + ha_1$ or $a_1 = \dfrac{f(a + h) - f(a)}{h} = \dfrac{\Delta f(a)}{h}$

For $x = a + 2h$, $\quad f(a + 2h) = a_0 + a_1 (2h) + a_2 (2h)(h)$

$$\Rightarrow \quad a_2 = \left[\frac{f(a + 2h) - a_0 - a_1 (2h)}{2h^2} \right]$$

$$= \frac{f(a + 2h) - f(a) - 2h \left\{ \dfrac{f(a + h) - f(a)}{h} \right\}}{2h^2}$$

$$= \frac{f(a + 2h) - 2f(a + h) + f(a)}{2h^2} = \frac{\Delta^2 f(a)}{2! h^2}$$

Similarly, $\qquad a_3 = \dfrac{\Delta^3 f(a)}{3! h^3}$ and so on.

Thus, $\qquad a_n = \dfrac{\Delta^n f(a)}{n! h^n}$

Putting these values in Eq. (4.1), we get

$$f(x) = f(a) + (x - a) \frac{\Delta f(a)}{h} + (x - a)(x - a - h) \frac{\Delta^2 f(a)}{2! h^2} + \dots +$$

$$(x - a)(x - a - h) \dots (x - a - \overline{n - 1}h) \frac{\Delta^n f(a)}{n! h^n}$$

Now putting $\qquad u = \dfrac{x - a}{h}$ or $x = a + uh$, we have

$$f(a + uh) = f(a) + \frac{uh \Delta f(a)}{h} + \frac{uh (uh - h) \Delta^2 f(a)}{2! h^2} + \dots +$$

$$\frac{uh (uh - h)(uh - 2h) \dots (uh - \overline{n - 1}h)}{n! h^n} \Delta^n f(a)$$

$$= f(a) + \frac{u \Delta f(a)}{1!} + \frac{u(u - 1) \Delta^2 f(a)}{2!} + \dots +$$

$$\frac{u(u - 1)(u - 2) \dots (u - n + 1)}{n!} \Delta^n f(a) \qquad (4.2)$$

Equation (4.2) is known as Gregory-Newton forward interpolation formula. This method is also known as Newton's forward interpolation formula.

This is used to interpolate the values of $f(x)$ nearer to the beginning values of the table.

4.3 GREGORY-NEWTON BACKWARD INTERPOLATION FORMULA

Let the function $y = f(x)$ takes the values $f(a), f(a + h), f(a + 2h), \dots, f(a + nh)$ corresponding to the values of x. Let the values of x viz. $a, a + h, \dots, a + nh$ are equispaced. That is the difference between any two consecutive arguments will remain same.

Let $f(x)$ be a polynomial of nth degree.

$$f(x) = a_0 + a_1(x - a - nh) + a_2(x - a - nh)(x - a - \overline{n-1}h) + \dots +$$
$$a_n(x - a - nh)(x - a - \overline{n-1}h)\dots(x - a - h) \qquad (4.3)$$

where $a_0, a_1, a_2, \dots, a_n$ are to be determined.

Put $x = a + nh$ in Eq. (4.3), we get

$$f(a + nh) = a_0 \implies a_0 = f(a + nh)$$

Put $x = a + \overline{n-1}h$ in Eq. (4.3), we get

$$f(a + \overline{n-1}h) = a_0 + a_1(-h)$$

$$= f(a + nh) - a_1 h \implies a_1 = \frac{f(a + nh) - f(a + \overline{n-1}h)}{h}$$

$\therefore \qquad\qquad a_1 = \dfrac{\nabla f(a + nh)}{1!\,h}$

Put $x = a + \overline{n-2}h$ in Eq. (4.3), we get

$$f(a + \overline{n-2}h) = a_0 + a_1(-2h) + a_0 + a_2(-2h)(-h)$$

$\implies \qquad 2h^2 a_2 = -a_0 + 2h a_1 + f(a + \overline{n-2}h)$

$$= -f(a + nh) + 2h\left\{\frac{f(a + nh) - f(a + \overline{n-1}h)}{h}\right\} + f(a + \overline{n-2}h)$$

$\implies \qquad a_2 = \dfrac{f(a + nh) - 2f(a + \overline{n-1}h) + f(a + \overline{n-2}h)}{2!\,h^2}$

$$= \frac{\nabla^2 f(a + nh)}{2!\,h^2}$$

Similarly, we get $\qquad a_3 = \dfrac{\nabla^3 f(a + nh)}{3!\,h^3}$

$$\dots \ \dots \ \dots$$
$$\dots \ \dots \ \dots$$

$$a_n = \frac{\nabla^n f(a + nh)}{n!\,h^n}$$

Substituting the values of constants in Eq. (4.3), we get

$$f(x) = f(a + nh) + (x - a - nh)\frac{\nabla f(a + nh)}{1!\,h} + (x - a - nh)$$

$$(x - a - \overline{n-1}h)\frac{\nabla^2 f(a + nh)}{2!\,h^2} + \dots + (x - a - nh)(x - a - \overline{n-1}h)$$

$$\dots (x - a - h)\frac{\nabla^n f(a + nh)}{n!\,h^n} \qquad (4.4)$$

If $x = a + nh + uh$, then

$$u = \frac{x - (a + nh)}{h}$$

So, Eq. (4.4) becomes,

$$f(a + nh + uh) = f(a + nh) + \frac{u\nabla f(a + nh)}{1!} + \frac{u(u + 1)\,\nabla^2 f(a + nh)}{2!} + \dots +$$

$$\frac{u(u + 1)\dots(u + \overline{n-1})}{n!}\nabla^n f(a + nh)$$

Equation (4.4) is known as Gregory-Newton backward interpolation formula. This is used to interpolate the value of $f(x)$ nearer to the end of the table.

Example 1: From the following table of half-yearly premium for policies maturing at different ages, estimate the premium for policies maturing at age of 46.

Age	45	50	55	60	65
Premium (in rupees)	114.84	96.16	83.32	74.48	68.48

Solution: Let us find the premium for policies maturing at age of 46.

Since, 5 data are given, so degree of $f(x)$ is 4.

Here $h = 5$, $a = 45$.

The difference table is:

Age (x)	Premium (in rupees)	$\Delta f(x)$	$\Delta^2 f(x)$	$\Delta^3 f(x)$	$\Delta^4 f(x)$
45	114.84				
		−18.68			
50	96.16		5.84		
		−12.84		−1.84	
55	83.32		4		0.68
		−8.84		−1.16	
60	74.48		2.84		
		−6			
65	68.48				

Now
$$u = \frac{x-a}{h} = \frac{46-45}{5} = 0.2$$

By Gregory-Newton forward difference formula

$$f(a + uh) = f(a) + \frac{u \Delta f(a)}{1!} + \frac{u(u-1)}{2!} \Delta^2 f(a) + \frac{u(u-1)(u-2)}{3!} \Delta^3 f(a)$$
$$+ \frac{u(u-1)(u-2)(u-3)}{4!} \Delta^4 f(a)$$

$\Rightarrow$
$$f(46) = 114.84 + 0.2(-18.68) + \frac{0.2(0.2-1)}{2}(5.84) +$$
$$\frac{0.2(0.2-1)(0.2-2)}{6}(-1.84) + \frac{0.2(0.2-1)(0.2-2)(0.2-3)}{24}(0.68)$$
$$= 110.525632$$

Hence the premium for policies maturing at the age of 46 is Rs. 110.52.

Example 2: From the data given below, find the number of students whose weight is between 40 to 45.

Weight in lbs	0–40	40–60	60–80	80–100	100–120
No. of students	250	120	100	70	50

Solution: First of all we shall prepare the cumulative frequency table as under.

Weight less than (x)	40	60	80	100	120
No. of students $f(x)$	250	370	470	540	590

The difference table is:

x	$f(x)$	$\Delta f(x)$	$\Delta^2 f(x)$	$\Delta^3 f(x)$	$\Delta^4 f(x)$
40	$\boxed{250}$				
		$\boxed{120}$			
60	370		$\boxed{-20}$		
		100		$\boxed{-10}$	
80	470		-30		$\boxed{20}$
		70		10	
100	540		-20		
		50			
120	590				

In order to obtain $f(45)$ (No. of students whose weight less than 45), we take $a = 40$, $h = 20$.

$$\therefore \qquad u = \frac{x - a}{h} = \frac{45 - 40}{10} = 0.5$$

By Gregory-Newton forward difference formula

$$f(a + uh) = f(a) + \frac{u\Delta f(a)}{1!} + \frac{u(u-1)}{2!}\Delta^2 f(a) + \frac{u(u-1)(u-2)}{3!}\Delta^3 f(a)$$
$$+ \frac{u(u-1)(u-2)(u-3)}{4!}\Delta^4 f(a)$$

$$\Rightarrow \qquad f(45) = 250 + 0.5(120) + \frac{0.5(0.5-1)}{2}(-20) + \frac{0.5(0.5-1)(0.5-2)}{6}(-10)$$
$$+ \frac{0.5(0.5-1)(0.5-2)(0.5-3)}{24}(20)$$
$$= 250 + 60 + 2.5 - 0.625 + 0.78125$$
$$= 312.656 \approx 313$$

Hence the number of students having weight less than 45 = 313.

By number of students having weight less than 40 = 250.

Hence number of students having weight between 40 and 45 = 313 – 250 = 63.

Example 3: The following table gives the values of density of saturated water vapours of various temperatures of saturated steam:

Temp °C	100	150	200	250	300
Density	958	917	865	799	712

Find the density when the temperature is 275°C.

Solution: Let us find the density for temperature at 275°C.

Here $h = 50$, $a + nh = 300$.

The difference table is:

Temp °C (x)	Density $f(x)$	$\Delta f(x)$	$\Delta^2 f(x)$	$\Delta^3 f(x)$	$\Delta^4 f(x)$
100	958				
		-41			
150	917		-11		
		-52		-3	
200	865		-14		$\boxed{-4}$
		-66		$\boxed{-7}$	
250	799		$\boxed{-21}$		
		$\boxed{-87}$			
300	$\boxed{712}$				

Now
$$u = \frac{x-(a+nh)}{h} = \frac{275-300}{50} = -0.5$$

By Gregory-Newton backward formula for interpolation, we have

$$f(x) = f(a+nh+uh) = f(a+nh) + \frac{u\nabla f(a+nh)}{1!} + \frac{u(u+1)}{2!}\nabla^2 f(a+nh)$$

$$+ \frac{u(u+1)(u+2)}{3!}\nabla^3 f(a+nh) + \frac{u(u+1)(u+2)(u+3)}{4!}\nabla^4 f(a+nh)$$

$$= 712 + (-0.5)\times(-87) + \frac{(-0.5)(-0.5+1)}{2}\times -21$$

$$+ \frac{(-0.5)(-0.5+1)(-0.5+2)}{6}\times -7$$

$$+ \frac{(-0.5)(-0.5+1)(-0.5+2)(-0.5+3)}{24}\times(-4)$$

$$= 712 + 43.5 + 2.625 + 0.4375 + 0.15625$$
$$= 758.76875 \approx 759.$$

EXERCISE 4.1

1. Given sin 45° = 0.7071, sin 50° = 0.7660, sin 55° = 0.8192, sin 60° = 0.8660, find sin 52° by using Newton forward formula for interpolation and estimate the error. [Ans. 0.7880, error = 0.0000392]

$$\left[\text{Hint: Error} = \frac{u(u-1)(u-2)\dots(u-n)}{(n+1)!}\Delta^{n+1}f(\xi), 45° < \xi < 60°\right]$$

2. Find the form of the function from the following data:

x	0	1	2	3	4
f(x)	3	6	11	18	27

[Ans. $x^2 + 2x + 3$]

3. The table gives the distance in nautical miles of the visible horizon from the given heights in feet above the earth's surface.

x (height)	100	150	200	250	300	350	400
y (distance)	10.63	13.03	15.04	16.81	18.42	19.90	21.27

Find the value of y when (i) x = 218 feet, (ii) x = 410 feet.

[Ans. (i) 15.7 (ii) 21.53 nautical miles]

4. The hourly declination of the moon on a day is given below. Find the declination at $3^h 35^m 15^s$.

Hour	0	1	2	3	4
December	8°29′53.7″	8°18′19.4″	8°6′43.5″	7°55′6.1″	7°43′27.2″

[Ans. 7°48′16″]

5. Find the value of $e^{1.85}$ given $e^{1.7} = 5.4739$, $e^{1.8} = 6.0496$, $e^{1.9} = 6.6859$, $e^{2.0} = 7.3891$, $e^{2.1} = 8.1662$, $e^{2.2} = 9.0250$, $e^{2.3} = 9.9742$. [Ans. 6.3598]

6. Calculate $\sqrt{5.5}$ given $\sqrt{5} = 2.236$, $\sqrt{6} = 2.449$, $\sqrt{7} = 2.646$, and $\sqrt{8} = 2.828$.

[Ans. 2.344]

7. From the following table, find tan (0.12) and tan (0.26).

x	0.10	0.15	0.20	0.25	0.30
tan x	0.1003	0.1511	0.2027	0.2553	0.3093

[Ans. 0.1205, 0.2662]

8. In the table below, the values of y are consecutive terms of a series of which 23.6 is the sixth term. Find the first and tenth terms of the series

x	3	4	5	6	7	8	9	
$f(x)$	4.8	8.4	14.5	23.6	36.2	52.8	73.9	[Ans. 3.1, 100]

9. Obtain the cubic polynomial which takes the values from the following table:

x	0	1	2	3	
$f(x)$	1	2	1	0	[Ans. $2x^2 - 7x^2 + 6x + 1, 3.5$]

and hence find $f(2.5)$.

10. The following table gives the amount of the chemical dissolved in water at different temperatures.

Temp.	10°	15°	20°	25°	30°	35°	
Solubility	19.97	21.51	22.47	23.52	24.65	25.89	[Ans. $x^2 + 2x + 3$]

Compute the amount dissolved at 8° and 22°. [Ans. 18.79, 22.87]

4.4 CENTRAL DIFFERENCE INTERPOLATION FORMULAE

Central difference formulae are used for interpolation near the middle of the tabulated values. Gregory-Newton forward and Gregory-Newton backward formula are not applicable to interpolate near the central value.

4.5 GAUSS'S FORWARD INTERPOLATION FORMULA

By Gregory-Newton forward formula for interpolation

$$f(a + uh) = f(a) + u\Delta f(a) + \frac{u(u-1)}{2!}\Delta^2 f(a) + \frac{u(u-1)(u-2)}{3!}\Delta^3 f(a)$$
$$+ \frac{u(u-1)(u-2)(u-3)}{4!}\Delta^4 f(a) + ... \qquad (4.5)$$

Put $a = 0, h = 1$, we get

$$f(\bar{u}) = f(0) + u\Delta f(0) + \frac{u(u-1)}{2!}\Delta^2 f(0) + \frac{u(u-1)(u-2)}{3!}\Delta^3 f(0)$$
$$+ \frac{u(u-1)(u-2)(u-3)}{4!}\Delta^4 f(0) + ... \qquad (4.6)$$

Now,
$$\Delta^3 f(-1) = \Delta^2 f(0) - \Delta^2 f(-1) \Rightarrow \Delta^2 f(0) = \Delta^3 f(-1) + \Delta^2 f(-1)$$
$$\Delta^4 f(-1) = \Delta^3 f(0) - \Delta^3 f(-1) \Rightarrow \Delta^3 f(0) = \Delta^4 f(-1) + \Delta^3 f(-1)$$
$$\Delta^5 f(-1) = \Delta^4 f(0) - \Delta^4 f(-1) \Rightarrow \Delta^4 f(0) = \Delta^5 f(-1) + \Delta^4 f(-1)$$

and so on.

So, from the Eq. (4.6), we have

$$f(u) = f(0) + u\Delta f(0) + \frac{u(u-1)}{2!}[\Delta^3 f(-1) + \Delta^2 f(-1)]$$
$$+ \frac{u(u-1)(u-2)}{3!}[\Delta^3 f(-1) + \Delta^4 f(-1)] + \frac{u(u-1)(u-2)(u-3)}{4!} \times$$
$$[\Delta^4 f(-1) + \Delta^5 f(-1)] + ...$$
$$= f(0) + u\Delta f(0) + \frac{u(u-1)}{2!}\Delta^2 f(-1) + \frac{u(u-1)}{2} \times$$

$$\left\{1+\frac{u-2}{3}\right\}\Delta^3 f(-1)+\frac{u(u-1)(u-2)}{6}\left\{1+\frac{u-3}{4}\right\}\Delta^4 f(-1)+$$

$$+\frac{u(u-1)(u-2)(u-3)}{4!}\Delta^5 f(-1)+...$$

$$= f(0)+u\Delta f(0)+\frac{u(u-1)}{2!}\Delta^2 f(-1)+\frac{(u+1)u(u-1)}{3!}\Delta^3 f(-1)$$

$$+\frac{(u+1)u(u-1)(u-2)}{4!}\Delta^4 f(-1)+\frac{u(u-1)(u-2)(u-3)}{4!}\Delta^5 f(-1)+... \quad (4.7)$$

But, $\qquad\qquad \Delta^5 f(-2) = \Delta^4 f(-1)-\Delta^4 f(-2)$

So, $\qquad\qquad \Delta^4 f(-1) = \Delta^4 f(-2)+\Delta^5 f(-2)$

Putting the above value in Eq. (4.7), we get

$$f(u) = f(0)+u\Delta f(0)+\frac{u(u-1)}{2!}\Delta^2 f(-1)+\frac{(u+1)u(u-1)}{3!}\Delta^3 f(-1)$$

$$+\frac{(u+1)u(u-1)(u-2)}{4!}\left\{\Delta^4 f(-2)+\Delta^5 f(-2)\right\}+\frac{u(u-1)(u-2)(u-3)}{4!}\Delta^5 f(-1)+...$$

$$= f(0)+u\Delta f(0)+\frac{u(u-1)}{2!}\Delta^2 f(-1)+\frac{(u+1)u(u-1)}{3!}\Delta^3 f(-1)$$

$$+\frac{(u+1)u(u-1)(u-2)}{4!}\Delta^4 f(-2)+... \quad (4.8)$$

Equation (4.8) is known as Gauss's forward interpolation formula.

Remarks: (i) This formula involves odd differences below the central line and even differences on the central line.

$$f(0)..................\Delta^2 f(-1) ....\,\Delta^4 f(-2)....................\Delta^6 f(-3) \qquad \text{Central line}$$
$$\searrow \Delta f(0)\nearrow \qquad \searrow \Delta^3 f(-1)\nearrow \qquad \searrow \Delta^5 f(-2)$$

(ii) This formula is used to evaluate the values of f for u between 0 and 1/2.

4.6 GAUSS'S BACKWARD INTERPOLATION FORMULA

By Gregory-Newton forward formula for interpolation

$$f(a+uh) = f(a)+u\Delta f(a)+\frac{u(u-1)}{2!}\Delta^2 f(a)+\frac{u(u-1)(u-2)}{3!}\Delta^3 f(a)+... \quad (4.9)$$

Put $a = 0$ and $h = 1$, we get

$$f(u) = f(0)+u\Delta f(0)+\frac{u(u-1)}{2!}\Delta^2 f(0)+\frac{u(u-1)(u-2)}{3!}\Delta^3 f(0)$$

$$+\frac{u(u-1)(u-2)(u-3)}{4!}\Delta^4 f(0)+... \quad (4.10)$$

Now, $\qquad \Delta f(0) = \Delta f(-1)+\Delta^2 f(-1)$

$$\Delta^2 f(0) = \Delta^2 f(-1)+\Delta^3 f(-1)$$

$$\Delta^3 f(0) = \Delta^3 f(-1) + \Delta^4 f(-1)$$

$$\Delta^4 f(0) = \Delta^4 f(-1) + \Delta^5 f(-1) \text{ and so on.}$$

So, from Eq. (4.10), we get

$$f(u) = f(0) + u[\Delta f(-1) + \Delta^2 f(-1)] + \frac{u(u-1)}{2!}[\Delta^2 f(-1) + \Delta^3 f(-1)]$$

$$+ \frac{u(u-1)(u-2)}{3!}[\Delta^3 f(-1) + \Delta^4 f(-1)]$$

$$+ \frac{u(u-1)(u-2)(u-3)}{4!}[\Delta^4 f(-1) + \Delta^5 f(-1)] + ... \tag{4.11}$$

$$= f(0) + u\Delta f(-1) + u\left(1 + \frac{u-1}{2}\right)\Delta^2 f(-1) + \frac{u(u-1)}{2}\left(1 + \frac{u-2}{3}\right)\Delta^3 f(-1)$$

$$+ \frac{u(u-1)(u-2)(u-3)}{4!} \times \Delta^4 f(-1) + ...$$

$$+ \frac{u(u-1)(u-2)(u-3)}{4!} \times \Delta^5 f(-1) + ...$$

$$= f(0) + u\Delta f(-1) + \frac{(u+1)u}{2!}\Delta^2 f(-1) + \frac{(u+1)u(u-1)}{3!}\Delta^3 f(-1)$$

$$+ \frac{(u+1)u(u-1)(u-2)}{4!}\Delta^4 f(-1) + ... \tag{4.12}$$

Now, $\Delta^3 f(-1) = \Delta^3 f(-2) + \Delta^4 f(-2)$

and $\Delta^4 f(-1) = \Delta^4 f(-2) + \Delta^5 f(-2)$ and so on.

So, Eq. (4.12) gives

$$f(u) = f(0) + u\Delta f(-1) + \frac{(u+1)u}{2!}\Delta^2 f(-1) + \frac{(u+1)u(u-1)}{3!} \times$$

$$\{\Delta^3 f(-2) + \Delta^4 f(-2)\} + \frac{(u+1)u(u-1)(u-2)}{4!}\{\Delta^4 f(-2) + \Delta^5 f(-2)\} + ...$$

$$\Rightarrow \quad f(u) = f(0) + u\Delta f(-1) + \frac{u(u+1)}{2!}\Delta^2 f(-1) + \frac{(u+1)u(u-1)}{3!}\Delta^3 f(-2)$$

$$+ \frac{(u+2)(u+1)u(u-1)}{4!}\Delta^4 f(-2) + ... \tag{4.13}$$

Equation (4.13) is known as Gauss's backward interpolation formula.

Remarks: (i) This formula involves odd differences above the central line and even differences on the central line.

$$f(0) \cdots\cdots \overset{\Delta f(-1)}{\underset{\Delta^2 f(-1)}{\nearrow\searrow}} \cdots\cdots \overset{\Delta^3 f(-2)}{\underset{\Delta^4 f(-2)}{\nearrow\searrow}} \cdots\cdots \overset{\Delta^5 f(-3)}{\underset{\Delta^6 f(-3)}{\nearrow\searrow}} \quad \text{Central line}$$

(ii) This formula is useful when u lies between $-1/2$ and 0.

4.7 STIRLING'S INTERPOLATION FORMULA

By Gauss's forward interpolation formula

$$f(u) = f(0) + u\Delta f(0) + \frac{u(u-1)}{2!}\Delta^2 f(-1) + \frac{(u+1)\,u(u-1)}{3!}\Delta^3 f(-1)$$

$$+ \frac{(u+1)\,u(u-1)\,(u-2)}{4!}\Delta^4 f(-2) + \dots \qquad (4.14)$$

By Gauss's backward interpolation formula

$$f(u) = f(0) + u\Delta f(-1) + \frac{(u+1)u}{2!}\Delta^2 f(-1) + \frac{(u+1)\,u(u-1)}{3!}\Delta^3 f(-2)$$

$$+ \frac{(u+2)\,(u+1)\,u(u-1)}{4!}\Delta^4 f(-2) + \dots \qquad (4.15)$$

Adding Eqs (4.14) and (4.15) and then dividing by 2, we have

$$f(u) = f(0) + u\left\{\frac{\Delta f(0) + \Delta f(-1)}{2}\right\} + \frac{u^2}{2!}\Delta^2 f(-1)$$

$$+ \frac{(u+1)\,u(u-1)}{3!}\left\{\frac{\Delta^3 f(-1) + \Delta^3 f(-2)}{2}\right\} + \frac{u^2(u^2-1)}{4!}\Delta^4 f(-2) + \dots \qquad (4.16)$$

The Eq. (4.16) is called the Stirling's formula. This formula is useful when $-\dfrac{1}{2} < u < \dfrac{1}{2}$. It gives best estimate when $-\dfrac{1}{4} < u < \dfrac{1}{4}$.

4.8 BESSEL'S INTERPOLATION FORMULA

By Gauss's forward interpolation formula

$$f(u) = f(0) + u\Delta f(0) + \frac{u(u-1)}{2!}\Delta^2 f(-1) + \frac{(u+1)\,u(u-1)}{3!}\Delta^3 f(-1)$$

$$+ \frac{(u+1)\,u(u-1)\,(u-2)}{4!}\Delta^4 f(-2) + \dots \qquad (4.17)$$

By Gauss's backward interpolation formula

$$f(u) = f(0) + u\Delta f(-1) + \frac{(u+1)u}{2!}\Delta^2 f(-1) + \frac{(u+1)\,u(u-1)}{3!}\Delta^3 f(-2)$$

$$+ \frac{(u+2)\,(u+1)\,u(u-1)}{4!}\Delta^4 f(-2) + \dots \qquad (4.18)$$

In Eq. (4.18), shift the origin to 1 by replacing u by $u-1$ and adding 1 to each argument 0, -1, -2, ..., we get

$$f(u) = f(1) + (u-1)\,\Delta f(0) + \frac{u(u-1)}{2!}\Delta^2 f(0) + \frac{u(u-1)\,(u-2)}{3!}\Delta^3 f(-1)$$

$$+ \frac{(u+1)\,u(u-1)\,(u-2)}{4!}\Delta^4 f(-1) + \dots \qquad (4.19)$$

Adding Eqs (4.17) and (4.19) and dividing it by 2, we get

$$f(u) = \left\{\frac{f(0)+f(1)}{2}\right\} + \left\{\frac{u+(u-1)}{2}\right\}\Delta f(0) + \frac{u(u-1)}{2!}\left\{\frac{\Delta^2 f(-1)+\Delta^2 f(0)}{2}\right\}$$

$$+ \frac{u(u-1)}{3!}(u+1+u-2)\frac{\Delta^3 f(-1)}{2} + \frac{(u+1)\,u(u-1)\,(u-2)}{4!} \times$$

$$\left\{\frac{\Delta^4 f(-2)+\Delta^4 f(-1)}{2}\right\} + \dots$$

Finally, we get

$$f(u) = \left\{\frac{f(0)+f(1)}{2}\right\} + \left(u-\frac{1}{2}\right)\Delta f(0) + \frac{u(u-1)}{2!}\left\{\frac{\Delta^2 f(-1)+\Delta^2 f(0)}{2}\right\}$$

$$+ \frac{(u-1)\left(u-\frac{1}{2}\right)u}{3!}\Delta^3 f(-1) + \frac{(u+1)\,u(u-1)\,(u-2)}{4!}\left\{\frac{\Delta^4 f(-2)+\Delta^4 f(-1)}{2}\right\} + \dots$$

$$(4.20)$$

The Eq. (4.20) is called the Bessel's formula.

Remarks: (i) This formula involves odd differences below the central line and means of even differences on and below the central line.

$$
\begin{array}{cccccc}
\text{Central line} & \left.f(0)\right) & & \left.\Delta^2 f(-1)\right) & & \left.\Delta^4 f(-2)\right) \\
\text{Next line} & \left.f(1)\right) & \Delta f(0) & \left.\Delta^2 f(0)\right) & \Delta^3 f(-1) & \left.\Delta^4 f(-1)\right) \\
& \uparrow & & \uparrow & & \uparrow \\
& \text{average} & & \text{average} & & \text{average}
\end{array}
$$

(ii) This formula gives accurate value u between 0.25 and 0.75.

4.9 LAPLACE-EVERETT'S INTERPOLATION FORMULA

By Gauss's forward interpolation formula

$$f(u) = f(0) + u\Delta f(0) + \frac{u(u-1)}{2!}\Delta^2 f(-1) + \frac{(u+1)\,u(u-1)}{3!}\Delta^3 f(-1)$$

$$+ \frac{(u+1)\,u(u-1)\,(u-2)}{4!}\Delta^4 f(-2) + \frac{(u+2)\,(u+1)\,u(u-1)\,(u-2)}{5!}\Delta^5 f(-2) + \dots$$

$$(4.21)$$

We have
$$\Delta f(0) = f(1) - f(0)$$
$$\Delta^3 f(-1) = \Delta^2 f(0) - \Delta^2 f(-1)$$
$$\Delta^5 f(-2) = \Delta^4 f(-1) - \Delta^4 f(-2)$$

So, from Eq. (4.21)

$$f(u) = f(0) + u\{f(1) - f(0)\} + \frac{u(u-1)}{2!}\Delta^2 f(-1) + \frac{(u+1)u(u-1)}{3!} \times$$

$$\{\Delta^2 f(0) - \Delta^2 f(-1)\} + \frac{(u+1)\,u(u-1)\,(u-2)}{4!}\Delta^4 f(-2)$$

$$\frac{(u+2)\,(u+1)\,u(u-1)\,(u-2)}{5!}\{\Delta^4 f(-1) - \Delta^4 f(-2)\} + \dots$$

$$= (1 - u)\, f(0) + uf(1) + \frac{(u + 1)\, u(u - 1)}{3!}\, \Delta^2 f(0) - \frac{u(u - 1)\,(u - 2)}{3!}\, \Delta^2 f(-1)$$

$$+ \frac{(u + 2)\,(u + 1)\, u(u - 1)\,(u - 2)}{5!}\, \Delta^4 f(-1) -$$

$$\frac{(u + 1)\, u(u - 1)\,(u - 2)\,(u - 3)}{5!}\, \Delta^4 f(-2) + \ldots$$

$$= \left\{ uf(1) + \frac{(u + 1)\, u(u - 1)}{3!}\, \Delta^2 f(0) + \frac{(u + 2)\,(u + 1)\, u(u - 1)\,(u - 2)}{5!}\, \Delta^4 f(-1) + \ldots \right\}$$

$$+ \left\{ (1 - u)\, f(0) + \frac{(1 - u + 1)\,(1 - u)\,(1 - u - 1)}{3!}\, \Delta^2 f(-1) \right.$$

$$\left. + \frac{(1 - u + 2)\,(1 - u + 1)\,(1 - u)\,(1 - u - 1)\,(1 - u - 2)}{5!}\, \Delta^4 f(-2) + \ldots \right\}$$

$$f(u) = \left\{ uf(1) + \frac{(u + 1)\, u(u - 1)}{3!}\, \Delta^2 f(0) + \frac{(u + 2)\,(u + 1)\, u(u - 1)\,(u - 2)}{5!}\, \Delta^4 f(-1) + \ldots \right\}$$

$$+ \left\{ vf(0) + \frac{(v + 1)\, v(v - 1)}{3!}\, \Delta^2 f(-1) \right.$$

$$\left. + \frac{(v + 2)\,(v + 1)\, v(v - 1)\,(v - 2)}{5!} \times \Delta^4 f(-2) + \ldots \right\} \tag{4.22}$$

where $v = 1 - u$.

The Eq. (4.22) is known as Laplace-Everett's formula. It gives best estimate when $u > 1/2$. It is used to compute any entry against any argument between 0 and 1. It is useful when intervening values in successive intervals are required.

Example 4: Apply Gauss's forward formula and Gauss's backward formula to find the value of $f(x)$ at $x = 3.75$ from the table:

x	2.5	3	3.5	4	4.5	5.0
$f(x)$	24.145	22.043	20.225	18.644	17.262	16.047

Solution: The difference is constructed as follows:

u	x	$f(x)$	$\Delta f(x)$	$\Delta^2 f(x)$	$\Delta^3 f(x)$	$\Delta^4 f(x)$	$\Delta^5 f(x)$
−2	2.5	24.145					
			−2.102				
−1	3.0	22.043		0.284			
			−1.818		−0.47		
0	3.5	20.225		0.237		0.009	
			−1.581		−0.38		−0.003
1	4.0	18.644		0.199		0.006	
			−1.382		−0.32		
2	4.5	17.262		0.167			
			−1.215				
3	5.0	16.047					

Here $h = 0.5$, $x = 3.75$. On taking the origin at 3.5 = a (say), we have

$$u = \frac{x - a}{h} = \frac{3.75 - 3.5}{0.5} = 0.5$$

(i) Using Gauss's forward formula

$$f(u) = f(0) + u\Delta f(0) + \frac{u(u-1)}{2!}\Delta^2 f(-1) + \frac{(u+1)\,u(u-1)}{3!}\Delta^3 f(-1)$$

$$+ \frac{(u+1)\,(u-1)\,(u-2)}{4!}\Delta^4 f(-2) + ...$$

We have $\quad f(0.5) = 20.225 + 0.5(-1.581) + \dfrac{0.5(-0.5)}{2} \times 0.237$

$$+ \frac{1.5 \times 0.5 \times (-0.5)}{6} \times -0.38 + \frac{1.5 \times 0.5 \times (-0.5) \times (-1.5)}{24} \times 0.009$$

$$= 20.225 - 0.7905 - 0.029625 + 0.002375 + 0.0002109375$$

$$= 19.407$$

(ii) Using Gauss's backward formula

$$f(u) = f(0) + u\Delta f(-1) + \frac{(u+1)u}{2!}\Delta^2 f(-1) + \frac{(u+1)u(u-1)}{3!}\Delta^2 f(-2) + ...$$

$$f(0.5) = 20.225 + 0.5 \times (-1.818) + \frac{(0.5+1)\,(0.5)}{2} \times 0.237$$

$$+ \frac{(1.5)\,(0.5)\,(-0.5)}{6} \times (-0.47) + \frac{(2.5)\,(1.5)\,(0.5)\,(-0.5)}{24} \times 0.009$$

$$= 19.407$$

Example 5: The following table gives the value of probability integral

$f(x) = \dfrac{2}{\sqrt{\pi}} \displaystyle\int_0^x e^{-x^2}\,dx$ for certain values of x. Find the value of this integral when $x =$ 0.5437 using (i) Stirling's formula (ii) Bessel's formula and (iii) Everett's formula.

x	0.51	0.52	0.53	0.54	0.55	0.56	0.57
$f(x)$	0.5292437	0.5378987	0.5464641	0.5549392	0.5633233	0.5716157	0.5798158

Solution: The difference is constructed as follows:

u	x	$f(x)$	$\Delta f(x)$	$\Delta^2 f(x)$	$\Delta^3 f(x)$	$\Delta^4 f(x)$
−3	0.51	0.5292437				
			0.0086550			
−2	0.52	0.5378987		−0.0000896		
			0.0085654		−0.0000007	
−1	0.53	0.5464641		−0.0000903		0
			0.0084751		−0.0000007	
0	0.54	0.5549392		−0.0000910		0
			0.0083841		−0.0000007	
1	0.55	0.5633233		−0.0000917		0.0000001
			0.0082924		−0.0000006	
2	0.56	0.5716157		−0.0000923		
			0.0082001			
3	0.57	0.5798158				

Here $h = 0.1$, $x = 0.5437$. Taking the origin at $0.54 = a$, we have

$$u = \frac{x-a}{h} = \frac{0.5437 - 0.54}{0.1} = 0.37$$

and

$$v = 1 - u = 1 - 0.37 = 0.63$$

(i) By Stirling's formula

$$f(x = 0.5437) = f(u = 0.37) = f(0) + u\left\{\frac{\Delta f(0) + \Delta f(-1)}{2}\right\} + \frac{u^2}{2!}\Delta^2 f(-1)$$

$$+ \frac{(u+1)u(u-1)}{3!}\left\{\frac{\Delta^3 f(-1) + \Delta^3 f(-2)}{2}\right\} + \frac{u^2(u^2-1)}{4!}\Delta^4 f(-2) + ...$$

$$= 0.5549392 + 0.37\left(\frac{0.0083841 + 0.0084751}{2}\right) + \frac{(0.37)^2}{2} \times (-0.0000910)$$

$$+ \frac{0.37[(0.37)^2 - 1]}{6}\left(\frac{-0.0000007 - 0.0000007}{2}\right)$$

$$= 0.5549392 + 0.003118952 - 0.00000623 + 0.00000004$$

$$= 0.55805196$$

(ii) By Bessel's formula

$$f(x = 0.5437) = \frac{1}{2}[f(0) + f(+1)] + \left(u - \frac{1}{2}\right)\Delta f(0) + \frac{u(u-1)}{2}\left[\frac{\Delta^2 f(-1) + \Delta^2 f(0)}{2}\right]$$

$$+ \frac{\left(u - \frac{1}{2}\right)u(u-1)}{3!}\Delta^3 f(-1) + ...$$

$$= \frac{1}{2}[0.5549392 + 0.5633233] + (0.37 - 0.5)(0.0083841)$$

$$+ \frac{0.37(-0.63)}{2}\left(\frac{-0.0000910 - 0.0000917}{2}\right)$$

$$+ \frac{0.37(-0.63)(0.37 - 0.5)}{6}(-0.0000007) + ...$$

$$= 0.55913125 - 0.001089933 + 0.0000106468 = 0.55805196$$

(iii) By Everett's formula

$$f(x = 0.5437) = \left[uf(1) + \frac{u(u^2 - 1)}{6}\Delta^2 f(0) + \frac{u(u^2 - 1)(u^2 - 4)}{120}\Delta^4 f(-1) + ...\right]$$

$$+ \left[vf(0) + \frac{v(v^2 - 1)}{6}\Delta^2 f(-1) + \frac{v(v^2 - 1)(v^2 - 4)}{120}\Delta^4 f(-2) + ...\right]$$

$$= (0.37)(0.5633233) + \frac{0.37[(0.37)^2 - 1]}{6}[-0.0000917] + ...$$

$$+ (0.63)(0.5549392) + (0.63)\frac{[(0.63)^2 - 1]}{6}[-0.0000910] + ...$$

$$= [0.208429621 + 0.00000488 + ...] + [0.349611696 + 0.00000576262 + ...]$$

$$= 0.55805195$$

Example 6: Given

θ	0°	5°	10°	15°	20°	25°	30°
tan θ	0	0.0875	0.1763	0.2679	0.364	0.4663	0.5774

Find the value of tan 16° using Stirling and Bessel's formula.

Solution: Construct the difference table as follows:

u	θ	$10^4f(\theta)$	$10^4\Delta f(\theta)$	$10^4\Delta^2 f(\theta)$	$10^4\Delta^3 f(\theta)$	$10^4\Delta^4 f(\theta)$	$10^4\Delta^5 f(\theta)$	$10^4\Delta^6 f(\theta)$
−3	0	0						
			875					
−2	5	875		13				
			888		15			
−1	10	1763		28		2		
			916		17		−2	
0	15	2679		45		0		11
			961		17		9	
1	20	3640		62		9		
			1023		26			
2	25	4663		88				
			1111					
3	30	5774						

Here $h = 5$, $x = 16°$. Taking the origin at $15° = a$

$$\therefore \qquad u = \frac{x-a}{h} = \frac{16-15}{5} = 0.2$$

(i) Using Stirling's formula

$$y_u = y_0 + \frac{u}{2}[\Delta y_0 + \Delta y_{-1}] + \frac{u^2}{2}\Delta^2 y_{-1}$$

$$+ \frac{u(u^2-1)}{3!}\frac{1}{2}[\Delta^3 y_{-1} + \Delta^3 y_{-3}] + \frac{u^2(u^2-1)}{4!}\Delta^4 y_{-2}$$

$$+ \frac{u(u^2-1)(u^2-4)}{5!}\frac{1}{2}[\Delta^5 y_{-2} + \Delta^5 y_{-3}] + \frac{u^2(u^2-1)(u^2-4)}{6!}\Delta^6 y_{-3}$$

Hence $\quad y_{0.2} = \dfrac{2679}{10^4} + \dfrac{0.2}{2}\dfrac{[961+916]}{10^4} + \dfrac{(0.2)^2}{2}\dfrac{45}{10^4} + \dfrac{(0.2)(0.04-1)}{6}\left[\dfrac{17+17}{10^4}\right]$

$$+ \frac{(0.04)(0.04-1)}{24}\left(\frac{0}{10^4}\right) + \frac{(0.2)(0.04-1)(0.04-4)}{120}\frac{(-2+9)}{2\times10^4}$$

$$+ \frac{(0.04)(0.04-1)(0.04-4)}{720}\frac{11}{10^4}$$

$$= 0.2679 + 0.01877 + 0.00009 + \text{negligible quantities}$$

$$= 0.28676$$

Thus the estimated value of tan 16° = 0.28676.

(ii) Using Bessel's formula, we have

$$y_u = \frac{1}{2}[y_0 + y_1] + \left(u - \frac{1}{2}\right)\Delta y_0 + \frac{u(u-1)}{2}\frac{[\Delta^2 y_{-1} + \Delta^2 y_0]}{2}$$

$$+ \frac{\left(u - \frac{1}{2}\right)u(u-1)}{3!}\Delta^3 y_{-1} + \ldots$$

$$y_{0.2} = \frac{1}{2}[0.2679 + 0.3640] + (0.2 - 0.5) \times 0.0961 +$$

$$\frac{(0.2)(0.2-1)}{2}\frac{[0.0045 + 0.0028]}{2} + \frac{(0.2-0.5)(0.2)(0.2-1)}{6}$$

$$\times 0.0017 + \frac{(0.2-0.5)(0.2)(0.2-1)(0.2-2)}{24} \times \left(\frac{0.001-1}{2}\right)$$

$$= 0.31595 - 0.02883 - 0.000584 + 0.0000136 + 0.00000648$$

$$= 0.28655608$$

EXERCISE 4.2

1. Use Gauss's formula to find y_{30} for the given data:
$y_{21} = 18.4708, y_{25} = 17.8144, y_{29} = 17.1070, y_{33} = 16.3432, y_{37} = 15.5154.$
[Ans. $y_{30} = 16.9216$]

2. Use Stirling's formula to find y_{28} for the data:
$y_{20} = 49225, y_{25} = 48360, y_{30} = 47236, y_{35} = 45926, y_{40} = 44306$ [Ans. $y_{28} = 47692$]

3. Use Bessel's formula to obtain y_{25} for the given data:
$y_{20} = 2854, y_{24} = 3162, y_{28} = 3544, y_{32} = 3992$ [Ans. $y_{25} = 3250.875$]

4. Using Gauss's backward formula obtain $\sqrt{12516}$ for the given data:

$\sqrt{12500} = 111.803399, \sqrt{12510} = 111.84811,$

$\sqrt{12520} = 111.892806, \sqrt{12530} = 111.937483$ [Ans. 111.874930]

4.10 LAGRANGE'S INTERPOLATION FORMULA

Let $y = f(x)$ be a function such that $f(x)$ takes the values $f(x_0), f(x_1), f(x_2), \ldots, f(x_n),$ corresponding to the arguments $x_0, x_1, x_2, \ldots, x_n$. That is $y_i = f(x_i), i = 0, 1, 2, \ldots, n.$

Now, there are $(n + 1)$ paired values $(x_i, y_i), i = 0, 1, 2, \ldots, n$ and hence $f(x)$ can be represented by a polynomial function of degree n in x. We will assume that $f(x)$ as follows

$$f(x) = A_0(x - x_1)(x - x_2) \ldots (x - x_n) + A_1(x - x_0)(x - x_2) \ldots (x - x_n) + \ldots + A_n(x - x_0)(x - x_1) \ldots (x - x_{n-1}) \tag{4.23}$$

where $A_0, A_1, A_2, \ldots, A_n$ are constants.

The relation in Eq. (4.23) is true for all values of x. To determine A_0 put $x = x_0$

$\therefore$ $f(x_0) = A_0(x_0 - x_1)(x_0 - x_2) \ldots (x_0 - x_n)$

or $A_0 = \dfrac{f(x_0)}{(x_0 - x_1)(x_0 - x_2) \ldots (x_0 - x_n)}$

Similarly, put $x = x_1$ in Eq. (4.23), we have

$$f(x_1) = A_1(x_1 - x_0)(x_1 - x_2) \ldots (x_1 - x_n)$$

or

$$A_1 = \frac{f(x_1)}{(x_1 - x_0)(x_1 - x_2)\ldots(x_1 - x_n)}$$

In the same way, we get

$$A_2 = \frac{f(x_2)}{(x_2 - x_0)(x_2 - x_1)\ldots(x_2 - x_n)}$$

$$\ldots \ldots \ldots$$

$$\ldots \ldots \ldots$$

$$A_n = \frac{f(x_n)}{(x_n - x_0)(x_n - x_1)\ldots(x_n - x_{n-1})}$$

Substituting these values of A's in Eq. (4.23), we get

$$f(x) = \frac{(x - x_1)(x - x_2)\ldots(x - x_n)}{(x_0 - x_1)(x_0 - x_2)\ldots(x_0 - x_n)} f(x_0) + \frac{(x - x_0)(x - x_2)\ldots(x - x_n)}{(x_1 - x_0)(x_1 - x_2)\ldots(x_1 - x_n)}$$

$$f(x_1) + \ldots + \frac{(x - x_0)(x - x_1)\ldots(x - x_{n-1})}{(x_n - x_0)(x_n - x_1)\ldots(x_n - x_{n-1})} f(x_n) \qquad (4.24)$$

which is Lagrange's interpolation formula for unequal intervals.

4.10.1 Algorithm for Lagrange's Interpolation Formula

1. Read x, n
2. For $i = 1$ to $(n + 1)$ in steps of 1 do read x_i, f_i end for
   Remarks: The above statements read x_is and the corresponding values of f_is.
3. Sum $\leftarrow 0$
4. For $i = 1$ to $(n + 1)$ in steps of 1 do
5. prodfunc $\leftarrow 1$
6. for $j = 1$ to $(n + 1)$ in steps of 1 do
7. if $(j \neq i)$ then
   prodfunc $\leftarrow$ prodfunc $\times (x - x_j)/(x - x_j)$
   end for
8. Sum $\leftarrow$ Sum $+ f_i \times$ prodfunc
   Remarks: Sum is the value of f at x
   end for
9. Write x, sum
10. Stop

4.10.2 Different Form of Lagrange's Interpolation Formula

The Lagrange's interpolation formula can also be expressed in the form

$$P_n(x) = \sum_{i=0}^{n} \frac{\pi_n(x)}{(x - x_i)\,\pi_n'(x_i)} f(x_i)$$

where
$$\pi_n(x) = (x - x_0)(x - x_1) \dots (x - x_n)$$

and
$$\pi'_n(x) = \frac{d}{dx}[\pi_n(x)]$$

$$\pi_n(x) = (x - x_0)(x - x_1) \dots (x - x_n)$$

Differentiating this and substituting $x = x_i$, we get

$$P_n(x) = \frac{\pi_n(x)}{x - x_0} \frac{f(x_0)}{\pi'_n(x_0)} + \frac{\pi_n(x)}{x - x_1} \frac{f(x_1)}{\pi'_n(x_1)} + \dots + \frac{\pi_n(x)}{x - x_n} \frac{f(x_n)}{\pi'_n(x_n)}$$

$$= \sum_{i=0}^{n} \frac{\pi_n(x)}{\pi'_n(x_i)} \frac{f(x_i)}{x - x_i}$$

Example 7: Find the unique polynomial $P(x)$ of degree 2 such that $P(1) = 1$, $P(3) = 27$, $P(4) = 64$ by using the Lagrange method of interpolation.

Solution: By Lagrange's interpolation formula, we have

$$f(x) = \frac{(x - x_1)(x - x_2)}{(x_0 - x_1)(x_0 - x_2)} f(x_0) + \frac{(x - x_0)(x - x_2)}{(x_1 - x_0)(x_1 - x_2)} f(x_1) + \frac{(x - x_0)(x - x_1)}{(x_2 - x_0)(x_2 - x_1)} f(x_2)$$

$$= \frac{(x - 3)(x - 4)}{(1 - 3)(1 - 4)} \times 1 + \frac{(x - 1)(x - 4)}{(3 - 1)(3 - 4)} \times 27 + \frac{(x - 1)(x - 3)}{(4 - 1)(4 - 3)} \times 64$$

$$= \frac{1}{6}(x^2 - 7x + 12) - \frac{1}{2}(x^2 - 5x + 4) \times 27 + \frac{1}{3}(x^2 - 4x + 3) \times 64$$

$$= 8x^2 - 19x + 12$$

Example 8: Find the form of function $f(x)$ of the following table using Lagrange's method.

x	0	1	4	5
f(x)	8	11	68	123

Solution: Here $x_0 = 0, x_1 = 1, x_2 = 4, x_3 = 5$ and $f(x_0) = 9, f(x_1) = 11, f(x_2) = 68, f(x_3) = 123$. Hence by Lagrange's formula

$$f(x) = \frac{(x - 1)(x - 4)(x - 5)}{(0 - 1)(0 - 4)(0 - 5)} \times 8 + \frac{(x - 0)(x - 4)(x - 5)}{(1 - 0)(1 - 4)(1 - 5)} \times 11 +$$

$$\frac{(x - 0)(x - 1)(x - 5)}{(4 - 0)(4 - 1)(4 - 5)} \times 68 + \frac{(x - 0)(x - 1)(x - 4)}{(5 - 0)(5 - 1)(5 - 4)} \times 123$$

$$= -\frac{8}{20}(x^3 - 10x^2 + 29x - 20) + \frac{11}{12}(x^3 - 9x^2 + 20x) -$$

$$\frac{68}{12}(x^3 - 6x^2 + 5x) + \frac{123}{60}(x^3 - 5x^2 + 4x)$$

$$= -3.1x^3 + 19.5x^2 - 13.4x + 8$$

Example 9: Find the parabola of the form $y = ax^2 + bx + c$ passing through the points $(0, 0)$, $(1, 1)$ and $(2, 20)$.

Solution: By Lagrange's interpolation formula

$$f(x) = \frac{(x - 1)(x - 2)}{(0 - 1)(0 - 2)} \times 0 + \frac{(x - 0)(x - 2)}{(1 - 0)(1 - 2)} \times 1 + \frac{(x - 0)(x - 1)}{(2 - 0)(2 - 1)} \times 20$$

$$= 9x^2 - 8x.$$

Example 10: Using Lagrange's formula, prove that

$$y_1 = y_3 - 0.3 (y_5 - y_{-3}) + 0.2 (y_{-3} - y_{-5}) \text{ nearby.}$$

Solution: For the arguments $-5, -3, 3, 5$, the Lagrange's formula is

$$y_x = \frac{(x+3)(x-3)(x-5)}{(-5+3)(-5-3)(-5-5)} y_{-5} + \frac{(x+5)(x-3)(x-5)}{(-3+5)(-3-3)(-3-5)} \cdot y_{-3}$$

$$+ \frac{(x+5)(x+3)(x-5)}{(3+5)(3+3)(3-5)} y_3 + \frac{(x+5)(x+3)(x-3)}{(5+5)(5+3)(5-3)} \cdot y_5$$

Put $x = 1$

$$y_1 = \frac{(1+3)(1-3)(1-5)}{(-2)(-8)(-10)} y_{-5} + \frac{(1+5)(1-3)(1-5)}{(2)(-6)(-8)} y_{-3} +$$

$$\frac{(1+5)(1+3)(1-5)}{(8)(6)(-2)} y_3 + \frac{(1+5)(1+3)(1-3)}{(10)(8)(2)} y_5$$

$$= -0.2y_{-5} + 0.5y_{-3} + y_3 - 0.3y_5$$

$$y_1 = y_3 - 0.3 (y_5 - y_{-3}) + 0.2 (y_{-3} - y_{-5})$$

Example 11: The mode of a certain frequency curve $y = f(x)$ is very near to $x = 9$ and the values of the frequency density $f(x)$ for $x = 8.9, 9, 9.3$ are respectively 0.30, 0.35 and 0.25. Calculate the approximate value of the mode.

Solution: We are given that

x	8.9	9.0	9.3
$f(x)$	0.3	0.35	0.25

By Lagrange's interpolation formula

$$f(x) = \frac{(x-9)(x-9.3)}{(8.9-9)(8.9-9.3)} \times 0.3 + \frac{(x-8.9)(x-9.3)}{(9-8.9)(9-9.3)} \times 0.35 +$$

$$\frac{(x-8.9)(x-9)}{(9.3-8.9)(9.3-9)} = \frac{1}{12}(-25x^2 + 453.5x - 2052.3)$$

To get the mode, $f'(x) = 0$ and $f''(x) = -ve$

$$\therefore \qquad f'(x) = 0 \Rightarrow \frac{1}{12}(-50x + 453.5) = 0$$

i.e. $\qquad x = 9.07$

$$f''(9.07) = \frac{1}{12} \times -50 = -ve$$

Hence $f(x)$ is maximum at $x = 9.07$.

Therefore, mode is 9.07.

Example 12: Certain corresponding values of x and $\log_{10}x$ are $(300, 2.4771)$, $(304, 2.4829)$, $(305, 2.4843)$, $(307, 2.4871)$. Find $\log_{10}301$.

Solution: Here $x_0 = 300, x_1 = 304, x_2 = 305, x_3 = 307$ and $x = 301$. $f(x_0) = 2.4771, f(x_1) = 2.4829, f(x_2) = 2.4843, f(x_3) = 2.4871$.

$$f(x) = \frac{(301-304)(301-305)(301-307)}{(300-304)(300-305)(300-307)} \times 2.4771$$

$$+ \frac{(301-300)(301-305)(301-307)}{(304-300)(304-305)(304-307)} \times 2.4829$$

$$+ \frac{(301-300)(301-304)(301-307)}{(305-300)(305-304)(305-307)} \times 2.4843$$

$$+ \frac{(301-300)(301-304)(301-305)}{(307-300)(307-304)(307-305)} \times 2.4871$$

$$= 1.2739 + 4.9658 - 4.4717 + 0.7106 = 2.4786.$$

EXERCISE 4.3

1. Find the cubic Lagrange's interpolating polynomial from the following data:

x	0	1	2	5
$f(x)$	2	3	12	147

[Ans. $x^3 + x^2 - x + 2$}

2. The values of $f(x)$ are given at a, b and c. Show that the maximum value is obtained by

$$x = \frac{(b^2 - c^2) f(a) + (c^2 - a^2) f(b) + (a^2 - b^2) f(c)}{2[(b-c) f(a) + (c-a) f(b) + (a-b) f(c)]}$$

3. Using Lagrange's formula, prove that

$$y_0 = \frac{1}{2}(y_1 + y_{-1}) - \frac{1}{8}\left[\frac{1}{2}(y_3 - y_1) - \frac{1}{2}(y_{-1} - y_{-3})\right]$$

4. Find U_3, given that $U_0 = 540$, $U_1 = 556$, $U_2 = 520$ and $U_4 = 385$. [Ans. 465.25]

5. Using Lagrange's formula, prove that

$$y_3 = 0.05 (y_0 + y_6) - 0.3 (y_1 + y_5) + 0.75 (y_2 + y_4)$$

6. Apply Lagrange's formula to find $f(4)$ given that:
$f(1) = 2, f(2) = 4, f(3) = 8, f(4) = 16, f(7) = 128$
Explain why the result differs from 2^5. [Ans. 32.93334]

7. For equidistant values U_{-1}, U_0, U_1 and U_2 being given, a value is interpolated by Lagrange's formula. Show that it may be written in the form

$$U_x = y U_0 + x U_1 + \frac{y(y^2 - 1)}{3!} \Delta^2 U_{-1} + x \frac{(x^2 - 1)}{3!} \Delta^2 U_0, \text{ where } x + y = 1.$$

8. Find the value of $\tan 33°$ by Lagrange's formula, if $\tan 30° = 0.5774$, $\tan 32° = 0.6249$, $\tan 35° = 0.7002$, $\tan 38° = 0.7813$. [Ans. 0.64942]

9. The following are measurements t made as a curve recorded by the oscillograph representing a change of current i due to a change in the conditions of an electric current

t	1.2	2.0	2.5	3.0
i	1.36	0.58	0.34	0.20

Using Lagrange's formula, find i at $t = 1.6$. [Ans. 0.8932]

10. By Lagrange's formula, find the form of $f(x)$ from the following data:

x	0	2	3	4
$f(x)$	659	705	729	804

Also find maximum value of $f(x)$.

$$\left[\text{Ans. } \frac{151}{24}x^3 - \frac{249}{8}x^2 + \frac{721}{12}x + 659, \text{ no real value exists} \right]$$

4.11 DIVIDED DIFFERENCE

Let the function $y = f(x)$, assume that $f(x_0), f(x_1), ..., f(x_n)$ corresponding to the arguments $x_0, x_1, ..., x_n$ respectively where the difference between two consecutive arguments need not be equal.

The first divided difference of $f(x)$ for the arguments x_0, x_1 is defined as

$$\frac{f(x_1) - f(x_0)}{x_1 - x_0}.$$

It is denoted by $f[x_0, x_1]$ or $[x_0, x_1]$ or $\underset{x_1}{\Delta} f(x_0) = \frac{f(x_1) - f(x_0)}{x_1 - x_0}$ \hfill (4.25)

In the same notatioon, we have

$$f[x_1, x_2] = \underset{x_2}{\Delta} f(x_1) = \frac{f(x_2) - f(x_1)}{x_2 - x_1} \text{ and so on.}$$

The second divided difference for x_0, x_1, x_2 is defined as

$$f[x_0, x_1, x_2] = \underset{x_1, x_2}{\Delta^2} f(x_0) = \frac{f[x_1, x_2] - f[x_0, x_1]}{x_2 - x_0} \hfill (4.26)$$

In the same way, we define the third divided difference of $f(x)$ for four arguments x_0, x_1, x_2, x_3 as

$$f[x_0, x_1, x_2, x_3] = \underset{x_1, x_2, x_3}{\Delta^3} f(x_0) = \frac{f[x_1, x_2, x_3] - f[x_0, x_1, x_2]}{x_3 - x_0} \hfill (4.27)$$

and so on.

Equations (4.25)–(4.27) refers to the divided differences of order one, two and three respectively.

4.11.1 Newton's Divided Difference Interpolation Formula

Let $f(x_0), f(x_1), f(x_2), ..., f(x_n)$ be the values of $f(x)$ corresponding to the arguments $x_0, x_1, x_2, ..., x_n$.

By the first divided difference

$$f[x, x_0] = \frac{f(x) - f(x_0)}{x - x_0}$$

we have $\qquad f(x) = f(x_0) + (x - x_0) f[x, x_0]$

Now, second divided difference is given by

$$f[x, x_0, x_1] = \frac{f[x, x_0] - f[x_0, x_1]}{x - x_1}$$

Therefore $\qquad f[x, x_0] = f[x_0, x_1] + (x - x_1) f[x, x_0, x_1]$

$\Rightarrow \qquad \dfrac{f(x) - f(x_0)}{x - x_0} = f[x_0, x_1] + (x - x_1) f[x, x_0, x_1]$

$\Rightarrow \qquad f(x) = f(x_0) + (x - x_0) f[x_0, x_1] + (x - x_0)(x - x_1) f[x, x_0, x_1]$

Similarly

$$f(x) = f(x_0) + (x - x_0) f[x_0, x_1] + (x - x_0) (x - x_1) f[x_0, x_1, x_2] +$$
$$(x - x_0) (x_0 - x_1) (x - x_2) f[x, x_0, x_1, x_2]$$

Proceding in the same way, we get

$$f(x) = f(x_0) + (x - x_0) f[x_0, x_1] + (x - x_0) (x - x_1) f[x_0, x_1, x_2] +$$
$$(x - x_0) (x_0 - x_1) (x - x_2) f[x_0, x_1, x_2, x_3] + ... +$$
$$(x - x_0) (x - x_1) ... (x - x_{n-1}) f[x_0, x_1, x_2, ..., x_n] +$$
$$(x - x_0) (x - x_1) ... (x - x_{n-1}) (x - x_n) f[x, x_0, x_1, ..., x_n]$$

If $f(x)$ is a polynomial of degree n, then

$$f(x, x_0, x_1, ..., x_n) = 0$$

Hence, we have

$$f(x) = f(x_0) + (x - x_0) f[x_0, x_1] + (x - x_0) (x - x_1) f[x_0, x_1, x_2] +$$
$$(x - x_0) (x - x_1) (x - x_2) f[x_0, x_1, x_2, x_3] + ... +$$
$$(x - x_0) (x - x_1) ... (x - x_{n-1}) f[x_0, x_1, x_2, ..., x_n]$$

This equation is called Newton's divided difference interpolation formula for unequal intervals. With a mild variation in notations, it is usually written as

$$f(x) = f(x_0) + (x - x_0) \Delta f(x_0) + (x - x_0) (x - x_1)$$
$$\Delta^2 f(x_0) + ... + (x - x_0) (x - x_1) ... (x - x_{n-1}) \Delta^n f(x_0)$$

4.11.2 Algorithm for Newton's Divided Difference Interpolation Formula

1. Read $n, x, x_0, ..., x_n, f(x_0), ..., f(x_n)$
2. For $i = 0$ to n do
3. $d_{i,0} = f(x_i)$
4. end for
5. For $i = 1$ to n do
6. For $j = 1$ to i do
7. $d_{i,j} = (d_{i,j-1} - d_{i-1,j-1})/(x_i - x_{i-1})$
8. end for
9. end for
10. Sum = do 0.0
11. Prod = 1.0
12. For $i = 1$ to n do
13. Prod = prod * $(x - x_{i-1})$
14. Sum = Sum + $d_{i,i}$ * prod
15. end for
16. Print 'Approximation at $x =$ ', x, 'is' sum.

Example 13: Find the polynomial of the lowest possible degree which assumes the values 3, 12, 15, –21 when x has the values 3, 2, 1, –1 respectively.

Solution: The divided difference table is as under:

x	$f(x)$	$\Delta f(x)$	$\Delta^2 f(x)$	$\Delta^3 f(x)$
-1	-21			
		$\dfrac{15+21}{1+1}=18$		
1	15		$\dfrac{-3-18}{2+1}=-7$	
		$\dfrac{12-15}{2-1}=-3$		$\dfrac{-3+7}{3+1}=1$
2	12		$\dfrac{-9+3}{3-1}=-3$	
		$\dfrac{3-12}{3-2}=-9$		
3	3			

Now applying Newton's divided diference formula, we get

$$f(x) = f(x_0) + (x - x_0)\,\Delta f(x_0) + (x - x_0)(x - x_1)\,\Delta^2 f(x_0)$$
$$\qquad + (x - x_0)(x - x_1)(x - x_2)\,\Delta^3 f(x_0)$$
$$= -21 + [x - (-1)] \times 18 + [x - (-1)](x - 1) \times (-7) + [x - (-1)](x - 1)(x - 2) \times 1$$
$$= -21 + (x + 1)(18) + (x + 1)(x - 1)(-7)(x + 1)(x - 1)(x - 2)$$
$$= x^3 - 9x^2 + 13x + 6.$$

Example 14: The following are the mean temperature (Fahrenheit) on three days, 30 days apart, round the period of summer and winter. Estimate the approximate dates and the value of the maximum and minimum temperature.

	Summer		Winter	
Days	Date	Temperature	Date	Temperature
0	15th June	58.8	16th December	40.7
30	15th July	63.4	15th January	38.1
60	14th August	62.5	14th February	39.3

Solution: Suppose 1 unit = 30 days

For summer, the divided difference table is as under

x	Temp $f(x)$	$\Delta f(x)$	$\Delta^2 f(x)$
0	58.8		
		$\dfrac{63.4-58.8}{1-0}=4.6$	
1	63.4		$\dfrac{-0.9-4.6}{2-0}=-2.75$
		$\dfrac{62.5-63.4}{2-1}=-0.9$	
2	62.5		

By Newton's divided difference formula, we have

$$f(x) = f(x_0) + (x - x_0)\,\Delta f(x_0) + (x - x_0)(x - x_1)\,\Delta^2 f(x_0)$$
$$= 58.8 + (x - 0)(4.6) + (x - 0)(x - 1)(-2.75)$$
$$= -2.75x^2 + 7.35x + 58.8$$

For maximum or minimum temperature, we have $f'(x) = 0$ which gives

$$-5.5x + 7.35 = 0$$
$$\Rightarrow \qquad x = 1.336$$

Further, at $x = 1.336$, we observe $f''(x) = -5.5 < 0$.

This implies that $x = 1.336$ is a point of maxima.

Now 1 unit = 30 days $\Rightarrow$ 1.336 units = 30 × 1.336 days = 40.08 days.

Required date of maximum temperature = 15th June + 40 days = 25th July and maximum temperature = $-2.75 (1.336)^2 + 7.35 (1.336) + 58.8 = 63.71°F$.

For winter, the divided difference table is as under:

x	Temp $f(x)$	$\Delta f(x)$	$\Delta^2 f(x)$
0	40.7		
		$\dfrac{38.1 - 40.7}{1 - 0} = -2.6$	
1	38.1		$\dfrac{1.2 + 2.6}{2 - 0} = 1.9$
		$\dfrac{39.3 - 38.1}{2 - 1} = 1.2$	
2	39.3		

By Newton's divided difference formula, we have

$$f(x) = f(x_0) + (x - x_0)\,\Delta f(x_0) + (x - x_0)(x - x_1)\,\Delta^2 f(x_0)$$
$$= 40.7 + (x - 0)(-2.6) + (x - 0)(x - 1)(1.9)$$
$$= 1.9x^2 - 4.5x + 40.7$$

For maximum or minimum temperature we have $f'(x) = 0$ which gives

$$3.8x - 4.5 = 0 \Rightarrow x = 1.184$$

Further at $x = 1.184$, we observe $f''(x) = 3.8 > 0$

$\Rightarrow$ The point $x = 1.184$ is a point of minima.

Again 1 unit = 30 days $\Rightarrow$ 1.184 units = 30 × 1.184 days = 35.52 days.

Required date of minimum temperature = 16th December + 36 days = 21st January and minimum temperature = $1.9 (1.184)^2 - 4.5 × 1.184 + 40.7 = 38.036°F$

Example 15: The following table gives same relation between steam pressure and temperature. Find the pressure at temperature 372.1°.

T	361°	367°	378°	387°	399°
P	154.9	167.9	191.0	212.5	244.2

Solution: We will form the divided difference table below:

T P	ΔP	$\Delta^2 P$	$\Delta^3 P$	$\Delta^4 P$
361°154.9				
	$\dfrac{167 - 154.9}{367 - 361} = 2.0166$			
367°167.0		$\dfrac{2.1818 - 2.0166}{378 - 361} = 0.0097$		
	$\dfrac{191 - 167}{378 - 367} = 2.1818$		$\dfrac{0.01035 - 0.0097}{387 - 361} = 0.000025$	
378°191.0		$\dfrac{2.3888 - 2.1818}{387 - 367} = 0.01035$		0.0000007
	$\dfrac{212.5 - 191}{387 - 378} = 2.3888$		$\dfrac{0.0120 - 0.01035}{399 - 367} = 0.000052$	
387°212.5		$\dfrac{2.6417 - 2.3888}{399 - 378} = 0.0120$		
	$\dfrac{244.2 - 212.5}{399 - 387} = 2.6417$			
399°244.2				

By Newton's divided difference formula

$$P(T = 372.1°) = 154.9 + (11.1) \times 2.0166 + (11.1)\,(5.1)\,(0.0097) +$$
$$(11.1)\,(5.1)\,(-5.9)\,(0.000025) + (11.1)\,(5.1)\,(-5.9)\,(-14.9) \times (0.0000007)$$
$$= 177.83$$

Example 16: Prove that $\Delta^3_{bcd}\left(\dfrac{1}{a}\right) = -\dfrac{1}{abcd}$.

Solution: Construct the following divided difference table:

x	$f(x)$	$\Delta f(x)$	$\Delta^2 f(x)$	$\Delta^3 f(x)$
a	$1/a$			
		$\dfrac{\frac{1}{b}-\frac{1}{a}}{b-a} = -\dfrac{1}{ab}$		
b	$1/b$		$(-1)\left[\dfrac{\frac{1}{bc}-\frac{1}{ba}}{c-a}\right] = \dfrac{1}{abc}$	
		$\dfrac{\frac{1}{c}-\frac{1}{b}}{c-b} = -\dfrac{1}{bc}$		$\dfrac{\frac{1}{bcd}-\frac{1}{abc}}{d-a} = -\dfrac{1}{abcd}$
c	$1/c$		$(-1)\left[\dfrac{\frac{1}{dc}-\frac{1}{bc}}{d-b}\right] = \dfrac{1}{bcd}$	
		$\dfrac{\frac{1}{d}-\frac{1}{c}}{d-c} = -\dfrac{1}{dc}$		
d	$1/d$			

From the table, we observe that

$$\Delta^3_{bcd}\frac{1}{a} = -\frac{1}{abcd}$$

Example 17: Use Newton's divided difference formula to find $f(7)$ if $f(3) = 24, f(5) = 120, f(8) = 504, f(9) = 720$ and $f(12) = 1716$.

Solution: Construct the following dividing difference table:

x	$f(x)$	$\Delta f(x)$	$\Delta^2 f(x)$	$\Delta^3 f(x)$	$\Delta^4 f(x)$
3	24				
		$\dfrac{120-24}{5-3} = 48$			
5	120		$\dfrac{128-48}{8-3} = 16$		
		$\dfrac{504-120}{8-5} = 128$		$\dfrac{22-16}{9-3} = 1$	
8	504		$\dfrac{216-128}{9-5} = 22$		$\dfrac{1-1}{12-3} = 0$
		$\dfrac{720-504}{9-8} = 216$		$\dfrac{29-22}{12-5} = 1$	
9	720		$\dfrac{332-216}{12-8} = 29$		
		$\dfrac{1716-720}{12-9} = 332$			
12	1716				

By Newton's divided difference formula

$$f(x = 7) = 24 + (7-3) \times 48 + (7-3)\,(7-5) \times 16 + (7-3)\,(7-5)\,(7-8) \times 1 +$$
$$(7-3)\,(7-5) \times (7-8)\,(7-9) \times 0$$
$$= 24 + 4 \times 48 + 4 \times 2 \times 16 + 4 \times 2 \times -1 \times 1 + 0$$
$$= 336$$

Example 18: Calculate $f(3)$ using Newton's divided difference from the following data:

x	0	1	2	4	5	6
f	1	14	15	5	6	19

Solution: Construct the following divided difference table:

x	$f(x)$	$\Delta f(x)$	$\Delta^2 f(x)$	$\Delta^3 f(x)$	$\Delta^4 f(x)$	$\Delta^5 f(x)$
0	1					
		$\dfrac{14-1}{1-0}=13$				
1	14		$\dfrac{1-13}{2-0}=-6$			
		$\dfrac{15-14}{2-1}=1$		$\dfrac{-2+6}{4-0}=1$		
2	15		$\dfrac{-5-1}{4-1}=-2$			$\dfrac{1-1}{5-0}=0$
		$\dfrac{5-15}{4-2}=-5$		$\dfrac{2+2}{5-1}=1$		0
4	5		$\dfrac{1+5}{5-2}=2$		$\dfrac{1-1}{6-1}=0$	
		$\dfrac{6-5}{5-1}=1$		$\dfrac{6-2}{6-2}=1$		
5	6		$\dfrac{13-1}{6-4}=6$			
		$\dfrac{19-6}{6-5}=13$				
6	19					

By Newton's divided differerence formula

$$f(3) = 1 + (x-0) \times 13 + (x-0)(x-1) \times -6 + (x-0)(x-1)(x-2) \times 1 +$$
$$(x-0)(x-1)(x-2)(x-4) \times 0 + (x-0)(x-1)(x-4)(x-5) \times 0$$
$$= 1 + 13x - 6x^2 + 6x + x^3 - 3x^2 + 2x$$
$$= x^3 - 9x^2 + 21x + 1$$

Example 19: For the following table, find $f(x)$ as a polynomial in powers of $(x-5)$:

x	0	2	3	4	7	9
f	4	26	58	112	466	922

Solution:

x	$f(x)$	$\Delta f(x)$	$\Delta^2 f(x)$	$\Delta^3 f(x)$
0	4			
		$\dfrac{26-4}{2-0}=11$		
2	26		$\dfrac{32-11}{3-0}=7$	
		$\dfrac{58-26}{3-2}=32$		$\dfrac{11-7}{4-0}=1$
3	58		$\dfrac{54-32}{4-2}=11$	
		$\dfrac{112-58}{4-3}=54$		$\dfrac{16-11}{7-2}=1$
4	112		$\dfrac{118-54}{7-3}=16$	
		$\dfrac{466-112}{7-4}=118$		$\dfrac{22-16}{9-3}=1$
7	466		$\dfrac{228-118}{9-4}=22$	
		$\dfrac{922-466}{9-7}=228$		
9	922			

By Newton's divided difference formula, we get

$$f(x) = 4 + (x - 0) \times 11 + (x - 0)(x - 2) \times 7 + (x - 0)(x - 2)(x - 3) \times 1$$
$$= x^3 + 2x^2 + 3x + 4$$

In order to express it in power of $(x - 5)$, we use synthetic division, we get:

5	1	2	3	4
		5	35	190
5	1	7	38	194
		5	60	
5	1	12	98	
		5		
	1	17		

∴ $x^3 + 2x^2 + 3x + 4 = (x - 5)^3 + 17(x - 5)^2 + 98(x - 5) + 194$

EXERCISE 4.4

1. Obtain the Newton's divided difference interpolating polynomial and hence find $f(6), f(5)$ and $f(8)$.

x	3	7	9	10
$f(x)$	168	120	72	63

   [Ans. $x^3 - 21x^2 + 119x - 27$; 147, 168, 93]

2. Use Newton's divided difference formula to find the interpolating polynomial and hence evaluate $y(9.5)$ from the given data:

x	7	8	9	10
$f(x)$	3	1	1	9

   [Ans. $x^3 - 23x^2 + 174x - 431$; 3.625]

3. Given the data $f(1) = 4, f(2) = 5, f(7) = 5, f(8) = 4$. Find the values of $f(6)$ and also the value of x for which $f(x)$ is maximum or minimum.   [Ans. 5.66, $x = 4.5$]

4. Compute $f'(3)$ from the following table:

x	1	2	4	8	10
$f(x)$	0	1	5	21	27

   [Ans. 1.97916]

5. Find the function u_x in powers of $x - 1$ given that $u_0 = 8, u_1 = 11, u_4 = 68, u_5 = 123$.
   [Ans. $(x - 1)^3 + 2(x - 1)^2 + 4(x - 1) + 11$]

6. Using divided difference table, find $f(x)$ which takes values 1, 4, 40, 85 as $x = 0, 1, 3, 4$.   [Ans. $x^3 + x^2 + x + 1$]

7. Using the following table, find $f(x)$ as a polynomial in powers of $(x - 6)$. Also find $f'(6), f''(6)$ and $f'''(6)$.   [Ans. $73 + 54(x - 6) + 13(x - 6)^2 + (x - 6)^3$]

8. Find $\log_{10} 323.5$ given

x	321.0	322.8	324.2	325.0
$\log_{10} x$	2.50651	2.50893	2.51081	2.51188

   [Ans. 2.5099063]

9. Find $y(x = 5.60275)$ from the table

x	5.600	5.602	5.605	5.607	4.608
y	0.77556588	0.77682686	0.77871250	0.77996571	0.78059114

   [Ans. 0.777298926]

10. Compute $f'(3)$ from the following table

x	1	2	4	8	10
$f(x)$	0	1	5	21	27

[Ans. 1.97916]

4.12 THE ERROR IN POLYNOMIAL INTERPOLATION: INTRODUCTION

We know that for a function $f(x)$ that has continuous derivatives up to and including the $(n + 1)$ th order, the Taylor's formula in the neighbourhood of the point $x = x_0$ and $x_0 \in [a, b]$ can be written as

$$f(x) = f(x_0) + (x - x_0) f'(x_0) + \frac{(x - x_0)^2}{2!} f''(x_0) + \dots + \frac{(x - x_0)^n}{n!} f^{(n)}(x_0) + R_{n+1}(x) \quad (4.28)$$

where the remainder term $R_{n+1}(x)$ is given as

$$R_{n+1}(x) = \frac{(x - x_0)^{n+1}}{(n+1)!} f^{(n+1)}(\xi), \; x_0 < \xi < x \quad (4.29)$$

when we neglect $R_{n+1}(x)$ in Eq. (4.28), we get a polynomial of degree n

$$P(x) = f(x_0) + \frac{(x - x_0)}{1!} f'(x_0) + \frac{(x - x_0)^2}{2!} f''(x_0) + \dots + \frac{(x - x_0)^n}{n!} f^{(n)}(x_0)$$
$$(4.30)$$

The polynomial $P(x)$ is said to be an interpolating polynomial satisfying the $(n + 1)$ conditions

$$f^{(k)}(x_0) = p^{(k)}(x_0), \, k = 0, 1, 2, \dots, n \quad (4.31)$$

which are said to be the interpolating conditions. The conditions in Eq. (4.31) may be replaced by more general conditions such as the value of $P(x)$ and/or its certain order derivatives coincide with the corresponding values of $f(x)$ and the same order derivatives, at one or more distinct tabular points, $a \le x_0 < x_1 < \dots < x_{n-1} < x_n \le b$. In general, the deviation or remainder due to replacement of a function $f(x)$ by another function $P(x)$ can be written as

$$E(f, x) = f(x) - P(x) \quad (4.32)$$

In approximation, we measure the deviation of the given function $f(x)$ from the approximating function $P(x)$, $\forall \, x \in [a, b]$.

Now we are giving some methods for constructing the interpolating polynomials and approximating functions for a given function $f(x)$.

4.12.1 Error in Taylor Series Interpolation

If we write the polynomial $P(x)$ as the Taylor series expansion, for the function $f(x)$ about a point x_0, $x_0 \in [a, b]$ in the form

$$P(x) = f(x_0) + \frac{(x - x_0)}{1!} f'(x_0) + \frac{(x - x_0)^2}{2!} f''(x_0) + \dots + \frac{1}{n!} (x - x_0)^n f^{(n)}(x_0)$$

then, $P(x)$ may be regarded as an interpolating polynomial of degree n, satisfying the conditions

$$P^{(k)}(x_0) = f^{(k)}(x_0), \, k = 0, 1, 2, \dots, n$$

The term
$$R_{n+1} = \frac{1}{(n+1)!}(x-x_0)^{n+1} f^{(n+1)}(\xi), \quad x_0 < \xi < x$$

that has been neglected in the Taylor expansion is called the remainder or truncation error.

The number of terms to be included in the Taylor expansion may be determined by the acceptable error. If this error is $\varepsilon > 0$ and the series is truncated at the term $f^{(n)}(x_0)$, then, we may write

$$\frac{1}{(n+1)!}|x-x_0|^{n+1}\left|f^{(n+1)}(\xi)\right| \le \varepsilon$$

or
$$\frac{1}{(n+1)!}|x-x_0|^{n+1} M_{n+1} \le \varepsilon$$

where, $M_{n+1} = \max_{x \in [a,b]}\left|f^{(n+1)}(x)\right|$.

4.12.2 Truncation Error or Remainder Term

Truncation error or the errors that result from using an approximation in place of an exact mathematical procedures, e.g.

$$e^x = 1 + x + \frac{x^2}{2!} + \frac{x^3}{3!} + \dots + \frac{x^n}{n!} + \frac{x^{n+1}}{(n+1)!}$$

is the exact mathematical formulation for e^x. Here $\dfrac{x^{n+1}}{(n+1)!}$ is the truncation error or

remainder term and the $e^x = 1 + x + \dfrac{x^2}{2!} + \dfrac{x^3}{3!} + \dots + \dfrac{x^n}{n!}$ is the approximation for e^x.

4.12.3 Error Bound

We know that $E(f, x) = f(x) - P(x)$. We are mostly interested in the maximum of $|E(f, x)|$ over the interval $[a, b]$. This maximum is called the errror bound. To prove the error in Lagrange interpolation polynomial, we need two basic theorems, one is Weierstrass approximation theorem and another one is Roll's theorem.

4.12.4 Weierstrass Approximation Theorem

Let $f \in C[a, b]$ and $\varepsilon > 0$. Then there exists a polynomial p of sufficiently high degree such that

$$|f(x) - p(x)| < \varepsilon$$

4.12.5 Roll's Theorem

If $f \in C[a, b]$ and f' exists on (a, b), and if $f(a) = f(b) = 0$, then there exists a number $\xi \in (a, b)$ such that $f'(\xi) = 0$.

4.12.6 Truncation Error in Lagrange Interpolation Polynomial

The truncation error in Lagrange interpolation polynomial is given by
$$E_n(f; x) = f(x) - P(x)$$

Since $E_n(f; x) = 0$ at $x = x_i$, $i = 0, 1, 3, ..., n$, then for $x \in [a, b]$ and $x \neq x_i$, we define a function $\phi(t)$ as

$$\phi(t) = f(t) - P(t) - [f(x) - P(x)] \frac{(t - x_0)(t - x_1)...(t - x_n)}{(x - x_0)(x - x_1)...(x - x_n)}$$

We see that $\phi(t) = 0$ at $t = x$ and $t = x_i$, $i = 0, 1, 2, ..., n$.

Using the Roll's theorem repeatedly for $\phi(t)$, $\phi'(t)$, ..., and $\phi^{(n)}(t)$, we get $\phi^{(n+1)}(\xi) = 0$ where ξ is some point such that

$$\min(x_0, x_1, ..., x_n, x) < \xi < \max(x_0, x_1, ..., x_n, x)$$

Differentiating $\phi(t)$, $(n + 1)$ times with respect to t, we obtain

$$\phi^{(n+1)}(t) = f^{(n+1)}(t) - \frac{(n+1)![f(x) - P(x)]}{(x - x_0)(x - x_1)..(x - x_n)}$$

Let $\phi^{(n+1)}(\xi) = 0$ and solving for $f(x)$, we get

$$f(x) = P(x) + \frac{w(x) f^{(n+1)}(\xi)}{(n+1)!}$$

where $w(x) = (x - x_0)(x - x_1) ... (x - x_n)$.

Hence the truncation error in Lagrange interpolation is given by

$$E_n(f; x) = \frac{w(x) f^{(n+1)}(\xi)}{(n+1)!}$$

where $\min(x_0, x_1, ..., x_n, x) < \xi < \max(x_0, x_1, ..., x_n, x)$.

4.12.7 Truncation Error in Newton's Divided Difference Interpolation Formula

Consider the first divided difference

$$f[x_0, x_1] = \frac{f(x_0) - f(x_1)}{x_0 - x_1}$$

$$\underset{x \to x_0}{\text{Lim}} \frac{f(x) - f(x_0)}{x - x_0} = f'(x_0)$$

or

$$= \frac{f(x_1) - f(x_0)}{x_1 - x_o}$$

We can assume without any loss of generality, that $x_0 < x_1$. If we assume, in addition, that $f(x)$ is continuously differentiable in the interval $[x_0, x_1]$, then this divided difference equals to the derivative of $f(x)$ at an intermediate point, i.e.

$$f[x_0, x_1] = f'(\xi), \quad x_0 < \xi < x_1.$$

In other words, the first order divided difference can be viewed as an approximation of the first derivative of $f(x)$ in the interval. This notion can be extended to divided differences of higher order.

We know that the divided difference formula for interpolation is

$$f(x) = f[x_0] + (x - x_0) f[x_0, x_1] + (x - x_0)(x - x_1) f[x_0, x_1, x_2] + ... +$$
$$(x - x_0)(x - x_1) ... (x - x_{n-1}) f[x_0, x_1, x_2, ..., x_n] +$$
$$(x - x_0)(x - x_1) ... (x - x_n) f[x, x_0, x_1, ..., x_n]$$

or

$$f(x) = f(x_0) + (x - x_0) f[x_0, x_1] + (x - x_0)(x - x_1) f[x_0, x_1, x_2] + ... +$$
$$(x - x_0)(x - x_1) ... (x - x_{n-1}) f[x_0, x_1, ..., x_n] + R_n(x)$$

with the error term as

$$R_n(x) = \frac{(x - x_0)(x - x_1)\dots(x - x_n)\, f^{(n+1)}(\xi)}{(n + 1)!}$$

for some $\xi \in (x_0, x_n)$.

This error term is same as in case of Lagrange formula.

Remarks: **Lemma:** Let $x, x_0, \dots, x_{n-1}$ be $(n + 1)$ distinct points. Let $a = \min(x, x_0, \dots, x_{n-1})$ and $b = \max(x, x_0, \dots x_{n-1})$. Assume that $f(y)$ has a continuous derivative of order n in the interval (a, b), then

$$f[x_0, x_1, \dots, x_{n-1}, x] = \frac{f^{(n)}(\xi)}{n!}$$

where $\xi \in (a, b)$.

Example 20: Let $f(x) = \ln(1 + x)$, $x_0 = 1$ and $x_1 = 1.1$. Use linear interpolation to calculate an approximate value of $f(1.04)$ and obtain a bound on the truncation error.

Solution: We have $f(x) = \ln(1 + x)$

$$f(1.0) = \ln(1 + 1) = \ln(2.0) = 0.693147$$
$$f(1.1) = \ln(1 + 1.1) = \ln(2.1) = 0.741937$$

The Lagrange interpolating polynomial is

$$P_1(x) = \frac{(x - 1.1)}{1 - 1.1}(0.693147) + \frac{(x - 1)}{1.1 - 1}(0.741937)$$

$$P_1(1.04) = \frac{(1.04 - 1.1)}{1 - 1.1}(0.693147) + \frac{(1.04 - 1)}{1.1 - 1}(0.741937)$$

$$= 0.712663$$

The error in linear interpolation is given by

$$\text{Truncation error} = \frac{1}{2!}(x - x_0)(x - x_1)\, f''(\xi), \, x_0 < \xi < x_1$$

Hence, we obtain the bound on the error as

$$|\,\text{Truncation error (TE)}\,| \le \frac{1}{2}\max_{1 \le x \le 1.1} |(x - x_0)(x - x_1)|\max_{1 \le x \le 1.1}|f''(x)|$$

Since the maximum of $(x - x_0)(x - x_1)$ is obtained at $x = \dfrac{x_0 + x_1}{2}$ and $f''(x) = -\dfrac{1}{(1 + x)^2}$.

We get $$|\,\text{TE}\,| \le \frac{1}{2}\frac{(x_1 - x_0)^2}{4}\max_{1 \le x \le 1.1}\left|\frac{1}{(1 + x)^2}\right|$$

$$= \frac{(0.1)^2}{8} \cdot \frac{1}{4} = 0.0003125$$

Example 21: Determine an appropriate step size to use, in the construction of a table of $f(x) = (1 + x)^6$ on $[0, 1]$. The truncation error for linear interpolation is to be bounded by 5×10^{-5}.

Solution: The maximum error in the linear interpolation is given by $\dfrac{h^2 M_2}{8}$, where $M_2 = \max\limits_{0 \le x \le 1} |f''(x)|$

$$M_2 = \max\limits_{0 \le x \le 1} |30(1+x)^4| = 480$$

We choose h so that

$$60h^2 \le 0.00005$$

which gives $h \le 0.00091$.

Example 22: i. Show that the truncation error of quadratic interpolation in an equidistant is bounded by $\left(\dfrac{h^3}{9\sqrt{3}}\right) \max |f'''(\xi)|$.

ii. We want to set up an equidistant table of the function $f(x) = x^2 \ln x$ in the interval $5 \le x \le 10$. The function values are rounded to 5 decimals. Give the step size h which is to be used to yield a total error less than 10^{-5} on quadratic interpolation in this table.

Solution: i. Let x_{i-1}, x_i, x_{i+1} denote three consecutive points with step size h.

The truncation error of the quadratic Lagrange interpolation is bounded by

$$\text{TE} = \frac{(x - x_{i-1})(x - x_i)(x - x_{i+1})}{3!} f'''(\xi), \quad x_{i-1} < \xi < x_{i+1}.$$

Substitute $\dfrac{x - x_i}{h} = u$, then

$$x - x_{i-1} = x - (x_i - h) = x - x_i + h = uh + h = (u+1)h$$
$$x - x_{i+1} = x - (x_i + h) = x - x_i - h = uh - h = (u-1)h$$

and $(x - x_{i-1})(x - x_i)(x - x_{i+1}) = (u+1)h\, uh\,(u-1)h = u(u-1)(u+1)h^3$

$$\text{TE} = \frac{u(u-1)(u+1)h^3}{6} f'''(\xi), \quad -1 < \xi < 1$$

The extreme values of

$$g(u) = u(u-1)(u+1) = u^3 - u$$

we have

$$g'(u) = 0 \Rightarrow 3u^2 - 1 = 0$$

$\Rightarrow$

$$u = \pm \frac{1}{\sqrt{3}}$$

$$\max |g(u)| = h^3 \max\limits_{-1 \le u \le 1} |u(u^2 - 1)| = \frac{2h^3}{3\sqrt{3}}$$

Hence the truncation error of the quadratic interpolation is given by

$$|\text{TE}| = \frac{h^3}{9\sqrt{3}} \max |f'''(\xi)|$$

ii. We have

$$f(x) = x^2 \ln(x)$$

$$f'(x) = \frac{x^2}{x} + 2x \ln(x) = x + 2x \ln(x)$$

$$f''(x) = 1 + 2x\frac{1}{x} + 2\ln(x) = 1 + 2 + 2\ln(x) = 3 + 2\ln(x)$$

$$f'''(x) = \frac{2}{x}$$

or $$\max_{5 \le x \le 10} f'''(x) = \max_{5 \le x \le 10} \frac{2}{x} = \frac{2}{5}$$

The truncation error of the quadratic interpolation is given by

$$|TE| \le \frac{h^3}{9\sqrt{3}} \max|f'''(\xi)|$$

Hence the step size h is given by

$$\frac{h^3}{9\sqrt{3}} M_3 \le 10^{-5}$$

$$\Rightarrow \qquad \frac{h^3}{9\sqrt{3}}\left(\frac{2}{5}\right) \le 0.000005$$

$$\Rightarrow \qquad h \le 0.0580.$$

Example 23: Determine the maximum step size that can be used in the tabulation of $f(x) = e^x$ in [0, 1] so that the error in the linear interpolation will be less than 5×10^{-4}. Find also the step size if quadratic interpolation is used.

Solution: We have $\quad f(x) = e^x$
$$f^{(r)}(x) = e^x, r = 1, 2, ...$$

Maximum error in linear interpolation is given by

$$\frac{h^2}{8} \max_{0 \le x \le 1} |e^x| = \frac{h^2 e}{8}$$

We choose h so that

$$\frac{h^2 e}{8} \le 0.0005$$

which gives $\qquad h \le 0.03836$

Maximum error in quadratic interpolation is given by

$$\frac{h^3}{9\sqrt{3}} \max_{0 \le x \le 1} |e^x| = \frac{h^3 e}{9\sqrt{3}}$$

We choose h so that

$$\frac{h^3 e}{9\sqrt{3}} \le 0.0005$$

which gives $h \le 0.1420$.

Example 24: For the function $f(x) = \ln(x + 1)$, construct interpolation polynomials of degree one and two to approximate $f(0.45)$ from the given nodes. Find the error bound and the actual error:

x	0	0.6	0.9
$f(x) = \ln(x+1)$	0	0.47000	0.64185

Solution: First degree polynomial

$$P_1(x) = \frac{x-0.6}{0-0.6}(0) + \frac{x-0}{0.6-0}(0.47) = 0.78333\,x$$

$$P_1(0.45) = \frac{0.45}{0.6}(0.47) = 0.3525$$

Error bound

$$TE = \frac{1}{2!}(x-x_0)(x-x_1)\,f''(\xi)$$

We have

$$f(x) = \ln(x+1)$$

$$f'(x) = \frac{1}{x+1}$$

$$f''(x) = -\frac{1}{(1+x)^2}$$

$$f'''(x) = \frac{1}{(x+1)^3}$$

Hence, we obtain the bound on the error as

$$|TE| \le \frac{1}{2}\max_{0 \le x \le 0.6}|(x-x_0)(x-x_1)|\max_{0 \le x \le 0.6}|f''(x)|$$

$$= \frac{1}{2}(0.45-0)(0.45-0.6)\cdot\frac{1}{(0+1)^2} = 3.375 \times 10^{-2}$$

$$\text{Actual error} = |\ln(1+0.45) - P_1(0.45)| = |0.37156 - 0.3525|$$
$$= 1.906 \times 10^{-2}$$

Second degree polynomial

$$P_2(x) = \frac{(x-0.6)(x-0.9)}{(0-0.6)(0-0.9)}(0) + \frac{x(x-0.9)}{(0.6-0)(0.6-0.9)}(0.47) +$$

$$\frac{x(x-0.6)}{(0.9-0)(0.9-0.6)}(0.64185)$$

$$= -0.2339\,x^2 + 0.4156\,x$$

$$P_2(0.45) = \frac{0.45\,(0.45-0.9)}{0.6(-0.3)}(0.47) + \frac{0.45(0.45-0.6)}{0.9 \times 0.3}(0.64185)$$

$$= 0.36829$$

Error bound

$$TE = \frac{1}{3!}(x-x_0)(x-x_1)(x-x_2)\,f'''(\xi)$$

Hence, we obtain the bound on the error as

$$| \text{TE} | \leq \frac{1}{3!} \max_{0 \leq x \leq 0.9} |(x - x_0)(x - x_1)(x - x_2)| \max_{0 \leq x \leq 0.9} \left| \frac{1}{(x + 1)^3} \right|$$

To find maximum of $g(x) = x(x - 0.6)(x - 0.9) = x^3 - 1.5x^2 + 0.54x$, we put $g'(x) = 0 \Rightarrow 3x^2 - 3x + 0.54 = 0 \Rightarrow x = 0.7645, 0.2354$. Now $g''(x) = 6x - 3$ give $[g''(x)]_{x = 0.7645} = +ve$ and $[g''(x)]_{x = 0.2354} = -ve$. Hence $g(x)$ is max. at $x = 0.2354$ and max. $g(x) = 0.057$.

Thus
$$| \text{TE} | \leq \frac{1}{6} \times 0.057 \left| \frac{1}{(1 + 0)^3} \right| = 9.5 \times 10^{-3}$$

Actual error $= |\ln(1 + 0.45) - 0.36829| = 3.2729 \times 10^{-3}$.

Example 25: Consider $y = f(x) = \cos x$ over $[0.0, 1.2]$. Determine the error bounds for the Lagrange polynomials $P_1(x)$, $P_2(x)$ and $P_3(x)$. For $P_1(x)$, use nodes $x_0 = 0.0, x_1 = 1.2$. For $P_2(x)$, use nodes $x_0 = 0.0, x_1 = 0.6, x_2 = 1.2$. For $P_3(x)$, use nodes $x_0 = 0, x_1 = 0.4, x_2 = 0.6, x_3 = 1.2$

Solution: First of all, we determine the bounds M_2, M_3 and M_4 for derivatives $|f''(x)|$, $|f'''(x)|$, $|f''''(x)|$, respectively, taken over the interval $[0.0, 1.2]$.

We have $f(x) = \cos x$, $f'(x) = -\sin x$, $f''(x) = -\cos x$, $f'''(x) = \sin x$, $f''''(x) = \cos x$.

So
$$M_2 = \max_{0 \leq x \leq 1.2} |f''(x)| = \max_{0 \leq x \leq 1.2} |-\cos x|$$
$$\leq |-\cos(0.0)| = 1.0$$

$$M_3 = \max_{0 \leq x \leq 1.2} |f'''(x)| = \max_{0 \leq x \leq 1.2} |\sin x| \leq |\sin(1.2)| = 0.932039$$

$$M_4 = \max_{0 \leq x \leq 1.2} |f''''(x)| = \max_{0 \leq x \leq 1.2} |\cos x| \leq |\cos 0| = 1$$

For $P_1(x)$, the error bound is

$$| \text{TE} | \leq \frac{h^2}{8} M_2 \leq \frac{(1.2)^2}{8} \times 1 = 0.18 \qquad h = 1.2$$

For $P_2(x)$, the error bound is

$$| \text{TE} | \leq \frac{h^3}{9\sqrt{3}} M_3 \geq \frac{(0.6)^2}{9\sqrt{3}} \times (0.932039) = 0.012915 \qquad h = 0.6$$

For $P_3(x)$, the error bound is

$$| \text{TE} | \leq \frac{h^4}{24} M_4 \leq \frac{(0.4)^4}{24} \times 1 = 0.001067 \qquad h = 0.4$$

Example 26: The error function erf(x) is defined by the integral

$$\text{erf}(x) = \frac{2}{\sqrt{\pi}} \int_0^x e^{-t^2} dt$$

Approximate erf(0.08) by linear interpolation in the given table of correctly rounded values. Estimate the total error.

x	0.05	0.10	0.15	0.20
erf(x)	0.05637	0.11246	0.16800	0.22270

Solution: Using linear interpolation based on point 0.05 and 0.10, we have

$$P_1(x) = \frac{(x - 0.1)}{0.05 - 0.1}(0.05637) + \frac{(x - 0.05)}{0.1 - 0.05}(0.11246)$$

$$\text{erf}(0.08) = P_1(0.08) = \frac{(0.08 - 0.1)}{0.05 - 0.1}(0.05637) + \frac{(0.08 - 0.05)}{0.1 - 0.05}(0.11246)$$

$$= 0.09002$$

The maximum error of interpolation is given by

$$|\text{TE}| = \frac{h^2}{8} M_2$$

where

$$M_2 = \max_{0.05 \le x \le 0.10} |f''(x)|$$

$$= \max_{0.05 \le x \le 0.10} \left| -\frac{4x}{\sqrt{\pi}} e^{-x^2} \right| = 0.2251$$

Hence

$$|\text{TE}| = \frac{(0.05)^2}{8} \times 0.2251 = 7.0 \times 10^{-5}$$

Example 27: Construct the Lagrange interpolating polynomial of degree two for $f(x) = \sin(\ln x)$ on the interval $[2, 2.6]$ with the points $x_0 = 2.0$, $x_1 = 2.4$ and $x_2 = 2.6$. Find a bound for the absolute error.

Solution: The table is given below:

x	2.0	2.4	2.6
$f(x)$	0.012097	0.015279	0.016676

So,

$$P_2(x) = \frac{(x - 2.4)(x - 2.6)}{(2 - 2.4)(2 - 2.6)}(0.012097) + \frac{(x - 2.0)(x - 2.6)}{(2.4 - 2.0)(2.4 - 2.6)}(0.015279)$$

$$+ \frac{(x - 2.0)(x - 2.4)}{(2.6 - 2.0)(2.6 - 2.4)}(0.016676)$$

$$= \frac{(x^2 - 5x + 6.24)}{0.24}(0.012097) + \frac{(x^2 - 4.6x + 5.2)}{(-0.08)}(0.015279)$$

$$+ \frac{(x^2 - 4.4x + 4.8)}{0.12}(0.016676)$$

$$= 0.0504(x^2 - 5x + 6.24) - 0.1910(x^2 - 4.6x + 5.2) +$$
$$0.1389(x^2 - 4.4x + 4.8)$$

$$= -0.00170x^2 + 0.01544x - 0.01198$$

Now the error is given by

$$\text{TE} = \frac{1}{3!} f'''(x)(x - 2)(x - 2.4)(x - 2.6)$$

We have

$$f(x) = \sin(\ln x)$$

$$f'(x) = \frac{\cos(\ln x)}{x}$$

$$f''(x) = \frac{\{-\sin(\ln x)\} \cdot \dfrac{1}{x} - \cos(\ln x)}{x^2} = \frac{-[\sin(\ln x) + \cos(\ln x)]}{x^2}$$

$$f'''(x) = \frac{3\sin(\ln x) + \cos(\ln x)}{x^3}$$

$$\max_{2 \le x \le 2.6} |f'''(x)| \le \left| \frac{3\sin(\ln x) + \cos(\ln x)}{x^3} \right|$$

$$\le \left| \frac{3\sin(\ln 2) + \cos(\ln 2)}{8} \right|$$

$$= 0.335765$$

Now we bound $|g(x)|$ say

$$g(x) = (x-2)(x-2.4)(x-2.6)$$

To find the maximum of $|(x-2)(x-2.4)(x-2.6)|$

We require the derivative

$$g'(x) = 3x^2 - 14x + 16.24 = 0$$

$$\Rightarrow \qquad x_1 = 2.157, x_2 = 2.5$$

Here $\qquad |g(2.157)| = 0.0169$ and $|g(2.5)| = 0.005$

So max. of $g(x)$ is 0.0169.

Now the error bound for quadratic polynomial is

$$|TE| = \frac{1}{3!} \max_{2.0 \le x \le 2.6} |f'''(x)| \max_{2.0 \le x \le 2.6} |(x-2)(x-2.4)(x-2.6)|$$

$$= \frac{1}{6} \times 0.335765 \times 0.0169 = 9.457 \times 10^{-4}$$

EXERCISE 4.5

1. Use Lagrange's interpolation formula, find the value of $\sin(\pi/6)$ from the following data

x	0	$\pi/4$	$\pi/2$
$y = \sin x$	0	0.70711	1.0

Also estimate the error in the solution. [Ans. 0.51743, 0.02392]

2. Determine the step size that can be used in the tabulation of $f(x) = \sin x$ in the interval $[0, \pi/4]$ at equally spaced nodal points so that the truncation error of the quadratic interpolation is less than 5×10^{-8}. [Ans. $h = 0.009$]

3. In a computer program, quick access to the function 2^x is needed, $0 \le x \le 1$. A table with step size h is stored into an array and the truncation value is calculated by interpolation in this table:

   i. Which is the maximal step size to be used when function values are wanted correct to 5 decimal places by linear interpolation?

   ii. The same question when quadratic interpolation is used.

4. For the function $f(x) = \cos x$. Let $x_0 = 0$, $x_1 = 0.6$, $x_2 = 0.9$:

  i. Construct the interpolation polynomial of degree at most two to approximate $f(0.45)$.

  ii. Find the actual error at 0.45 and find the error bound.

  [Ans. Actual error = 0.0023, Error bound = 0.0074468]

5. Let $f(x) = e^{2x} - x$, $x_0 = 1$, $x_4 = 1.25$, and $x_2 = 1.6$. Construct interpolation polynomials of degree at most one and two to approximate $f(1.4)$, and find an error bound for the approximation. [Ans. 1.461899, 0.246838]

6. Determine the spacing h in a table of equally spaced values of the function $f(x) = \sqrt{x}$ between 1 and 2, so that interpolation with a quadratic polynomial will yield an accuracy of 5×10^{-8}. [Ans. $h = 0.01028$]

Chapter

5

Piecewise and Spline Interpolation

5.1 SPLINE INTERPOLATION

In engineering applications, piecewise polynomial approximations have been used. It was found that a low order polynomial approximation in every subintervals supplied a better approximation to the tabulated function instead of fitting a single higher-order polynomial to the entire interval. These connecting piecewise polynomials are called spline functions. The name was derived from a draftman's device to draw a curve through a given point. The connecting polynomials could be of any degree and so we have different types of spline functions, e.g. linear, quadratic, cubic etc. Out of these we have studied only linear and cubic spline.

5.2 PIECEWISE LINEAR INTERPOLATION

If we divide the interval $[a, b]$, where $a = x_0 < x_1 < x_2 ... < x_n = b$, into a number of nonoverlapping subintervals, then, we construct the corresponding linear interpolating polynomials fitting the given data on each subintervals. These polynomials define the piecewise linear interpolating polynomials.

For $x \in [x_0, x_1]$, by using Lagrange linear interpolating polynomials

$$F_1(x) = \frac{x - x_1}{x_0 - x_1} f(x_0) + \frac{x - x_0}{x_1 - x_0} f(x_1) \tag{5.1}$$

Similarly for $x \in [x_1, x_2]$, we get

$$F_1(x) = \frac{x - x_2}{x_1 - x_2} f(x_1) + \frac{x - x_1}{x_2 - x_1} f(x_2) \tag{5.2}$$

In general, for $x \in [x_{i-1}, x_i]$, the piecewise linear interpolating polynomial will be

$$F_1(x) = P_{i,1^*}(x) = \frac{x - x_i}{x_{i-1} - x_i} f(x_{i-1}) + \frac{x - x_{i-1}}{x_i - x_{i-1}} f(x_i) \tag{5.3}$$

where $i = 1, 2, 3, ..., n$. In this case we see that $P_{i,1}(x)$ is a continuous function of x, but its derivative is not continuous at the node x_i.

Example 1: Obtain the piecewise linear interpolating polynomials for the function $f(x)$ defined by the data

x	1	2	4	8
$f(x)$	3	7	21	73

Hence estimate the value of $f(3.1)$ and $f(7.5)$.

* Here it represents linear interpolation.

Solution: For $x \in [1, 2]$

$$F_1(x) = \frac{x - x_1}{x_0 - x_1} f(x_1) + \frac{x - x_0}{x_1 - x_0} f(x_2) = \frac{x - 2}{1 - 2} \times 3 + \frac{x - 1}{2 - 1} \times 7$$

$$= -3(x - 2) + 7(x - 1) = -3x + 6 + 7x - 7 = 4x - 1$$

For $x \in [2, 4]$

$$F_1(x) = \frac{x - 4}{2 - 4} \times 7 + \frac{x - 2}{4 - 2} \times 21 = -\frac{7}{2}(x - 4) + \frac{21}{2}(x - 2)$$

$$= -\frac{7}{2} x + 14 + \frac{21x}{2} - 21 = 7x - 7$$

For $x \in [4, 8]$

$$F_1(x) = \frac{x - 8}{4 - 8} \times 21 + \frac{x - 4}{8 - 4} \times 73 = -\frac{21}{4}(x - 8) + \frac{(x - 4)73}{4}$$

$$= 13x - 31$$

Now we get

$$f(3.1) = F_1(3.1) = 7 \times 3.1 - 7 = 14.7$$

and

$$f(7.5) = F_1(7.5) = 13 \times 7.5 - 31 = 66.5$$

Example 2: The function $y = f(x)$ is given at the points $(7, 3)$, $(8, 1)$, $(9, 1)$ and $(10, 9)$. Find the value of y for $x = 9.5$. Obtain piecewise linear interpolation polynomial.

Solution: For $x \in [7, 8]$

$$F_1(x) = \frac{x - x_1}{x_0 - x_1} f(x_0) + \frac{x - x_0}{x_1 - x_0} f(x_1) = \frac{x - 8}{7 - 8} \times 3 + \frac{x - 7}{8 - 7} \times 1$$

$$= -3(x - 8) + (x - 7) = -3x + 24 + x - 7 = -2x + 17$$

For $x \in [8, 9]$

$$F_1(x) = \frac{x - 9}{8 - 9} \times 1 + \frac{x - 8}{9 - 8} \times 1 = -(x - 9) + (x - 8) = 1$$

For $x \in [9, 10]$

$$F_1(x) = \frac{x - 10}{9 - 10} \times 1 + \frac{x - 9}{10 - 9} \times 9 = -(x - 10) + 9(x - 9) = 8x - 71$$

Now we get

$$f(9.5) = F_1(9.5) = 8 \times 9.5 - 71 = 5.$$

Example 3: For the data points $(0.82, 2.2705)$ and $(0.83, 2.2933)$, find $F_1(x)$ and evaluate $F_1(0.82)$.

Solution: For $x \in [0.82, 0.83]$

$$F_1(x) = \frac{x - x_1}{x_0 - x_1} f(x_0) + \frac{x - x_0}{x_1 - x_0} f(x_1)$$

$$= \frac{x - 0.83}{0.82 - 0.83} \times 2.2705 + \frac{x - 0.82}{0.83 - 0.82} \times 2.2933$$

$$= -227.05(x - 0.83) + 229.33(x - 0.82)$$

$$= -227.05x + 188.4515 + 229.33x - 188.0506$$

$$= 2.28x + 0.4009$$

$$F_1(0.82) = 2.28 \times 0.82 + 0.4009 = 2.2705$$

5.3 CUBIC SPLINE INTERPOLATION

A cubic spline satisfies the following properties:

i. $F(x_i) = f_i, i = 0, 1, ..., n$.

ii. On each subinterval $[x_{i-1}, x_i], 1 \le i \le n, F(x)$ is a third degree polynomial.

iii. $F(x), F'(x)$ and $F''(x)$ are continuous on $[a, b]$.

iv. $F(x)$ is a polynomial of degree one for $x < x_0$ and $x > x_n$.

Since $F(x)$ is piecewise cubic, polynomial $F''(x)$ is a linear function of x in the interval $x_{i-1} \le x \le x_i$ and hence can be written as

$$F''(x) = \frac{x_i - x}{x_i - x_{i-1}} F''(x_{i-1}) + \frac{x - x_{i-1}}{x_i - x_{i-1}} F''(x_i) \tag{5.4}$$

Now space is equal between two intervals, i.e.

$$x_i - x_{i-1} = h, \quad 1 \le i \le n$$

Then Eq. (5.4) becomes

$$F''(x) = \frac{1}{h} [(x_i - x) F''(x_{i-1}) + (x - x_{i-1}) F''(x_i)] \tag{5.5}$$

Now integrating Eq. (5.5) two times, we get

$$F(x) = \frac{(x_i - x)^3}{6h} F''(x_{i-1}) + \frac{(x - x_{i-1})^3}{6h} F''(x_i) + c_1 x + c_2 \tag{5.6}$$

Now using the conditions

$$F(x_{i-1}) = f(x_{i-1}) \text{ and } F(x_i) = f(x_i), \text{ we have}$$

$$f(x_{i-1}) = F(x_{i-1}) = \frac{(x_i - x_{i-1})^3}{6h} F''(x_{i-1}) + c_1 x_{i-1} + c_2$$

$$\Rightarrow \qquad f_{i-1} = f(x_{i-1}) = \frac{h^3}{6h} F''(x_{i-1}) + c_1 x_{i-1} + c_2$$

$$= \frac{h^2}{6} F''(x_{i-1}) + c_1 x_{i-1} + c_2 \tag{5.7}$$

and

$$f_i = \frac{h^2}{6} F''(x_i) + c_1 x_i + c_2 \tag{5.8}$$

On subtracting Eq. (5.7) from Eq. (5.8), we get

$$f_i - f_{i-1} = \frac{1}{6} h^2 [F''(x_i) - F''(x_{i-1})] + c_1 (x_i - x_{i-1})$$

$$= \frac{1}{6} h^2 [F''(x_i) - F''(x_{i-1})] + c_1 h$$

From where

$$c_1 = \frac{f_i - f_{i-1}}{h} - \frac{1}{6} h [F''(x_i) - F''(x_{i-1})] \tag{5.9}$$

Multiplying Eq. (5.7) by x_i and Eq. (5.8) by x_{i-1} and then subtracting, we get

$$x_i f_{i-1} - x_{i-1} f_i = \frac{h^2}{6} [x_i F''(x_{i-1}) - x_{i-1} F''(x_i)] + c_2 (x_i - x_{i-1})$$

$$c_2 h = (x_i \, f_{i-1} - x_{i-1} \, f_i) - \frac{h^2}{6} [x_i \, F''(x_{i-1}) - x_{i-1} \, F''(x_i)]$$

$$c_2 = \frac{(x_i \, f_{i-1} - x_{i-1} \, f_i)}{h} - \frac{h}{6} [x_i \, F''(x_{i-1}) - x_{i-1} \, F''(x_i)] \qquad (5.10)$$

Substituting the value of Eq. (5.9) and (5.10) in Eq. (5.6), we get

$$F(x) = \frac{(x_i - x)^3}{6h} F''(x_{i-1}) + \frac{(x - x_{i-1})^3}{6h} F''(x_i) +$$

$$x \left[\frac{f_i - f_{i-1}}{h} - \frac{1}{6} h \{ F''(x_i) - F''(x_{i-1}) \} \right] +$$

$$\frac{x_i \cdot f_{i-1} - x_{i-1} f_i}{h} - \frac{h}{6} [x_i \cdot F''(x_{i-1}) - x_{i-1} \, F''(x_i)]$$

Denoting $F''(x_i) = M_i$ and simplifying the above equation, we get

$$F(x) = \frac{1}{6h} [(x_i - x)^3 M_{i-1} + (x - x_{i-1})^3 M_i + (x_i - x) \{ 6 f_{i-1} - h^2 M_{i-1} \}$$

$$+ (x - x_{i-1}) \{ 6 f_i - h^2 M_i \}] \qquad (5.11)$$

Now differentiating Eq. (5.11), we get

$$F'(x) = \frac{1}{6h} [-3 (x_i - x)^2 M_{i-1} + 3 (x - x_{i-1})^2 M_i - (6 f_{i-1} - h^2 M_{i-1}) + (6 f_i - h^2 M_i)]$$

$$= \frac{1}{6h} [-3 (x_i - x)^2 M_{i-1} + 3 (x - x_{i-1})^2 M_i + 6 (f_i - f_{i-1}) + h^2 (M_{i-1} - M_i)] \quad (5.12)$$

Now we require that the derivative $F'(x)$ be continuous at $x = x_i \pm \eta$ as $\eta \to 0$

i. $F'(x_{i-1} + 0) = \frac{1}{6h} [-3h^2 M_{i-1} + h^2 M_{i-1} - h^2 M_i + 6 (f_i - f_{i-1})]$

$$= \frac{1}{6h} [-h^2 M_i - 2h^2 M_{i-1} + 6 (f_i - f_{i-1})] \qquad (5.13)$$

Again in the interval $[x_{i-2}, x_{i-1}]$

$$F'(x) = \frac{1}{6h} [-3 (x_{i-1} - x)^2 M_{i-2} + 3 (x - x_{i-2})^2 M_{i-1} + 6 (f_{i-1} - f_{i-2})$$

$$+ h^2 M_{i-2} - h^2 M_{i-1}] \qquad (5.14)$$

ii. From Eq. (5.14)

$$F'(x_{i-1} - 0) = \frac{1}{6h} [3h^2 M_{i-1} + 6 f_{i-1} - 6 f_{i-2} + h^2 M_{i-2} - h^2 M_{i-1}]$$

$$= \frac{1}{6h} [2h^2 M_{i-1} + h^2 M_{i-2} + 6 f_{i-1} - 6 f_{i-2}] \qquad (5.15)$$

As $F'(x)$ is continuous at x_{i-1}

$\therefore$ $F'(x_{i-1} - 0) = F'(x_{i-1} + 0)$

$\Rightarrow 2h^2 M_{i-1} + h^2 M_{i-2} + 6 f_{i-1} - 6 f_{i-2} = -h^2 M_i - 2h^2 M_{i-1} + 6 f_i - 6 f_{i-1}$

$\Rightarrow$ $h^2 [M_i + 4 M_{i-1} + M_{i-2}] = 6 (f_i - 2 f_{i-1} + f_{i-2}) \qquad (5.16)$

For the interval $[x_{i-1}, x_i]$, we have

$$h^2[M_{i+1} + 4M_i + M_{i-1}] = 6[f_{i+1} - 2f_i + f_{i-1}] \tag{5.17}$$

where $i = 1, 2, 3, ..., n-1$.

Further, in view of fourth property, that $F(x)$ is linear for $x < x_0$ and $x > x_n$, we have $F''(x) = 0$ at $x = x_0$ and $x = x_n$.

Hence $\quad M_0 = 0, M_n = 0 \tag{5.18}$

Equations (5.17) and (5.18) give $(n + 1)$ equations in $(n + 1)$ unknowns, $M_0, M_1, ...$ M_n. Hence, we can solve for $M_0, M_1, M_2, ..., M_n$. Substituting in Eq. (5.11), we get the cubic spline in each interval.

If $M_0 = M_n = 0$, we call this cubic spline as natural spline.

Example 4: From the following table compute $y(1.5)$ using cubic spline.

x	1	2	3
y	−8	−1	18

Solution: Here $h = 1$ and $n = 2$. We consider that $M_0 = M_2 = 0$

We have

$$h^2[M_{i-1} + 4M_i + M_{i+1}] = 6[f_{i-1} - 2f_i + f_{i+1}]$$

Putting $i = 1$, we have

$$1(M_0 + 4M_1 + M_2) = 6(f_0 - 2f_1 + f_2) = 6(-8 - 2 \times -1 + 18)$$

$$\Rightarrow \qquad M_0 + 4M_1 + M_2 = 72$$

$$\Rightarrow \qquad 0 + 4M_1 + 0 = 72 \quad \Rightarrow \; M_1 = 18$$

Now $\quad F(x) = \dfrac{1}{6h}[(x_i - x)^3 M_{i-1} + (x - x_{i-1})^3 M_i + (x_i - x)(6f_{i-1} - h^2 M_{i-1})$

$$+ (x - x_{i-1})(6f_i - h^2 M_i)]$$

For $x \in [1, 2]$ and putting $i = 1$, we get

$$F(x) = \frac{1}{6}[(x_1 - x)^3 M_0 + (x - x_0)^3 M_1 + (x_1 - x)(6f_0 - h^2 M_0)$$
$$+ (x - x_0)(6f_1 - h^2 M_1)]$$

$$= \frac{1}{6}[(2 - x)^3 \times 0 + (x - 1)^3 \times 18 + (2 - x)(6 \times -8 - 1^2 \times 0)$$

$$+ (x - 1)(6 \times -1 - 1^2 \times 18)]$$

$$= \frac{1}{6}[18(x - 1)^3 - 48(2 - x) - 24(x - 1)] = 3(x - 1)^3 + 4x - 12$$

$$= 3x^3 - 9x^2 + 13x - 15, \quad 1 \le x \le 2$$

Thus $y(1.5) \approx F(1.5) = 3 \times (1.5)^3 - 9 \times (1.5)^2 + 13 \times 1.5 - 15 = -5.625$

Example 5: Find natural cubic spline to every subinterval for the following data

x	0	1	2	3
y	2	−6	−8	2

Hence compute $y(2.5)$.

Solution: We have the cubic spline for the interval (x_{i-1}, x_i) is

$$F(x) = \frac{1}{6h}[(x_i - x)^3 M_{i-1} + (x - x_{i-1})^3 M_i + (x_i - x)(6f_{i-1} - h^2 M_{i-1})$$
$$+ (x - x_{i-1})(6f_i - h^2 M_i)]$$

Here $h = 1$ and $n = 3$. We consider that $M_0 = M_3 = 0$.

Again we have

$$h^2[M_{i-1} + 4M_i + M_{i+1}] = 6[f_{i-1} - 2f_i + f_{i+1}]$$

For $x \in [0, 1]$ and putting $i = 1$

$$M_0 + 4M_1 + M_2 = 6[f_0 - 2f_1 + f_2]$$

and for $x \in [1, 2]$ $\quad M_1 + 4M_2 + M_3 = 6(f_1 - 2f_2 + f_3)$

Here $\qquad\qquad\qquad 4M_1 + M_2 = 6(2 - 2 \times -6 - 8) = 36$

$$M_1 + 4M_2 = 6(-6 - 2 \times -8 + 2) = 72$$

On solving, we get $\quad M_1 = 4.8, M_2 = 16.8$.

Hence for $x \in [0, 1]$

$$F(x) = \frac{1}{6}[(1 - x)^3 M_0 + (x - 0)^3 M_1 + (1 - x)(6f_0 - M_0) + (x - 0)(6f_1 - M_1)$$

$$= \frac{1}{6}[4.8x^3 + 12(1 - x) - 40.8x]$$

$$= 0.8x^3 - 8.8x + 2$$

For $x \in [1, 2]$

$$F(x) = \frac{1}{6}[(2 - x)^3 M_1 + (x - 1)^3 M_2 + (2 - x)(6f_1 - M_1) + (x - 1)(6f_2 - M_2)$$

$$= \frac{1}{6}[4.8(2 - x)^3 + 16.8(x - 1)^3 - 40.8(2 - x) - 64.8(x - 1)]$$

$$= 2x^3 - 3.6x^2 - 5.2x + 0.8$$

For $x \in [2, 3]$

$$F(x) = \frac{1}{6}[(3 - x)^3 M_2 + (x - 2)^3 M_3 + (3 - x)(6f_2 - h^2 M_2) + (x - 2)(6f_3 - h^2 M_3)$$

$$= \frac{1}{6}[(3 - x)^3 (16.8) - 64.8(3 - x) + 12(x - 2)]$$

$$= -2.8x^3 + 25.2x^2 - 62.8x + 39.2$$

Therefore cubic spline in different intervals are enlisted as follows:

Interval	Cubic spline
[0, 1]	$0.8x^3 - 8.8x + 2$
[1, 2]	$2x^3 - 3.6x^2 - 5.2x + 0.8$
[2, 3]	$-2.8x^3 + 25.2x^2 - 62.8x + 39.2$

Thus $y(2.5) = -2.8 \times (2.5)^3 + 25.2 \times (2.5)^2 - 62.8 \times 2.5 + 39.2 = -4.05$

Example 6: Test whether the following functions are cubic splines or not:

i. $P_1(x) = x^2 - x + 1$, $1 \leq x \leq 2$
   $P_2(x) = 3x - 3$, $2 \leq x \leq 3$

ii. $P_1(x) = -2x^2 + x^3$, $-1 \leq x \leq 0$
    $P_2(x) = x^2 - 2x^3$, $0 \leq x \leq 1$

Solution: Each polynomial is at most of degree three in each subinterval.

i. $P_1(2) = 3 = P_2(2)$
   $P_1'(2) = 2 \times 2 - 1 = 3,\ P_2'(2) = 3$
   $P_1''(2) = 2,\ P_2''(2) = 0$
   $\therefore$ Not a cubic spline since $F''(x)$ is not continuous at $x = 2$.

ii. $P_1(0) = 0 = P_2(0)$
    $P_1'(0) = 0 = P_2'(0)$
    $P_1''(0) = -4,\ P_2''(0) = 2$
    $\therefore$ Not a cubic spline since $F''(x)$ is not continuous at $x = 0$.

Example 7: Find the cubic spline given in the table

x	0	2	4	6
y	1	9	41	41

and $M_0 = 0$, $M_3 = -12$. Hence compute $y\,(2.5)$.

Solution: We have the cubic spline for the interval (x_{i-1}, x_i) is

$$F(x) = \frac{1}{6h}[(x_i - x)^3 M_{i-1} + (x - x_{i-1})^3 M_i + (x_i - x)(6f_{i-1} - h^2 M_{i-1})$$
$$+ (x - x_{i-1})(6f_i - h^2 M_i)]$$

Here $h = 2$ and $n = 3$.

We have $M_0 = 0$ and $M_3 = -12$

$\therefore$ $h^2[M_{i-1} + 4M_i + M_{i+1}] = 6[f_{i-1} - 2f_i + f_{i+1}]$

For $x \in [0, 2]$ and putting $i = 1$

$\qquad 4[M_0 + 4M_1 + M_2] = 6[f_0 - 2f_1 + f_2]$

$\Rightarrow \qquad 4[0 + 4M_1 + M_2] = 6(1 - 2 \times 9 + 41) = 144$

$\Rightarrow \qquad\qquad 4M_1 + M_2 = 36$

and for $x \in [2, 4]$

$\qquad 4[M_1 + 4M_2 + M_3] = 6[f_1 - 2f_2 + f_3]$

$\Rightarrow \qquad M_1 + 4M_2 - 12 = \frac{6}{4}[9 - 2 \times 41 + 41] = -48$

$\Rightarrow \qquad\qquad M_1 + 4M_2 = -36$

On solving, we get

$$M_1 = 12,\ M_2 = -12$$

Hence for $x \in [0, 2]$

$$F(x) = \frac{1}{6 \times 2} [(2-x)^3 \times 0 + (x-0)^3 \times 12 + (2-x)(6 \times 1 - 2^2 \times 0)$$
$$+ (x-0)(6 \times 9 - 2^2 \times 12)]$$

$$= \frac{1}{12} \{12x^3 + 6(2-x) + 6x\} = \frac{1}{12} [12x^3 + 12] = x^3 + 1$$

Similarly, for $x \in [2, 4]$

$$F(x) = 25 - 36x + 18x^2 - 2x^3$$

and for $x \in [4, 6]$

$$F(x) = -103 + 60x - 6x^2$$

EXERCISE 5.1

1. For the data points (2, 3) and (5, 7), find $P_1(x)$ by using piecewise linear interpolation. $\left[\text{Ans.} \dfrac{2}{3} x + \dfrac{5}{3} \right]$

2. Compute ln 9.2 from ln 9.0 = 2.1972 and ln 9.5 = 2.2513 by the piecewise linear interpolation. [Ans. 2.2188]

3. Obtain the piecewise linear interpolating polynomials for the function $f(x)$ defined by the data

x	1.1	1.3
$f(x)$	0.8912	0.9636

   Hence estimate the value of $P(1.2)$. [Ans. $P(1.2) = 0.9274$]

4. Find the cubic spline approximation for the function given below

x	0	1	2	3
$y = f(x)$	1	2	33	244

   with the end conditions $M_0 = M_3 = 0$. Also find $y(2.5)$

   [Ans. $-4x^3 + 5x + 1, \quad 0 \le x \le 1$
   $50x^3 - 162x^2 + 167x - 53, \quad 1 \le x \le 2$
   $-46x^3 + 414x^2 - 985x + 715, \quad 2 \le x \le 3$

   and $f(2.5) = 121.25$]

5. Using cubic spline, find $y(0.5)$ and $y'(1)$ given $M_0 = M_2 = 0$ and the table

x	0	1	2
y	-5	-4	3

   $\left[\text{Ans.} F(0.5) = -\dfrac{81}{16} = y(0.5) \; F'(1) = y'(1) = 4 \right]$

6. Find the cubic spline valid in the interval [3, 4] for the function given by the following table under the conditions $M(1) = 0, M(4) = 0$

x	1	2	3	4
y	3	10	29	65

   $\left[\text{Ans. } M_1 = \dfrac{62}{5}, M_2 = \dfrac{112}{5}, P_3(x) = \dfrac{1}{15}(-56x^3 + 672x^2 - 2092 + 2175) \text{ for } 3 \le x \le 4 \right]$

7. Test whether the following functions are cubic spline or not
$$P_1(x) = -x^2 + 20x^3, \qquad -1 \le x \le 0$$
$$P_2(x) = -x^2 + 6x^3, \qquad 0 \le x \le 1 \qquad\qquad \text{[Ans. Yes, it is a cubic spline]}$$

8. Evaluate $I = \int_0^1 \dfrac{1}{1+x}\,dx$ using the cubic spline method.

$$\left[\text{Hint: Use the table of values} \begin{array}{ccccc} x & 0 & 1/3 & 2/3 & 1 \\ y & 1 & 3/4 & 3/5 & 1/2 \end{array}\right]$$

$$\left[\text{Ans. } F(x) = \begin{cases} 0.63x^3 - 0.82x + 1; & 0 \le x \le \dfrac{1}{3} \\ -0.45x^3 + 1.08x^2 - 1.18x + 1.04; & \dfrac{1}{3} \le x \le \dfrac{2}{3} \\ -0.18x^3 + 0.54x^2 - 0.82x + 0.96, & \dfrac{2}{3} \le x \le 1 \end{cases}, I = 0.695 \right]$$

9. Obtain the cubic spline fit for the data:
$$\begin{array}{ccccc} x & 0 & 1 & 2 & 3 \\ y & 1 & 4 & 10 & 8 \end{array}$$
Under the end conditions $f''(0) = 0 = f''(3)$ and valid in the interval [1, 2]. Hence obtain the estimate of $f(1.5)$

$$\left[\text{Ans. } M_1 = 8, M_2 = -14, F(x) = \dfrac{-11x^3 + 45x^2 - 40x + 18}{3}, F(1.5) = 7.375 \right]$$

10. Find the natural cubic spline to fit the data in [0, 1]:
$$\begin{array}{cccc} x & 0 & 1 & 2 \\ y & 0 & 2 & 6 \end{array}$$
Also find $y(0.5)$ and $y'(1)$.

$$\left[\text{Ans. } M_1 = 1 \text{ in } [0,1], P_1(x) = \dfrac{1}{2}(3x + x^3), y'(1) = 3, y(0.5) = \dfrac{13}{16} \right]$$

Approximation of Functions

6.1 INTRODUCTION

The problem of approximating a function is a significant problem in numerical analysis because of its various application in the software development for digital computers. The functions which are generally used for approximating given functions are polynomials.

The Weierstrass approximation theorem stated below guarantees the existence of a polynomial function $P(x)$ which approximates any continuous function $f(x)$ defined on a finite interval $[a, b]$.

6.2 WEIERSTRASS APPROXIMATION THEOREM

If the function $f(x)$ is continuous on a finite interval $[a, b]$, then for any given $\varepsilon > 0$, there exists a positive integer $n = n(\varepsilon)$ and a polynomial $P(x)$ of degree n such that

$$|f(x) - P(x)| < \varepsilon \ \forall \ x \in [a, b]$$

In order to approximate a given function $f(x)$, we assume that

$$f(x) \simeq P(x, c_0, c_1, c_2, ..., c_n)$$
$$= c_0\phi_0(x) + c_1\phi_1(x) + ... + c_n\phi_n(x)$$

where $\phi_0(x), \phi_1(x), ..., \phi_n(x)$, are n suitably choosen linearly independent functions called coordinate functions and $c_0, c_1, ..., c_n$ are n parameters to be determined. For polynomial approximation we usually take $\phi_i(x) = x^i, i = 0, 1, 2, ..., n$.

The error of approximation (also called residual) is defined as

$$e(f; c_0, c_1, ... c_n) = \left\| f(x) - [c_0\phi_0(x) + c_1\phi_1(x) + ... + c_n\phi_n(x)] \right\| \tag{6.1}$$

where $\|\cdot\|$ is a well defined norm. The problem of approximation is to determine the parameter $c_0, c_1, ..., c_n$ such that this error is as small as possible in some sense.

We obtain different types of approximations by using different norms. Once a criterion for approximation is fixed, i.e. a particular norm is fixed, the functions which minimize the error of approximation for this criterion is called the best approximation.

The norms which are commonly used as given below.

Discrete Data

l^p**-norm:** It is defined as

$$\|x\| = \left(\sum_{i=1}^{n} |x_i|^p \right)^{1/p}, p \geq 1$$

where $\{x_i\}$ is a sequence of real or complex numbers.

Euclidean norm or square norm: It is defined as

$$\|x\| = \left(\sum_{i=1}^{n} |x_i|^2\right)^{1/2}$$

and it is particular case of L^P-norm for $p = 2$.

Uniform norm: It is defined as

$$\|x\| = \max_{1 \leq i \leq n} |x_i|$$

Continuous data: If the function $f(x)$ is continuous on $[a, b]$ and $|f(x)|^P$ is integrable on $[a, b]$, then the L^P norm is defined as

$$\|f\| = \left(\int_a^b w(x)|f(x)|^P \, dx\right)^{1/P}, p \geq 1$$

where $w(x) > 0$ is the weight function.

Euclidean norm or square norm for continuous data: It is defined as

$$\|f\| = \left(\int_a^b w(x)|f(x)|^2 \, dx\right)^{1/2}$$

and it is particular case of L^P-norm for $p = 2$.

Uniform norm for continuous data: It is defined as

$$\|f\| = \max_{a \leq x \leq b} |f(x)|$$

Note: When the Euclidean norm or square norm is used, we obtain the least square approximation and when uniform norm is used we get the uniform approximation.

6.3 WEIGHT FUNCTIONS

An integrable function w is called weight function on the interval I if $w(x) \geq 0 \; \forall x$ in I but $w(x) \equiv 0$ on any subinterval of I.

The aim of a weight function is to assign varying degree of importance to approximation on certain portion of the interval

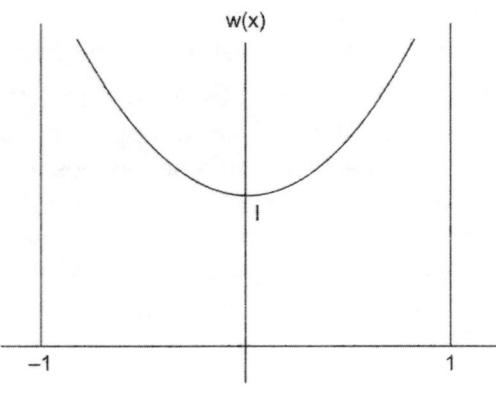

Fig. 6.1

For example consider the weight function $w(x) = \dfrac{1}{\sqrt{1-x^2}}$, which places less emphasis near the centre of the interval $(-1, 1)$ and more emphasis when $|x|$ is near to 1.

6.4 LINEARLY INDEPENDENT AND DEPENDENT FUNCTIONS

The set of functions $(\phi_0, \phi_1, \phi_2, ..., \phi_n)$ is said to be linearly independent on $[a, b]$ if whenever

$$c_0\phi_0(x) + c_1\phi_1(x) + c_2\phi_2(x) + ... + c_n\phi_n(x) = 0, \forall\ x \in [a, b]$$

then

$$c_0, c_1, c_2, ..., c_n = 0$$

Otherwise the set of functions is said to be linearly dependent.

6.5 LEAST SQUARE APPROXIMATION

The method of least squares is the most commonly used systematic procedure to fit a unique curve through given data points which may be given in tabular form or known explicitly over a given interval.

Let the given function

$$f(x) \approx c_0\phi_0(x) + c_1\phi_1(x) + c_2\phi_2(x) + ... + c_n\phi_n(x) \qquad (6.2)$$

where $\phi_0(x), \phi_1(x), ... \phi_n(x)$ are $n + 1$ appropriately choosen linearly independent functions and $c_0, c_1, c_2, ..., c_n$ are parameters to be determined.

The error of approximation E_k at the value of $x = x_k$ of the variable x is defined as

$$E_k = f(x_k) - [c_0\phi_0(x_k) + c_1\phi_1(x_k) + ... + c_n\phi_n(x_k)]$$

The best approximation in the least square sense is defined as that for which the constants $c_0, c_1, c_2, ..., c_n$ are so determined that the aggregate of $w(x)\ E^2$ over a given domain D is minimum, where $w(x) > 0$ is the weight function.

For function whose values are given at N points $x_1, x_2, ..., x_N$, we have

$I(c_0, c_1, ..., c_n)$ = Sum of square of the errors or residuals

$$I(c_0, c_1, ..., c_n) = \sum_{K=1}^{N} w(x_k) \left[f(x_k) - \sum_{i=0}^{n} c_i\phi_i(x_k) \right]^2$$

$$= \text{minimum} \qquad (6.3)$$

For functions which are continuous on $[a, b]$, we determine $c_0, c_1, c_2, ..., c_n$ such that

$$I(c_0, c_1, c_2, ..., c_n) = \int_a^b w(x) \left[f(x) - \sum_{i=0}^{n} c_i\phi_i(x) \right]^2 dx$$

$$= \text{minimum} \qquad (6.4)$$

The necessary condition for Eq. (6.3) or (6.4) to have a minimum value is that

$$\frac{\partial I}{\partial c_i} = 0, \ i = 0, 1, 2, ..., n \qquad (6.5)$$

which gives a system of $(n + 1)$ linear equations in $(n + 1)$ unknowns $c_0, c_1, c_2, ... c_n$ in the form

$$\sum_{K=1}^{N} w(x_k) \left[f(x_k) - \sum_{i=0}^{n} c_i\phi_i(x_k) \right] \phi_j(x_k) = 0, j = 0, 1, 2, ..., n \qquad (6.6)$$

or

$$\int_a^b w(x) \left[f(x) - \sum_{i=0}^{n} c_i\phi_i(x) \right] \phi_j(x) \, dx = 0, j = 0, 1, 2, ..., n \qquad (6.7)$$

The Eqs (6.6) or (6.7) are called normal equations. On solving these normal equations, we get the values of $c_0, c_1, c_2, ..., c_n$ and these values are substituted in Eq. (6.2) to obtain the required least square approximation for $f(x)$.

Note: For polynomial approximation, we usually set

$$\phi_i(x) = x^i, i = 0, 1, 2, ..., n \text{ and take } w(x) = 1$$

6.6 ALGORITHM FOR FITTING A STRAIGHT LINE OF THE FORM $y = a + bx$ FOR A GIVEN SET OF DATA POINTS

1. Start the program
2. Input number of terms observe
3. Input the array ax
4. Input the array ay
5. for $i = 0$ to observe
6. Sum $1 + = x[i]$
7. Sum $2 + = y[i]$
8. $xy[i] = x[i] * y[i]$
9. Sum $3 + = xy[i]$
10. End loop i
11. for $i = 0$ to observe
12. $x2[i] = x[i] * x[i]$
13. Sum $4 + = x2[i]$
14. End of loop i
15. temp $1 = $ (Sum 2 * Sum 4) – (Sum 3 * Sum 1)
16. $a = $ temp $1/$(observe * Sum 4) – (Sum 1 * Sum 1)
17. $b = $ (Sum $2 – $ observe * a)/Sum 1
18. Print out a, b
19. Print "line is: $y = a + bx$"
20. End of the program

Example 1: Obtain an approximation in the sense of the principle of least square in the form of a polynomial of degree 2 to the function $\dfrac{1}{1+x^2}$ in the range $[-1, 1]$.

Solution: Consider a polynomial of degree two is

$$P_2(x) = a_0 + a_1 x + a_2 x^2, \text{ such that}$$

$$I = \int_{-1}^{1} \left(\frac{1}{1+x^2} - a_0 - a_1 x - a_2 x^2 \right)^2 dx = \text{minimum}$$

The normal equations are

$$\frac{\partial I}{\partial a_0} = 0 = -2 \int_{-1}^{1} \left(\frac{1}{1+x^2} - a_0 - a_1 x - a_2 x^2 \right) dx$$

$$\int_{-1}^{1} \frac{1}{1+x^2} dx = a_0 \int_{-1}^{1} dx + a_1 \int_{-1}^{1} x \, dx + a_2 \int_{-1}^{1} x^2 dx$$

$$(\tan^{-1} x)^1_{-1} = a_0 (x)^1_{-1} + a_1 \left(\frac{x^2}{2}\right)^1_{-1} + a_2 \left(\frac{x^3}{3}\right)^1_{-1}$$

$$\frac{\pi}{4} + \frac{\pi}{4} = 2a_0 + 0 + \frac{2}{3}a_2 \implies 2a_0 + \frac{2}{3}a_2 = \frac{\pi}{2} \qquad \text{(i)}$$

Similarly,
$$\frac{\partial I}{\partial a_1} = 0 = -2 \int_{-1}^{1} \left(\frac{1}{1+x^2} - a_0 - a_1 x - a_2 x^2\right) x\, dx$$

$$\frac{\partial I}{\partial a_2} = 0 = -2 \int_{-1}^{1} \left(\frac{1}{1+x^2} - a_0 - a_1 x - a_2 x^2\right) x^2\, dx$$

Integrating the above equation, we get

$$\frac{2a_1}{3} = 0 \implies a_1 = 0 \qquad \text{(ii)}$$

$$\frac{2a_0}{3} + \frac{2a_2}{5} = 2 - \frac{\pi}{2} \qquad \text{(iii)}$$

On solving Eqs (i) and (iii), we get

$$a_0 = \frac{3}{4}(2\pi - 5), \quad a_2 = \frac{15}{4}(3 - \pi)$$

The least square approximation is

$$P_2(x) = \frac{1}{4}[3(2\pi - 5) + 15(3 - \pi) x^2]$$

Example 2: Obtain a linear polynomial approximation to the function $f(x) = x^3$ on the interval $[0, 1]$ using least square approximation with $w(x) = 1$.

Also obtain the linear polynomial approximation for the given function through the origin.

Solution: Let the linear polynomial approximation is
$$P(x) = a_0 + a_1 x$$
where a_0 and a_1 are arbitrary parameters.

We have
$$I(a_0, a_1) = \text{Sum of the squares of the errors}$$
$$= \int_{0}^{1}[x^3 - (a_0 + a_1 x)]^2\, dx = \text{minimum}$$

The normal equations are

$$\frac{\partial I}{\partial a_0} = 0 = -2 \int_{-1}^{1}[x^3 - (a_0 + a_1 x)]dx = 0$$

$$\implies \int_{0}^{1}[x^3 - (a_0 + a_1 x)]dx = 0$$

$$\implies \int_{0}^{1} x^3\, dx = a_0 \int_{0}^{1} dx + a_1 \int_{0}^{1} x\, dx$$

$$\implies \left(\frac{x^4}{4}\right)^1_0 = a_0 (x)^1_0 + a_1 \left(\frac{x^2}{2}\right)^1_0$$

$$\Rightarrow \qquad \frac{1}{4} = a_0 + \frac{a_1}{2} \qquad\qquad (i)$$

$$\frac{\partial I}{\partial a_1} = 0$$

$$\Rightarrow \quad -2\int_0^1 [x^3 - (a_0 + a_1 x)]\, x\, dx = 0$$

$$\Rightarrow \quad \int_0^1 [x^3 - (a_0 + a_1 x)]\, x\, dx = 0$$

$$\Rightarrow \quad \int_0^1 x^4 dx = a_0 \int_0^1 x\, dx + a_1 \int_0^1 x^2 dx$$

$$\Rightarrow \quad \left(\frac{x^5}{5}\right)_0^1 = a_0 \left(\frac{x^2}{2}\right)_0^1 + a_1 \left(\frac{x^3}{3}\right)_0^1$$

$$\Rightarrow \qquad \frac{1}{5} = \frac{a_0}{2} + \frac{a_1}{3} \qquad\qquad (ii)$$

On solving Eq. (i) and (ii), we get

$$a_0 = -\frac{1}{5} \text{ and } a_1 = \frac{9}{10}$$

Hence the desired linear polynomial approximation is

$$P_1(x) = a_0 + a_1 x = -\frac{1}{5} + \frac{9}{10} x$$

Linear polynomial approximation through the origin.

Let the linear polynomial approximation through the origin be $P(x) = ax$.

$$I(a) = \int_0^1 (x^3 - ax)^2 \, dx = \text{minimum}$$

The normal equation is

$$\frac{\partial I}{\partial a} = -2\int_0^1 (x^3 - ax)\, x\, dx = 0$$

$$\Rightarrow \quad \int_0^1 x^4 dx = a \int_0^1 x^2 dx$$

$$\Rightarrow \quad \left(\frac{x^5}{5}\right)_0^1 = a\left(\frac{x^3}{3}\right)_0^1$$

$$\Rightarrow \qquad \frac{1}{5} = \frac{a}{3} \Rightarrow a = \frac{3}{5}$$

Hence the desired polynomial approximation through the origin is

$$P(x) = \frac{3}{5} x$$

Example 3: Find the least square approximation polynomial of degree two for the function $f(x) = \sin \pi x$ on the interval $[0, 1]$.

Solution: Let the second order polynomial is

$$P_2(x) = a_0 + a_1 x + a_2 x^2 \qquad\qquad (i)$$

We have

$$I(a_0, a_1, a_2) = \text{Sum of the squares of the errors}$$

$$= \int_0^1 [\sin \pi x - (a_0 + a_1 x + a_x x^2)]^2 \, dx = \text{minimum}$$

The normal equations are

$$\int_0^1 [\sin \pi x - (a_0 + a_1 x + a_2 x^2)] \, dx = 0 \qquad \text{(ii)}$$

$$\int_0^1 [\sin \pi x - (a_0 + a_1 x + a_2 x^2)] \, x \, dx = 0 \qquad \text{(iii)}$$

$$\int_0^1 [\sin \pi x - (a_0 + a_1 x + a_2 x^2)] \, x^2 \, dx = 0 \qquad \text{(iv)}$$

From Eq. (ii)

$$\int_0^1 \sin \pi x \, dx = a_0 \int_0^1 dx + a_1 \int_0^1 x \, dx + a_2 \int_0^1 x^2 \, dx$$

$$\Rightarrow \qquad -\left(\frac{\cos \pi x}{\pi} \right)_0^1 = a_0 (x)_0^1 + a_1 \left(\frac{x^2}{2} \right)_0^1 + a_2 \left(\frac{x^3}{3} \right)_0^1$$

$$\left(-\frac{\cos \pi}{\pi} + \frac{1}{\pi} \right) = a_0 + \frac{a_1}{2} + \frac{a_2}{3}$$

$$\Rightarrow \qquad a_0 + \frac{a_1}{2} + \frac{a_2}{3} = \frac{2}{\pi} \qquad \text{(v)}$$

From Eq. (iii)

$$\int_0^1 x \sin \pi x \, dx = a_0 \int_0^1 x \, dx + a_1 \int_0^1 x^2 \, dx + a_2 \int_0^1 x^3 \, dx$$

$$\left[x \left(\frac{-\cos \pi x}{\pi} \right) \right]_0^1 + \left(\frac{\sin \pi x}{\pi^2} \right)_0^1 = a_0 \left(\frac{x^2}{2} \right)_0^1 + a_1 \left(\frac{x^3}{3} \right)_0^1 + a_2 \left(\frac{x^4}{4} \right)_0^1$$

$$\Rightarrow \qquad -\frac{\cos \pi}{\pi} = \frac{1}{\pi} = \frac{a_0}{2} + \frac{a_1}{3} + \frac{a_2}{4}$$

$$\Rightarrow \qquad \frac{a_0}{2} + \frac{a_1}{3} + \frac{a_2}{4} = \frac{1}{\pi} \qquad \text{(vi)}$$

Similarly from Eq. (iv), we have

$$\frac{a_0}{3} + \frac{a_1}{4} + \frac{a_2}{5} = \frac{\pi^2 - 4}{\pi^3} \qquad \text{(vii)}$$

On solving Eqs (v), (vi) and (vii), we get

$$a_0 = 12\pi^2 - \frac{120}{\pi^3} \approx -0.050465$$

$$a_1 = -a_2 = \frac{(720 - 60\pi^2)}{\pi^3} \approx 4.12251$$

Hence the desired second order polynomial approximation is

$$P_2(x) = -0.050465 + 4.12251x - 4.12251x^2$$

Example 4: Find the least square approximating polynomial of degree two for the function $f(x) = x^{3/2}$ on the interval $[0, 1]$.

Solution: Consider the second degree polynomial is

$$P_2(x) = a_0 + a_1 x + a_2 x^2 \tag{i}$$

We have $I(a_0, a_1, a_2) = $ Sum of the square of the errors

$$= \int_0^1 [x^{3/2} - (a_0 + a_1 x + a_2 x^2)]^2 \, dx$$

Then the normal equations are

$$\int_0^1 [x^{3/2} - (a_0 + a_1 x + a_2 x)] \, dx = 0 \tag{ii}$$

$$\int_0^1 [x^{3/2} - (a_0 + a_1 x + a_2 x^2)] \, x dx = 0 \tag{iii}$$

$$\int_0^1 [x^{3/2} - (a_0 + a_1 x + a_2 x^2)] \, x^2 dx = 0 \tag{iv}$$

From Eq. (ii)

$$\int_0^1 x^{3/2} \, dx = a_0 \int_0^1 dx + a_1 \int_0^1 x dx + a_2 \int_0^1 x^2 dx$$

$$\Rightarrow \qquad \left(\frac{x^{5/2}}{5/2}\right)_0^1 = a_0 \, (x)_0^1 + a_1 \left(\frac{x^2}{2}\right)_0^1 + a_2 \left(\frac{x^3}{3}\right)_0^1$$

$$\Rightarrow \qquad \frac{2}{5} = a_0 + \frac{a_1}{2} + \frac{a_2}{3} \tag{v}$$

From Eq. (iii)

$$\int_0^1 x^{5/2} \, dx = a_0 \int_0^1 x dx + a_1 \int_0^1 x^2 dx + a_2 \int_0^1 x^3 dx$$

$$\Rightarrow \qquad \left(\frac{x^{7/2}}{7/2}\right)_0^1 = a_0 \left(\frac{x^2}{2}\right)_0^1 + a_1 \left(\frac{x^3}{3}\right)_0^1 + a_2 \left(\frac{x^4}{4}\right)_0^1$$

$$\Rightarrow \qquad \frac{2}{7} = \frac{a_0}{2} + \frac{a_1}{3} + \frac{a_2}{4} \tag{vi}$$

From Eq. (iv)

$$\int_0^1 x^{7/2} \, dx = a_0 \int_0^1 x^2 dx + a_1 \int_0^1 x^3 dx + a_2 \int_0^1 x^4 dx$$

$$\Rightarrow \qquad \left(\frac{x^{9/2}}{9/2}\right)_0^1 = a_0 \left(\frac{x^3}{3}\right)_0^1 + a_1 \left(\frac{x^4}{4}\right)_0^1 + a_2 \left(\frac{x^5}{5}\right)_0^1$$

$$\Rightarrow \qquad \frac{2}{9} = \frac{a_0}{3} + \frac{a_1}{4} + \frac{a_2}{5}$$

On solving Eqs (v) to (ix), we get

$$a_0 = -0.019, \, a_1 = 0.457, \, a_2 = 0.571$$

Hence the desired second order polynomial approximation is

$$P_2(x) = -0.019 + 0.457x + 0.571x^2$$

6.7 DERIVATION OF NORMAL EQUATIONS FOR LEAST SQUARE APPROXIMATION FOR DISCRETE DATA

Consider we have N observations $(x_1, y_1), (x_2, y_2), ..., (x_N, y_N)$ of two variables x and y and it is required to fit these values to a polynomial curve of the form

$$y = P(x) = a_0 + a_1 x + a_2 x^2 + ... + a_n x^n \qquad (6.8)$$

by using principle of least square.

At the point $x = x_k$ of the variable x the expected value Y_k of y is given by

$$Y_k = a_0 + a_1 x_k + a_2 x_k^2 + ... + a_n x_k^n \qquad (6.9)$$

Then error is given by

$$E_k = y_k - Y_K = y_k - (a_0 + a_1 x_k + a_2 x_k^2 + ... + a_n x_k^n) \qquad (6.10)$$

Let $\quad I(a_0, a_1, a_2, ..., a_n) = $ Sum of the squares of the errors

$$= \sum_{K=1}^{N} [y_k - (a_0 + a_1 x_k + a_2 x_k^2 + ... + a_n x_k^n)]^2 \qquad (6.11)$$

By the principle of least square expression in Eq. (6.11) should be minimum. For a minimum of I, we have

$$\frac{\partial I}{\partial a_0} = 0, \frac{\partial I}{\partial a_1} = 0, \frac{\partial I}{\partial a_2} = 0, ..., \frac{\partial I}{\partial a_n} = 0$$

After simplifying these relations, we have

$$\Sigma y = N a_0 + a_1 \Sigma x + a_2 \Sigma x^2 + ... + a_n \Sigma x^n$$
$$\Sigma xy = a_0 \Sigma x + a_1 \Sigma x^2 + a_2 \Sigma x^3 + ... + a_n \Sigma x^{n+1}$$
$$\Sigma x^2 y = a_0 \Sigma x^2 + a_1 \Sigma x^3 + a_2 \Sigma x^4 + ... + a_n \Sigma x^{n+2}$$

$$... \quad \text{..}$$

$$... \quad \text{..}$$

$$\Sigma x^n y = a_0 \Sigma x^n + a_1 \Sigma x^{n+1} + a_2 \Sigma x^{n+2} + ... + a_n \Sigma x^{2n}$$

These equations are called normal equations and can be solved as simultaneous equations to give the values of constants.

Note: If $n = 1$, i.e. polynomial of degree one then curve to be fitted is a straight line $y = P_1(x) = a_0 + a_1 x$.

For $n = 2$, i.e. polynomial of degree two then the curve to be fitted is a parabola $y = P_2(x) = a_0 + a_1 x + a_2 x^2$.

Change of origin:

If N is odd then $\qquad u = \dfrac{x - \text{middle term}}{\text{interval }(h)}$

If N is even then $\qquad u = \dfrac{x - (\text{mean of two middle terms})}{\text{interval}/2}$

6.7.1 Transformation of Data to Linear Form

When original data is not in a linear form then we reduce it to linear form by some simple transformation of variables.

1. **Fitting of an exponential curve $y = ae^{bx}$:** To fit the given data to the exponential curve, first we take log of both sides as

$$\log_{10} y = \log_{10} a + bx \log_{10} e$$

or
$$Y = A + Bx, \quad \text{where } A = \log_{10} a, Y = \log_{10} y \text{ and } B = b\log_{10} e$$

Then the normal equations are
$$\Sigma Y = NA + B\Sigma x \text{ and } \Sigma xY = A\Sigma x + B\Sigma x^2$$

Solving these equations, we get A and B and consequently, we have $a = $ antilog (A) and $b = \dfrac{B}{\log_{10} e}$.

2. **Fitting of the curve of the type $y = ab^x$:** Taking log of both sides, we get

$$\log_{10} y = \log_{10} a + x \log_{10} b$$

or
$$Y = A + Bx, \quad \text{where } Y = \log_{10} y, A = \log_{10} a \text{ and } B = \log_{10} b$$

Then the normal equations are
$$\Sigma Y = NA + B\Sigma x \text{ and } \Sigma xY = A\Sigma x + B\Sigma x^2$$

Solving these equations, we get A and B and consequently, we have $a = $ antilog (A) and $b = $ antilog (B).

3. **Fitting of the curve of the type $y = ax^b$:** Taking log of both sides, we get

$$\log_{10} y = \log_{10} a + b\log_{10} x$$

or
$$Y = A + bX, \quad \text{where } Y = \log_{10} y, A = \log_{10} a \text{ and } X = \log_{10} x$$

Then the normal equations are
$$\Sigma Y = NA + b\Sigma X \text{ and } \Sigma XY = A\Sigma X + b\Sigma X^2$$

Solving these equations, we get A and b and consequently, we have $a = $ antilog (A).

Example 5: Fit a straight line approximate to the data:

x	1	2	3	4
y	3	7	13	21

Solution: Let the straight line to be fitted to the given data be $y = a + bx$. Then the normal equations are

$$\Sigma y = NA + b\Sigma x \text{ and } \Sigma xy = A\Sigma x + b\Sigma x^2$$

x	y	xy	x^2
1	3	3	1
2	7	14	4
3	13	39	9
4	21	84	16
10	44	140	30

In this case $N = 4$, $\Sigma x = 10$, $\Sigma y = 44$, $\Sigma xy = 140$, $\Sigma x^2 = 30$

Substituting these values in normal equations, we have

$$44 = 4a + 10b$$

and
$$140 = 10a + 30b$$

Solving these equations, we get $a = -4$, $b = 6$

Hence the fitted line is $y = -4 + 6x$.

Example 6: If P is the pull required to lift a load W by means of a pulley block, find a linear law of the form $P = m\,W + C$ connecting P and W, using the data:

P	12	15	21	25
W	50	70	100	120

where P and W are taken in kg·wt.

Solution: Let the straight line to be fitted to the given data be $P = m\,W + C$. Then the normal equations are

$$\Sigma P = m\Sigma\,W + NC$$

$$\Sigma P\,W = m\Sigma\,W^2 + C\Sigma\,W$$

P	W	PW	W²
12	50	600	2500
15	70	1050	4900
21	100	2100	10000
25	120	3000	14400
73	340	6750	31800

In this case $N = 4$, $\Sigma P = 73$, $\Sigma\,W = 340$, $\Sigma PW = 6750$, $\Sigma\,W^2 = 31800$.

Substituting these values in normal equations, we have

$$73 = 340m + 4C$$

and

$$6750 = 31800m + 340C$$

Solving these equations, we get $m = 0.1879$, $C = 2.2785$

Hence the fitted line is $P = 0.1879m + 2.2785$.

Example 7: Show that the line of fit to the following data is given by $y = 0.7x + 11.285$:

x	0	5	10	15	20	25
y	12	15	17	22	24	30

Solution: Since N is even, let $x_0 = 12.5$, $h = 5$, $y_0 = 20$ (say). Then let

$$u = \frac{x - 12.5}{2.5} \text{ and } v = y - 20$$

x	y	u	v	uv	u²
0	12	-5	-8	40	25
5	15	-3	-5	15	9
10	17	-1	-3	3	1
15	22	1	2	2	1
20	24	3	4	12	9
25	30	5	10	50	25
		$\Sigma u = 0$	$\Sigma v = 0$	$\Sigma uv = 122$	$\Sigma u^2 = 70$

Normal equations are $0 = 6a$ and $122 = 70b$

$\Rightarrow \qquad\qquad\qquad a = 0,\, b = 1.743$

Line of fit is $\qquad v = 1.743\,u$

Put $\qquad\qquad u = \dfrac{x - 12.5}{2.5}$ and $v = y - 20$, we get

$$y = 0.7x + 11.285$$

Example 8: Fit a second degree parabola to the following data:

x	0	1	2	3	4
y	1	5	10	22	38

Solution: Let the parabola to be fitted to the given data be $y = a + bx + cx^2$. Then normal equations are

$$\Sigma y = Na + b\Sigma x + c\Sigma x^2$$
$$\Sigma xy = a\Sigma x + b\Sigma x^2 + c\Sigma x^3$$
$$\Sigma x^2 y = a\Sigma x^2 + b\Sigma x^3 + c\Sigma x^4$$

x	y	x^2	x^3	x^4	xy	$x^2 y$
0	1	0	0	0	0	0
1	5	1	1	1	5	5
2	10	4	8	16	20	40
3	22	9	27	81	66	198
4	38	16	64	256	152	608
$\Sigma x = 10$	$\Sigma y = 76$	$\Sigma x^2 = 30$	$\Sigma x^3 = 100$	$\Sigma x^4 = 354$	$\Sigma xy = 243$	$\Sigma x^2 y = 851$

Substituting these values in normal equations, we have

$$76 = 5a + 10b + 30c$$
$$243 = 10a + 30b + 100c$$
$$851 = 30a + 100b + 354c$$

Solving these equations simultaneously, we get $a = 1.43$, $b = 0.24$, $c = 2.21$
Hence the fitted parabola is

$$y = 1.43 + 0.24x + 2.21x^2$$

Example 9: Fit a parabolic curve to the following data:

x	1	2	3	4	5
y	1090	1220	1390	1625	1915

Solution: Changing the origin by $u = x - 3$, $v = \dfrac{(y - 1450)}{5}$, let the curve of best fit be $v = a + bu + cu^2$.

Then the normal equations are

$$\Sigma v = Na + b\Sigma u + c\Sigma u^2$$
$$\Sigma uv = a\Sigma u + b\Sigma u^2 + c\Sigma u^3$$
$$\Sigma u^2 v = a\Sigma u^2 + b\Sigma u^3 + c\Sigma u^4$$

x	y	u	v	u^2	u^3	u^4	uv	u^2v
1	1090	-2	-72	4	-8	16	144	-288
2	1220	-1	-46	1	-1	1	46	-46
3	1390	0	-12	0	0	0	0	0
4	1625	1	35	1	1	1	35	35
5	1915	2	93	4	-8	16	186	372
		$\Sigma u = 0$	$\Sigma v = -2$	$\Sigma u^2 = 10$	$\Sigma u^3 = 0$	$\Sigma u^4 = 34$	$\Sigma uv = 411$	$\Sigma u^2v = 73$

Putting the values in the normal equations, we get

$$-2 = 5a + b\cdot 0 + 10c \Rightarrow -2 = 5a + 10c$$
$$411 = a\cdot 0 + 10b + c\cdot 0 \Rightarrow 411 = 10b$$
$$73 = 10a + b\cdot 0 + 34c \Rightarrow 73 = 10a + 34c$$

Solving these, we get $a = -11.4$, $b = 41.1$, $c = 5.5$

Hence the curve of best fit is

$$v = -11.4 + 41.4u + 5.5u^2$$

Changing the origin back, i.e. putting $u = x - 3$ and $v = \dfrac{y - 1450}{5}$, the required curve of best fit is

$$\frac{1}{5}(y - 1450) = -11.4 + 41.1(x - 3) + 5.5(x - 3)^2$$
$$\Rightarrow \qquad y = 1024 + 40.5x + 27.5x^2$$

Example 10: The pressure and volume of a gas are related by the equation $pv^a = b$ where a and b are constants. Fit the equation to the following set of data:

p (kg/cm^2)	0.5	1.0	1.5	2	2.5	3
v (litres)	1.62	1	0.75	0.62	0.52	0.46

Solution: $Pv^a = b \Rightarrow P = bv^{-a}$

Taking log on both sides, we get

$$\log_{10} P = \log_{10} b - a \log v \Rightarrow y = A + BX$$

where $y = \log_{10} P$, $A = \log_{10} b$, $B = -a$ and $X = \log_{10} v$

Normal equations are

$$\Sigma Y = NA + B\Sigma X$$
$$\Sigma XY = A\Sigma X + B\Sigma X^2, \qquad \text{here } N = 6$$

p	v	$X = \log_{10} v$	$Y = \log_{10} P$	XY	X^2
0.5	1.62	0.2095	-0.3010	-0.0631	0.0439
1	1	0	0	0	0
1.5	0.75	-0.1249	0.1761	-0.0220	0.0156
2	0.62	-0.2076	0.3010	-0.0625	0.0431
2.5	0.52	-0.2840	0.3979	-0.1130	0.0806
3	0.46	-0.3372	0.4771	-0.1609	0.1137
		$\Sigma X = -0.7442$	$\Sigma Y = 1.0511$	$\Sigma XY = -0.4215$	$\Sigma X^2 = 0.2969$

Putting these values in the normal equations, we get
$$1.0511 = 6A - 0.7442B$$
$$-0.4215 = -7.442A + 0.2969B$$

Solving these, we get
$$A = -1.3 \times 10^{-3} = -0.0013$$
$$B = -1.4229$$
$$a = -B = 1.4229$$
$$b = \text{antilog } (A) = \text{antilog } (-0.0013) = \text{antilog } (\bar{1}.9987) = 0.9970$$

Hence the required equation of curve is
$$pv^{1.4229} = 0.9970$$

Example 11: Determine the constants a and b by the method of least squares such that $y = ae^{bx}$ fits the following data:

x	2	4	6	8	10
y	4.077	11.084	30.128	81.897	222.62

Solution: $y = ae^{bx}$

Taking log on both sides
$$\log_{10} y = \log_{10} a + bx \log_{10} e$$
$$\Rightarrow \qquad Y = A + Bx$$
where $Y = \log_{10} y$, $A = \log_{10} a$, $B = b \log_{10} e$

Normal equations are
$$\Sigma Y = NA + B\Sigma x$$
$$\Sigma xY = A\Sigma x + B\Sigma x^2, \quad \text{Here } N = 5$$

x	y	$Y = \log_{10} y$	xY	x^2
2	4.077	0.61034	1.22068	4
4	11.084	1.04469	4.17876	16
6	30.128	1.47897	8.87382	36
8	81.897	1.91326	15.30608	64
10	222.62	2.347564	23.47564	100
$\Sigma x = 30$		$\Sigma Y = 7.394824$	$\Sigma xY = 53.05498$	$\Sigma x^2 = 220$

Substituting these values in normal equations, we have
$$7.394824 = 5A + 30B$$
$$53.05498 = 30A + 220B$$

Solving these equations, we get $A = 0.1760594$, $B = 0.2171509$
$$\therefore \qquad a = \text{antilog } (A) = \text{antilog } (0.1760594) = 1.49989$$

$$b = \frac{B}{\log_{10} e} = \frac{0.2171509}{0.4342945} = 0.50001$$

Hence the required equation of curve is
$$y = 1.49989 e^{0.50001x}$$

Example 12: Use the method of least square to fit the curve $y = \dfrac{c_0}{x} + c_1\sqrt{x}$ to the table of values:

x	0.1	0.2	0.4	0.5	1	2
y	21	11	7	6	5	6

Solution: $y = \dfrac{c_0}{x} + c_1\sqrt{x}$

Normal equations are

$$\Sigma \frac{y}{x} = c_0 \Sigma \frac{1}{x^2} + c_1 \Sigma \frac{1}{\sqrt{x}}$$

$$\Sigma y\sqrt{x} = c_0 \Sigma \frac{1}{\sqrt{x}} + c_1 \Sigma x$$

Table is as below

x	y	y/x	$y\sqrt{x}$	$1/\sqrt{x}$	$1/x^2$
0.1	21	210	6.64078	3.16228	100
0.2	11	55	4.91935	2.23607	25
0.4	7	17.5	4.42719	1.58114	6.25
0.5	6	12	4.24264	1.41421	4
1	5	5	5	1	1
2	6	3	8.48528	0.70711	0.25
$\Sigma x = 4.2$		$\Sigma y/x = 302.5$	$\Sigma y\sqrt{x} = 33.71524$	$\Sigma \dfrac{1}{\sqrt{x}} = 10.10081$	$\Sigma \dfrac{1}{x^2} = 136.5$

Putting these values in the normal equations, we get

$$302.5 = 136.5c_0 + 10.10081c_1$$
$$33.71524 = 10.10081c_0 + 4.2c_1$$

Solving these, we get

$$c_0 = 1.97327, c_1 = 3.28182$$

Hence the required equation of curve is

$$y = \frac{1.97327}{x} + 3.28182\sqrt{x}$$

Example 13: The following measurements of a function f were made:

x	-2	-1	0	1	3
$f(x)$	7.0	4.8	2.3	2	13.8

Fit a third degree polynomial $P_3(x)$ to the data by the least square method. As the value for $x = 1$ is known to be exact and $f'(1) = 1$, we demand that $P_3(1) = 2$ and $P_3'(1) = 1$.

Solution: Let the polynomial
$$P_3(x) = a_0 + a_1(x - 1) + a_2(x - 1)^2 + a_3(x - 1)^3$$

Since $P_3(1) = 2$ and $P_3'(1) = 1$, we get $a_0 = 2, a_1 = 1$.

Hence by the method of least squares approximation, we determine a_2 and a_3 such that

$$I(a_2, a_3) = \sum_{i=1}^{5} [f(x_i) - 2 - (x_i - 1) - a_2(x_i - 1)^2 - a_3(x_i - 1)^3]^2 = \text{minimum}$$

The normal equations are given by $\dfrac{\partial I}{\partial a_2} = 0$

$$\Rightarrow \quad \sum_{i=1}^{5} 2[(f(x_i) - 2 - (x_i - 1) - a_2(x_i - 1)^2 - a_3(x_i - 1)^3] \cdot [-(x_i - 1)^2] = 0$$

$$\Rightarrow \quad \sum_{i=1}^{5}(x_i - 1)^2 f(x_i) - 2\sum_{i=1}^{5}(x_i - 1)^2 - \sum_{i=1}^{5}(x_i - 1)^3 - a_2\sum_{i=1}^{5}(x_i - 1)^4 - a_3\sum_{i=1}^{5}(x_i - 1)^5 = 0 \quad \text{(i)}$$

and $\dfrac{\partial I}{\partial a_3} = 0$

$$\Rightarrow \quad \sum_{i=1}^{5} 2[f(x_i) - 2 - (x_i - 1) - a_2(x_i - 1)^2 - a_3(x_i - 1)^3] \cdot [-(x_i - 1)^3] = 0$$

$$\Rightarrow \quad \sum_{i=1}^{5}(x_i - 1)^3 f(x_i) - 2\sum_{i=1}^{5}(x_i - 1)^3 - \sum_{i=1}^{5}(x_i - 1)^4 - a_2\sum_{i=1}^{5}(x_i - 1)^5 - a_3\sum_{i=1}^{5}(x_i - 1)^6 = 0 \quad \text{(ii)}$$

Using the given data values, Eqs (i and ii) become

$$-114a_2 + 244a_3 = -131.7$$

and

$$244a_2 - 858a_3 = 177.3$$

Solving these equations, we get

$$a_2 = 1.8220 \text{ and } a_3 = 0.3115$$

Hence the required least square approximation is

$$P_3(x) = 2 + (x - 1) + 1.822(x - 1)^2 + 0.3115(x - 1)^3$$

Example 14: Experiments with a periodic process gave the following data:

$t°$	0	50	100	150	200	250	300	350
y	0.754	1.762	2.041	1.412	0.303	−0.484	−0.380	0.520

Estimate the parameters a and b in the model $y = b + a\sin t$, using the least square approximation.

Solution: We have equations $y = b + a\sin t$

The normal equations are:

$$\Sigma y = Nb + a\Sigma \sin t$$
$$\Sigma y \sin t = b\Sigma \sin t + a\Sigma \sin^2 t$$

The table is as below

$t°$	y	$\sin t$	$y \sin t$	$\sin^2 t$
0	0.753	0	0	0
50	1.762	0.7660	1.3497	0.5868
100	2.041	0.9848	2.0100	0.9698
150	1.412	0.5	0.706	0.25
200	0.303	−0.3420	−0.1036	0.1170
250	−0.484	−0.9397	0.4548	0.8830
300	−0.380	−0.8660	0.3291	0.7499
350	0.520	−0.1736	0.0903	0.0301
	$\Sigma y = 5.928$	$\Sigma \sin t = -0.0705$	$\Sigma y \sin t = 4.8363$	$\Sigma \sin^2 t = 3.5866$

Putting these values in normal equation, we get

$$5.928 = 8b - 0.0705a$$
$$4.8363 = -0.0705b + 3.5866a$$

On Solving, we get $b = 0.7530$, $a = 1.3632$

Hence the least square approximation is

$$y = 1.3632 \sin t + 0.7530$$

EXERCISE 6.1

1. Show that the line of fit to the following data is given by $y = -0.5x + 8$:

x	6	7	7	8	8	8	9	9	10
y	5	5	4	5	4	3	4	3	3

2. Find the line of fit to the following data:

x	0	5	10	15	20	25
y	12	15	17	22	24	30

[Ans. $y = 0.7x + 11.28$]

3. In the following table y in the weight of potassium bromide which will dissolve in 100 gm of water at temperature $x°$. Find a linear law between x and y.

$x°$(C)	0	10	20	30	40	50	60	70
y gm		53.5	59.5	65.2	70.6	75.5	80.2	85.5	90

[Ans. $y = 54.35 + 0.5184\, x°$]

4. Find the least square line for the data points:

$(-1, 10), (0, 9), (1, 7), (2, 5), (3, 4), (4, 3), (5, 0)$ and $(6, -1)$

[Ans. $y = -1.6071429x + 8.6428571$]

5. The profit of a certain company in the xth year of its life are given by

x	1	2	3	4	5
y	1250	1400	1650	1950	2300

taking $u = x - 3$ and $v = \dfrac{y - 1650}{50}$ show that the parabola of the second degree of v on u is

$$v + 0.086 = 5.30u + 0.643u^2$$

and deduce that the parabola of second degree of y on x is
$$y = 1140 + 72x + 32.15x^2$$

6. Employ the method of least squares to fit a parabola $y = a + bx + cx^2$ in the following data:
   $(x, y) : (-1, 2), (0, 0), (0, 1), (1, 2)$ [Ans. $y = 0.5 + 1.5x^2$]

7. Determine the least squares approximation of the type $ax^2 + bx + c$ to the function 2^x at the points $x_i = 0, 1, 2, 3, 4$.   [Ans. $1.143x^2 - 0.971x + 1.286$]

8. A person runs the same race track for five consecutive days and is timed as follows:

day (x)	1	2	3	4	5
time (y)	15.3	15.1	15.0	14.5	14.0

   Make a least square fit to the above data using a function $a + \dfrac{b}{x} + \dfrac{c}{x^2}$.

            [Ans. $a = 13.0065, b = 6.7512, c = -4.4738$]

9. We are given the following values of a function $f(t)$ of the variable t:

t	0.1	0.2	0.3	0.4
$f(t)$	0.76	0.58	0.44	0.35

   Obtain a least square fit of the form
   $$f = ae^{-3t} + be^{-2t}$$ [Ans. $f = 0.6853e^{-3t} + 0.3058e^{-2t}$]

10. Fit an equation of the form $y = ab^x$ to the following data:

x	2	3	4	5	6
y	144	172.8	207.4	248.8	298.6

11. Fit a curve $y = ax^b$ to the following data:

x	1	2	3	4	5	6
y	2.98	4.26	5.21	6.10	6.80	7.50

             [Ans. $y = 2.978x^{-0.5444}$]

12. A physicist wants to approximate the following data:

x	0.0	0.5	1.0	2.0
$f(x)$	0.0	0.57	1.46	5.05

    using a function $ae^{bx} + c$, he believes that $b \approx 1$:
    i. Compute the values of a and c that give the best least squares approximation assuming that $b = 1$.   [Ans. $a = 0.784976, c = -0.733298$]
    ii. Use these values of a and c to obtain a better value of b.   [Ans. $b = 0.9995$]

13. Derive the least square equations for fitting a curve of the type $y = ax^2 + \dfrac{b}{x}$ to set of n points. Hence fit a curve of this type to the data:

x	1	2	3	4
y	-1.51	0.99	3.88	7.66

    $$\left[\text{Ans. } y = 0.509x^2 - \frac{2.04}{x} \right]$$

14. Obtain an approximation in least square sense in the form of polynomials of degree 2 to the function: $f(x) = x^3$ in the interval $[0, 2]$.
             [Ans. $P(x) = 0.4 - 2.4x + 3x^2$]

15. Obtain the least square polynomials of degree one and two for $f(x) = x^{1/2}$ on interval $[0, 1]$.   [Ans. $4/15 (1 + 3x), 1/35 (6 + 48x - 20x^2)$]

16. Obtain a linear polynomial approximation to the function $f(x) = x^3$ on the interval $[0, 2]$ using the least square approximation with $w(x) = 1$. [Ans. $-1.6 + 3.6x$]

17. Construct a least square quadratic approximation to the function $y(x) = \sin x$ on interval $[0, \pi/2]$ with respect to the wegiht function $w(x) = 1$.

$$\left[\text{Ans.}\left(\frac{18}{\pi} + \frac{96}{\pi^2} - \frac{480}{\pi^3}\right) + \left(\frac{-144}{\pi^2} - \frac{1344}{\pi^3} + \frac{5760}{\pi^4}\right)x + \left(\frac{240}{\pi^3} + \frac{2880}{\pi^4} - \frac{11520}{\pi^5}\right)x^2\right]$$

6.8 ILL CONDITIONED SYSTEM AND USE OF ORTHOGONAL FUNCTIONS

We have seen that the method of determining a least square approximation to a continuous functions give satisfactory results. However, this method possesses the disadvantage of solving a large linear system of equations. Besides, while fitting a polynomial of degree n to a given data by the method of least squares, the normal equations become ill conditioned, if n is large. This means that small change in any of its parameters c_i, $i = 0, 1, 2, ..., n$ introduce large errors in the solution – the degree of ill conditioning increasing with the order of system.

These difficulties can be avoided if the coordinate functions $\phi_i(x)$ are so choosen that they are orthogonal with respect to the weight function $w(x)$ on an interval $[a, b]$.

6.8.1 Orthogonal Functions

1. **For discrete data:** A set of real functions $\phi_1(x), \phi_2(x), \phi_3(x), ..., \phi_n(x)$ is said to be orthogonal over a set of points $x_1, x_2, ..., x_N$ with respect to the weight function $w(x)$ if

$$\sum_{K=1}^{N} w(x_k)\, \phi_i(x_k)\, \phi_j(x_k) = 0, \text{ whenever } i \neq j$$

2. **For continuous data:** A set of real functions $\phi_1(x), \phi_2(x), \phi_3(x), ..., \phi_n(x)$ defined on an interval $[a, b]$ is said to be orthogonal on this interval with respect to the weight function $w(x)$ if

$$\int_a^b w(x)\, \phi_i(x)\, \phi_j(x)\, dx = 0 \text{ whenever } i \neq j$$

If the coordinate function $\phi_i(x)$, $i = 0, 1, 2, ..., n$ are orthogonal, then from the equation (6.6) of section of least square approximation in the same chapter, we obtain

$$\sum_{K=1}^{N} w(x_k)\left[f(x_k) - \sum_{i=0}^{n} c_i\, \phi_i(x_k)\right]\phi_j(x_k) = 0$$

From where

$$\sum_{K=1}^{N} w(x_k)\, f(x_k)\, \phi_i(x_k) = \sum_{K=1}^{N} c_i\, w(x_k)\, \phi_i^2(x_k)$$

Hence

$$c_i = \frac{\displaystyle\sum_{K=1}^{N} w(x_k)\, f(x_k)\, \phi_i(x_k)}{\displaystyle\sum_{K=1}^{N} w(x_k)\, \phi_i^2(x_k)} \cdot i = 0, 1, 2, ..., n$$

For the functions which are continuous on $[a, b]$ and the expression for c_i's are given explicitly as we have the approximation of the form

$$f(x) = c_0 \phi_0(x) + c_1 \phi_1(x) + c_2 \phi_2(x) + \ldots + c_n \phi_n(x) \tag{6.12}$$

where $\phi_i(x)$ is a polynomial in x of degree i. Then, we write

$$I(c_0, c_1, c_2, \ldots, c_n) = \int_a^b w(x) [f(x) - \{c_0 \phi_0(x) + c_1 \phi_1(x) + c_2 \phi_2(x) + \ldots + c_n \phi_n(x)\}]^2 \, dx$$

$$= \text{minimum}$$

Now $\quad \dfrac{\partial I}{\partial c_0} = 0 = -2 \int_a^b w(x) [f(x) - \{c_0 \phi_0(x) + c_1 \phi_1(x) + c_2 \phi_2(x) + \ldots + c_n \phi_n(x)\}] \phi_0(x) \cdot dx$

$$\dfrac{\partial I}{\partial c_1} = 0 = -2 \int_a^b w(x) [f(x) - \{c_0 \phi_0(x) + c_1 \phi_1(x) + c_2 \phi_2(x) + \ldots + c_n \phi_n(x)\}] \phi_1(x) \cdot dx$$

$$\dfrac{\partial I}{\partial c_2} = 0 = -2 \int_a^b w(x) [f(x) - \{c_0 \phi_0(x) + c_1 \phi_1(x) + c_2 \phi_2(x) + \ldots + c_n \phi_n(x)\}] \phi_2(x) \, dx$$

... .

... .

$$\dfrac{\partial I}{\partial c_n} = 0 = -2 \int_a^b w(x) [f(x) - \{c_0 \phi_0(x) + c_1 \phi_1(x) + c_2 \phi_2(x) + \ldots + c_n \phi_n(x)\}] \phi_n(x) \, dx$$

The normal equations are

$$\int_a^b w(x) f(x) \phi_0(x) \, dx = c_0 \int_a^b w(x) \phi_0^2(x) \, dx + c_1 \int_a^b \phi_0(x) \phi_1(x) \, dx + \ldots$$

$$+ c_n \int_a^b w(x) \phi_0(x) \phi_n(x) \, dx$$

$$\int_a^b w(x) f(x) \phi_1(x) \, dx = c_0 \int_a^b w(x) \phi_0(x) \phi_1(x) \, dx + c_1 \int_a^b \phi_1^2(x) \, dx + \ldots$$

$$+ c_n \int_a^b w(x) \phi_1(x) \phi_n(x) \, dx \tag{6.13}$$

... .

... .

$$\int_a^b w(x) f(x) \phi_n(x) \, dx = c_0 \int_a^b w(x) \phi_0(x) \phi_n(x) \, dx + c_1 \int_a^b w(x) \phi_1(x) \phi_n(x) \, dx + \ldots$$

$$+ c_n \int_a^b w(x) \phi_n^2(x) \, dx$$

The above system can be written more simply as

$$\int_a^b w(x) f(x) \phi_i(x) \, dx = c_0 \int_a^b w(x) \phi_0(x) \phi_i(x) \, dx + c_1 \int_a^b w(x) \phi_1(x) \phi_i(x) \, dx + \ldots +$$

$$c_n \int_a^b w(x) \phi_n(x) \phi_i(x) \, dx \tag{6.14}$$

In Eq. (6.13), we find the product of the type $\phi_p(x)\,\phi_q(x)$ in the integrands, and if we assume that

$$\int_a^b w(x)\,\phi_p(x)\,\phi_q(x)\,dx = \begin{cases} 0 & p \ne q \\ \int_a^b w(x)\,\phi_p^2(x)\,dx & p = q \end{cases} \tag{6.15}$$

then the system in Eq. (6.13) reduces to

$$c_0 \int_a^b w(x)\,\phi_0^2(x)\,dx = \int_a^b w(x)\,f(x)\,\phi_0(x)\,dx$$

$$c_1 \int_a^b w(x)\,\phi_1^2(x)\,dx = \int_a^b w(x)\,f(x)\,\phi_1(x)\,dx$$

$$\cdots\cdots\cdots\cdots\cdots\cdots \quad \cdots\cdots\cdots\cdots\cdots\cdots$$

$$\cdots\cdots\cdots\cdots\cdots\cdots \quad \cdots\cdots\cdots\cdots\cdots\cdots$$

$$c_n \int_a^b w(x)\,\phi_n^2(x)\,dx = \int_a^b w(x)\,f(x)\,\phi_n(x)\,dx$$

From the above equation, we obtain

$$c_i = \frac{\int_a^b w(x)\,f(x)\,\phi_i(x)\,dx}{\int_a^b w(x)\,\phi_i^2(x)\,dx}, \quad i = 0, 1, 2, \ldots, n \tag{6.16}$$

Substituting the value of $c_0, c_1, \ldots, c_n$ in Eq. (6.12), then yield the required least square approximation, but the functions $\phi_0(x)$, $\phi_1(x)$, ... $\phi_n(x)$ are still not known. The $\phi_i(x)$, which are polynomial in x satisfying the conditions in Eq. (6.15) are called orthogonal polynomials and are said to be orthogonal with respect to $w(x)$. As we mentioned earlier, the function $\phi_i(x)$ are to be determined. These are obtained by using the Gram–Schmidt's orthogonalization process, which has important application in numerical analysis.

6.9 GRAM–SCHMIDT PROCESS OF ORTHOGONALIZATION

Suppose that the orthogonal polynomial $\phi_i(x)$, valid on the interval $[a, b]$, has the leading term x^i. Then starting with

$$\phi_0(x) = x^0 = 1 \tag{6.17}$$

We find that the linear polynomial $\phi_1(x)$, with leading term x, can be written as

$$\phi_1(x) = x + K_{1,0}\,\phi_0(x) \tag{6.18}$$

where $K_{1,0}$ is a constant to be determined. Since $\phi_0(x)$, $\phi_1(x)$ are orthogonal, we have from Eq. (6.18)

$$\int_a^b w(x)\,\phi_0(x)\,\phi_1(x)\,dx = \int_a^b w(x)\,x\,\phi_0(x)\,dx + K_{1,0} \int_a^b w(x)\,\phi_0^2(x)\,dx$$

by using Eq. (6.15), we get

$$0 = \int_a^b w(x)\,x\,\phi_0(x)\,dx + K_{1,0} \int_a^b w(x)\,\phi_0^2(x)\,dx$$

$$K_{1,0} = -\frac{\int_a^b w(x)\,x\,\phi_0(x)\,dx}{\int_a^b w(x)\,\phi_0^2(x)\,dx} \tag{6.19}$$

Then from Eq. (6.18), we have

$$\phi_1(x) = x - \frac{\int_a^b w(x)\, x\, \phi_0(x)\, dx}{\int_a^b w(x)\, \phi_0^2(x)\, dx}$$

Now, the polynomial $\phi_2(x)$ of degree two in x with leading term x^2 may be written as

$$\phi_2(x) = x^2 + K_{2,0} f_0(x) + K_{2,1} f_1(x) \tag{6.20}$$

where $K_{1,0}$ and $K_{2,0}$ are to be determined by using the orthogonality conditions in Eq. (6.15). Since $\phi_2(x)$ is orthogonal to $\phi_0(x)$, we have

$$\int_a^b w(x)\, \phi_0(x)\, [x^2 + K_{2,0}\, \phi_0(x) + K_{2,1}\, \phi_1(x)]\, dx = 0$$

$$\int_a^b w(x)\, x^2 \phi_0(x)\, dx + K_{2,0} \int_a^b w(x)\, \phi_0^2(x)\, dx + K_{2,1} \int_a^b w(x)\, \phi_0(x)\, \phi_1(x)\, dx = 0$$

After using condition in Eq. (6.15), we get

$$K_{2,0} = -\frac{\int_a^b x^2 w(x)\, \phi_0(x)\, dx}{\int_a^b w(x)\, \phi_0^2(x)\, dx}$$

But $\phi_0(x) = 1$

$$\therefore \qquad K_{2,0} = -\frac{\int_a^b x^2 w(x)\, dx}{\int_a^b w(x)\, \phi_0^2(x)\, dx} \tag{6.21}$$

Again, since $\phi_2(x)$ is orthogonal to $\phi_1(x)$, we have

$$\int_a^b w(x)\, \phi_1(x)\, [x^2 + K_{2,0}\, \phi_0(x) + K_{2,1}\, \phi_1(x)]\, dx = 0$$

Using the condition that $\int_a^b w(x)\, \phi_0(x)\, \phi_1(x)\, dx = 0$, we have

$$K_{2,1} = -\frac{\int_a^b x^2 w(x)\, \phi_1(x)\, dx}{\int_a^b w(x)\, \phi_1^2(x)\, dx} \tag{6.22}$$

Since $K_{2,0}$ and $K_{2,1}$ are known, Eq. (6.20) determines $\phi_2(x)$. Procedding in this way

$$\phi_j(x) = x^j + K_{j,0}\, \phi_0(x) + K_{j,1}\, \phi_1(x) + \dots + K_{j,j-1}\, \phi_{j-1}(x) \tag{6.23}$$

where the constants $K_{j,1}$ are so choosen that $\phi_{j(x)}$ is orthogonal to $\phi_0(x), \phi_1(x), \dots, \phi_{j-1}(x)$. These conditions yield

$$K_{j,i} = -\frac{\int_a^b x^j w(x)\, \phi_i(x)\, dx}{\int_a^b w(x)\, \phi_i^2(x)\, dx} \tag{6.24}$$

Since the c_i and $\phi_i(x)$ in Eq. (6.12) are known, the approximation $f(x)$ can now be determined.

Example 15: Using the Gram–Schmidt orthogonalization process, compute the first three orthogonal polynomials $P_0(x)$, $P_1(x)$, $P_2(x)$ which are orthogonal on $[0, 1]$ with respect to the weight function $w(x) = 1$. Use these polynomials to obtain the least square approximation of second degree for $f(x) = x^{1/2}$ on $[0, 1]$.

Solution: Let $P_0(x) = \phi_0(x) = 1$ and $w(x) = 1$

Then
$$K_{1,0} = -\frac{\int_0^1 x \times 1 \times 1\, dx}{\int_0^1 1 \times 1^2\, dx} = -\frac{\left(\dfrac{x^2}{2}\right)_0^1}{(x)_0^1} = -\frac{1}{2}$$

Now
$$P_1(x) = \phi_1(x) = x + K_{1,0}\,\phi_0(x) = x - \frac{1}{2}$$

Again
$$K_{2,0} = -\frac{\int_0^1 x^2\, dx}{\int_0^1 dx} = -\frac{\left(\dfrac{x^3}{3}\right)_0^1}{(x)_0^1} = -\frac{1}{3}$$

and
$$K_{2,1} = -\frac{\int_0^1 x^2\left(x - \dfrac{1}{2}\right)dx}{\int_0^1\left(x - \dfrac{1}{2}\right)^2 dx} = -\frac{\left(\dfrac{x^4}{4} - \dfrac{x^3}{6}\right)_0^1}{\left(\dfrac{x^3}{3} + \dfrac{1}{4}x - \dfrac{x^2}{2}\right)_0^1}$$

$$= -\frac{\left(\dfrac{1}{4} - \dfrac{1}{6}\right)}{\left(\dfrac{1}{3} + \dfrac{1}{4} - \dfrac{1}{2}\right)} = -1$$

Now
$$P_2(x) = \phi_2(x) = x^2 + K_{2,0}\,\phi_0(x) + K_{2,1}\,\phi_1(x)$$

$$= x^2 - \frac{1}{3} - \left(x - \frac{1}{2}\right) = x^2 - x + \frac{1}{6}$$

Using these polynomials, we have for $n = 2$

$$I(c_0, c_1, c_2) = \int_0^1 [x^{1/2} - \{c_0 P_0(x) + c_1 P_1(x) + c_2 P_2(x)\}]^2\, dx = \text{minimum}$$

The normal equations are given by

$$\frac{\partial I}{\partial c_0} = 0 \Rightarrow \int_0^1 [x^{1/2} - \{c_0 P_0(x) + c_1 P_1(x) + c_2 P_2(x)\}] P_0(x)\, dx = 0$$

$$\frac{\partial I}{\partial c_1} = 0 \Rightarrow \int_0^1 [x^{1/2} - \{c_0 P_0(x) - c_1 P_1(x) - c_2 P_2(x)\}] P_1(x)\, dx = 0$$

$$\frac{\partial I}{\partial c_2} = 0 \Rightarrow \int_0^1 [x^{1/2} - \{c_0 P_0(x) + c_1 P_1(x) + c_2 P_2(x)\}] P_2(x)\, dx = 0$$

Using the orthogonality conditions, we get

$$c_0 = \frac{\int_0^1 x^{1/2} P_0(x)\,dx}{\int_0^1 P_0^2(x)\,dx} = \frac{\int_0^1 x^{1/2}\,dx}{\int_0^1 dx} = \frac{\left(\dfrac{x^{3/2}}{3/2}\right)_0^1}{(x)_0^1} = \frac{2}{3}$$

$$c_1 = \frac{\int_0^1 x^{1/2} P_1(x)\,dx}{\int_0^1 P_1^2(x)\,dx} = \frac{\int_0^1 x^{1/2}\left(x-\dfrac{1}{2}\right)dx}{\int_0^1 \left(x-\dfrac{1}{2}\right)^2 dx} = \frac{\left(\dfrac{x^{5/2}}{5/2} - \dfrac{1}{2}\dfrac{x^{3/2}}{3/2}\right)_0^1}{\left(\dfrac{x^3}{3} + \dfrac{1}{4}x - \dfrac{x^2}{2}\right)_0^1}$$

$$= \frac{\left(\dfrac{2}{5} - \dfrac{1}{3}\right)}{\left(\dfrac{1}{3} + \dfrac{1}{4} - \dfrac{1}{2}\right)} = \frac{4}{5}$$

$$c_2 = \frac{\int_0^1 x^{1/2} P_2(x)\,dx}{\int_0^1 P_2^2(x)\,dx} = \frac{\int_0^1 x^{1/2}\left(x^2-x+\dfrac{1}{6}\right)dx}{\int_0^1 \left(x^2-x+\dfrac{1}{6}\right)^2 dx} = -\frac{4}{7}$$

Hence the required least square approximation is

$$y = P(x) = \frac{2}{3}P_0(x) + \frac{4}{5}P_1(x) - \frac{4}{7}P_2(x)$$

$$= \frac{2}{3} + \frac{4}{5}\left(x-\frac{1}{2}\right) - \frac{4}{7}\left(x^2-x+\frac{1}{6}\right)$$

$$= \frac{1}{35}(6 + 48x - 20x^2)$$

6.10 CHEBYSHEV POLYNOMIALS

In the theory of approximation of functions we often use the well known orthogonal polynomials, Chebyshev polynomials, as the coordinate functions while applying the method of least squares. Chebyshev polynomials are also used in the economization of power series which we shall discuss later. We shall now define this polynomial and give their important properties.

If n is a non negative integer, the Chebyshev polynomial of first kind is denoted as

$$T_n(x) = \cos(n\cos^{-1}x)$$

$T_n(x)$ is a solution of Chebyshev's equation

$$(1-x^2)\frac{d^2y}{dx^2} - x\frac{dy}{dx} + n^2y = 0$$

The set $\{T_0(x), T_1(x), T_2(x), ..., T_n(x), ...\}$ of Chebyshev polynomials for the interval $[-1, 1]$ can be derived by Gram–Schmidt process using the weight function

$$w(x) = \frac{1}{\sqrt{1-x^2}}.$$

We will derive the Chebyshev polynomials differently here and then we will show that they satisfy the required orthogonal property. For $x \in [-1, 1]$, define

$$T_n(x) = \cos(n \cos^{-1} x) \quad \text{for each } n \geq 0$$

Let $\theta = \cos^{-1} x$, then

$$T_n(x) = \cos n\theta \quad \text{where } \theta \in (0, \pi) \tag{6.25}$$

From Eq. (6.25), we have

$$\begin{aligned} T_0(x) &= \cos 0 = 1 \\ T_1(x) &= \cos\theta = x \end{aligned} \quad \bigg| \quad \because \ \theta = \cos^{-1} x \Rightarrow \cos\theta = x$$

Using the trignometric identity

$$\cos(n-1)\theta + \cos(n+1)\theta = 2\cos n\theta \cos\theta \tag{6.26}$$

From Eq. (6.25), we get

$$\begin{aligned} T_{n-1}(x) &= \cos(n-1)\theta \\ T_{n+1}(x) &= \cos(n+1)\theta \end{aligned}$$

Substituting the above values in Eq. (6.26), we get

$$2x\,T_n(x) = T_{n-1}(x) + T_{n+1}(x)$$

or

$$T_{n+1}(x) = 2x\,T_n(x) - T_{n-1}(x), \quad \forall \ n \geq 1$$

This is the recurrence relation satisfying the Chebyshev polynomials and can be used to compute successively all $T_n(x)$.

From recurrence relation

$$T_2(x) = 2xT_1(x) - T_0(x) = 2x \times x - 1 = 2x^2 - 1$$
$$T_3(x) = 2xT_2(x) - T_1(x) = 2x(2x^2 - 1) - x = 4x^3 - 3x$$
$$T_4(x) = 2xT_3(x) - T_2(x) = 2x(4x^3 - 3x) - (2x^2 - 1) = 8x^4 - 8x^2 + 1$$
$$T_5(x) = 2xT_4(x) - T_3(x) = 2x(8x^4 - 8x^2 + 1) - (4x^3 - 3x) = 16x^5 - 20x^3 + 5x$$
$$T_6(x) = 2xT_5(x) - T_4(x) = 2x(16x^5 - 20x^3 + 5x) - (8x^4 - 8x^2 + 1) = 32x^6 - 48x^4 + 18x^2 - 1$$

It is also possible to express powers of x in terms of Chebyshev polynomials, we find

$$1 = T_0(x)$$
$$x = T_1(x)$$
$$x^2 = \frac{1}{2}(T_0 + T_2)$$
$$x^3 = \frac{1}{4}(3T_1 + T_3)$$
$$x^4 = \frac{1}{8}(3T_0 + 4T_2 + T_4)$$
$$x^5 = \frac{1}{16}(10T_1 + 5T_3 + T_5)$$
$$x^6 = \frac{1}{32}(10T_0 + 15T_2 + 6T_4 + T_6)$$

and so on.

6.10.1 Orthogonal Properties

In order to show the orthogonality, consider

$$\int_{-1}^{1} \frac{T_n(x)\,T_m(x)}{\sqrt{1-x^2}}\,dx \;=\; \int_{-1}^{1} \frac{\cos(n\cos^{-1}x)\cdot\cos(m\cos^{-1}x)}{\sqrt{1-x^2}}\,dx$$

We reintroduce the substitution $\theta = \cos^{-1}x$, due to which, we have

$$d\theta = -\frac{1}{\sqrt{1-x^2}}\,dx$$

and for $m \neq n$

$$\int_{-1}^{1} \frac{T_n(x)\,T_m(x)\,dx}{\sqrt{1-x^2}} = \int_{0}^{\pi} \cos n\theta \cdot \cos m\theta\, d\theta$$

$$= \frac{1}{2}\int_{0}^{\pi}[\cos(n+m)\theta + \cos(m-n)\theta]\,d\theta = 0$$

and for $m = n$

$$\int_{-1}^{1} \frac{T_n(x)\,T_m(x)\,dx}{\sqrt{1-x^2}} = \int_{0}^{\pi} \cos n\theta \cdot \cos n\theta\, d\theta$$

$$= \frac{1}{2}\int_{0}^{\pi}(1 + \cos 2n\theta)\,d\theta = \frac{\pi}{2}$$

Similarly for $m = n = 0$

$$\int_{-1}^{1} \frac{T_0(x)\,T_0(x)\,dx}{\sqrt{1-x^2}} = \int_{0}^{\pi} d\theta = \pi$$

Thus, we can write

$$\int_{-1}^{1} \frac{T_n(x)\,T_m(x)\,dx}{\sqrt{1-x^2}} = \begin{cases} 0; & m \neq n \\ \dfrac{\pi}{2} & m = n \neq 0 \\ \pi; & m = n = 0 \end{cases}$$

6.10.2 Properties of Chebyshev Polynomials

Chebyshev polynomial $T_n(x)$ possesses the following properties:

1. $T_n(x)$ is a polynomial of degree n. We have $T_n(-x) = (-1)^n\, T_n(x)$ so that $T_n(x)$ is an even function of x if n is even and it is odd function of x if n is odd.

2. $T_n(x)$ has n simple zeros at $x_k = \cos\left(\dfrac{2k-1}{2n}\pi\right)$, $k = 1, 2, 3, \ldots, n$ on the interval.

3. $T_n(x)$ assumes extreme values at $(n+1)$ points $x_k = \cos\left(\dfrac{k\pi}{n}\right)$, $k = 0, 1, 2, \ldots, n$ and the extreme value at x_k is $(-1)^k$.

4. $|T_n(x)| \leq 1, x \in [-1, 1]$.

5. $T_n(x)$ is orthogonal with respect to the weight function

$$w(x) = (1-x^2)^{-1/2} \text{ and } \int_{-1}^{1} \frac{T_n(x)\,T_m(x)}{\sqrt{1-x^2}}\,dx = \begin{cases} 0; & m \neq n \\ \dfrac{\pi}{2}; & m = n \neq 0 \\ \pi; & m = n = 0 \end{cases}$$

6. If $P_n(x)$ is any monic polynomial of degree n i.e. polynomial of degree n in which the coefficient of x^n is 1 and $\tilde{T}_n(x) = \dfrac{T_n(x)}{2^{n-1}}$ is the monic Chebyshev polynomial, then $\max\limits_{x \in [-1,1]} |\tilde{T}_n(x)| \leq \max\limits_{x \in [-1,1]} |P_n(x)|$

This remarkable property of Chebyshev polynomial is known as minimax property and it can also be stated as below:

Of all monic polynomial $P_n(x)$ of degree n, the polynomial $2^{1-n}T_n(x)$ has the smallest least upper bound for its absolute value in the range $|x| \leq 1$ i.e. $-1 \leq x \leq 1$. Since $|T_n(x)| \leq 1$, the upper bound referred to above is 2^{1-n}.

Example 16: Using the Chebyshev polynomial, obtain the least squares approximation of second degree for $f(x) = x^4$ on $[-1, 1]$.

Solution: Let $\qquad f(x) \approx P(x) = C_0 T_0(x) + C_1 T_1(x) + C_2 T_2(x)$ $\qquad$ (i)

We have

$$I(c_0, c_1, c_2) = \int_{-1}^{1} \frac{1}{\sqrt{1-x^2}} [x^4 - c_0 T_0 - c_1 T_1 - c_2 T_2]^2\,dx = \text{minimum}$$

The normal equations are given by

$$\frac{\partial I}{\partial c_0} = 0 \Rightarrow \int_{-1}^{1} [x^4 - c_0 T_0 - c_1 T_1 - c_2 T_2]\frac{T_0}{\sqrt{1-x^2}}\,dx = 0 \qquad \text{(ii)}$$

$$\frac{\partial I}{\partial c_1} = 0 \Rightarrow \int_{-1}^{1} [x^4 - c_0 T_0 - c_1 T_1 - c_2 T_2]\frac{T_1}{\sqrt{1-x^2}}\,dx = 0 \qquad \text{(iii)}$$

$$\frac{\partial I}{\partial c_2} = 0 \Rightarrow \int_{-1}^{1} [x^4 - c_0 T_0 - c_1 T_1 - c_2 T_2]\frac{T_2}{\sqrt{1-x^2}}\,dx = 0 \qquad \text{(iv)}$$

By using orthogonal property in Eq. (ii), we have

$$\int_{-1}^{1} \frac{x^4 T_0\,dx}{\sqrt{1-x^2}} = c_0 \int_{-1}^{1} \frac{T_0^2\,dx}{\sqrt{1-x^2}} = \pi c_0$$

$$\Rightarrow \qquad c_0 = \frac{1}{\pi}\int_{-1}^{1} \frac{x^4\,dx}{\sqrt{1-x^2}} \qquad \qquad \because T_0(x) = 1$$

$$= \frac{2}{\pi}\int_{0}^{1} \frac{x^4\,dx}{\sqrt{1-x^2}}$$

We set $x = \sin \theta$, we get

$$c_0 = \frac{2}{\pi} \int_0^{\pi/2} \frac{\sin^4 \theta \, \cos \theta \, d\theta}{\sqrt{1 - \sin^2 \theta}} = \frac{2}{\pi} \int_0^{\pi/2} \sin^4 \theta \, dv$$

$$= \frac{2}{\pi} \frac{\left|\dfrac{4+1}{2}\right.\left|\dfrac{0+1}{2}\right.}{2\left|\dfrac{4+0+2}{2}\right.} = \frac{2}{\pi} \frac{\left|\dfrac{5}{2}\right.\left|\dfrac{1}{2}\right.}{2\left|3\right.}$$

$$= \frac{2}{\pi} \frac{\dfrac{3}{2}\dfrac{1}{2}\left|\dfrac{1}{2}\right.\left|\dfrac{1}{2}\right.}{2 \times 2\left|1\right.}$$

$$= \frac{2}{\pi} \frac{\dfrac{3}{2}\dfrac{1}{2}\pi}{2 \times 2} = \frac{3}{8} \quad \text{since} \left|\dfrac{1}{2}\right. = \sqrt{\pi}$$

Again, using orthogonal property in Eq. (iii), we get

$$c_1 \frac{\pi}{2} = \int_{-1}^{1} \frac{x^4 T_1}{\sqrt{1 - x^2}} \, dx$$

$\Rightarrow$

$$c_1 = \frac{2}{\pi} \int_{-1}^{1} \frac{x^4 \, x \, dx}{\sqrt{1 - x^2}} = \frac{2}{\pi} \int_{-1}^{1} \frac{x^5 \, dx}{\sqrt{1 - x^2}} = 0$$

by the property of definite integral.

Using the orthogonal property in Eq. (iv), we get

$$c_2 \frac{\pi}{2} = \int_{-1}^{1} \frac{x^4 T_2 \, dx}{\sqrt{1 - x^2}} \Rightarrow c_2 = \frac{2}{\pi} \int_{-1}^{1} \frac{x^4 (2x^2 - 1)}{\sqrt{1 - x^2}} \, dx$$

$$= \frac{2}{\pi} \left[2\int_{-1}^{1} \frac{x^6 dx}{\sqrt{1 - x^2}} - \int_{-1}^{1} \frac{x^4 \, dx}{\sqrt{1 - x^2}} \right] = \frac{4}{\pi} \int_{-1}^{1} \frac{x^6 \, dx}{\sqrt{1 - x^2}} - \frac{2}{\pi} \int_{-1}^{1} \frac{x^4 \, dx}{\sqrt{1 - x^2}}$$

$$= \frac{4}{\pi} \int_{-1}^{1} \frac{x^6 dx}{\sqrt{1 - x^2}} - \frac{3 \times 2}{8} = \frac{8}{\pi} \int_{-1}^{1} \frac{x^6 \, dx}{\sqrt{1 - x^2}} - \frac{6}{8}$$

$$= \frac{8}{\pi} \int_0^{\frac{\pi}{2}} \sin^6 \theta \, d\theta - \frac{6}{8}; \quad \text{putting } x = \sin \theta, \, dx = \cos \theta \, d\theta$$

$$= \frac{8}{\pi} \frac{\left|\dfrac{6+1}{2}\right.\left|\dfrac{0+1}{2}\right.}{2\left|\dfrac{6+0+2}{2}\right.} - \frac{6}{8} = \frac{8}{\pi} \frac{\left|\dfrac{7}{2}\right.\left|\dfrac{1}{2}\right.}{2\left|4\right.} - \frac{6}{8}$$

$$= \frac{8}{\pi} \frac{\dfrac{5}{2}\dfrac{3}{2}\dfrac{1}{2}\pi}{2 \times 3 \times 2} - \frac{6}{8} = \frac{15}{12} - \frac{6}{8} = \frac{30 - 18}{24} = \frac{12}{24} = \frac{1}{2}$$

Hence the required approximation is

$$f(x) = \frac{3}{8} T_0(x) + 0 + \frac{1}{2} T_2(x) = \frac{3}{8} + \frac{1}{2}(2x^2 - 1)$$

$$= x^2 - \frac{1}{8}$$

Example 17: Using Chebyshev polynomial, obtain the least square approximation of second degree for $f(x) = 3x^4 + 2x^3 + x + 2$ on $[-1, 1]$.

Solution: Let $\quad f(x) \approx P(x) = C_0 T_0(x) + C_1 T_1(x) + C_2 T_2(x)$ \hfill (i)

We have $\quad I(c_0, c_1, c_2) = \int_{-1}^{1} \frac{1}{\sqrt{1-x^2}} [(3x^4 + 2x^3 + x + 2) - c_0 T_0(x) - c_1 T_1(x) - c_2 T_2(x)]^2 \, dx$

The normal equations are given by

$$\frac{\partial I}{\partial c_0} = 0 \Rightarrow \int_{-1}^{1} [3x^4 + 2x^3 + x + 2 - c_0 T_0 - c_1 T_1 - c_2 T_2] \frac{T_0}{\sqrt{1-x^2}} \, dx = 0 \quad \text{(ii)}$$

$$\frac{\partial I}{\partial c_1} = 0 \Rightarrow \int_{-1}^{1} [3x^4 + 2x^3 + x + 2 - c_0 T_0 - c_1 T_1 - c_2 T_2] \frac{T_1}{\sqrt{1-x^2}} \, dx = 0 \quad \text{(iii)}$$

$$\frac{\partial I}{\partial c_2} = 0 \Rightarrow \int_{-1}^{1} [3x^4 + 2x^3 + x + 2 - c_0 T_0 - c_1 T_1 - c_2 T_2] \frac{T_2}{\sqrt{1-x^2}} \, dx = 0 \quad \text{(iv)}$$

By using orthogonal property in Eq. (ii), we have

$$\int_{-1}^{1} \frac{(3x^4 + 2x^3 + x + 2) T_0}{\sqrt{1-x^2}} \, dx = c_0 \int_{-1}^{1} \frac{T_0^2 \, dx}{\sqrt{1-x^2}} = \pi c_0$$

$$\Rightarrow \qquad c_0 = \frac{1}{\pi} \left[\int_{-1}^{1} \frac{3x^4 T_0 \, dx}{\sqrt{1-x^2}} + 2 \int_{-1}^{1} \frac{x^3 T_0 \, dx}{\sqrt{1-x^2}} + \int_{-1}^{1} \frac{x T_0 \, dx}{\sqrt{1-x^2}} + 2 \int_{-1}^{1} \frac{T_0 \, dx}{\sqrt{1-x^2}} \right]$$

$$= 3 \cdot \frac{1}{\pi} \int_{-1}^{1} \frac{x^4 T_0 \, dx}{\sqrt{1-x^2}} + 2 \cdot \frac{1}{\pi} \int_{-1}^{1} \frac{x^3 T_0 \, dx}{\sqrt{1-x^2}} + \frac{1}{\pi} \int_{-1}^{1} \frac{x T_0 \, dx}{\sqrt{1-x^2}} + 2 \cdot \frac{1}{\pi} \int_{-1}^{1} \frac{T_0 \, dx}{\sqrt{1-x^2}}$$

$$= 3 \cdot \frac{3}{8} + 0 + 0 + 2 \cdot \frac{1}{\pi} (\sin^{-1} x)_{-1}^{1} = \frac{9}{8} + \frac{2}{\pi} \left(\frac{\pi}{2} + \frac{\pi}{2} \right)$$

$$= \frac{9}{8} + \frac{2}{\pi} \cdot \pi = \frac{9}{8} + 2 = \frac{25}{8}$$

Again, using orthogonal property in Eq. (iii), we get

$$\int_{-1}^{1} \frac{(3x^4 + 2x^3 + x + 2)}{\sqrt{1-x^2}} T_1 \, dx = c_1 \int_{-1}^{1} \frac{T_1^2 \, dx}{\sqrt{1-x^2}} = c_1 \frac{\pi}{2}$$

$$\Rightarrow \qquad c_1 = \frac{2}{\pi} \int_{-1}^{1} \frac{3x^4 T_1 \, dx}{\sqrt{1-x^2}} + \frac{2}{\pi} \int_{-1}^{1} \frac{2x^3 T_1 \, dx}{\sqrt{1-x^2}} + \frac{2}{\pi} \int_{-1}^{1} \frac{x T_1 \, dx}{\sqrt{1-x^2}} + \frac{2}{\pi} \int_{-1}^{1} \frac{2 T_1 \, dx}{\sqrt{1-x^2}}$$

$$= 6 \cdot \frac{1}{\pi} \int_{-1}^{1} \frac{x^5 \, dx}{\sqrt{1 - x^2}} + 4 \cdot \frac{1}{\pi} \int_{-1}^{1} \frac{x^4 \, dx}{\sqrt{1 - x^2}} + 2 \cdot \frac{1}{\pi} \int_{-1}^{1} \frac{x^2 \, dx}{\sqrt{1 - x^2}} + 4 \cdot \frac{1}{\pi} \int_{-1}^{1} \frac{x \, dx}{\sqrt{1 - x^2}}$$

$$= 0 + 4 \cdot \frac{3}{8} + 2 \cdot \frac{2}{\pi} \int_{0}^{1} \frac{x^2 \, dx}{\sqrt{1 - x^2}} + 0 = \frac{3}{2} + \frac{4}{\pi} \int_{0}^{\pi/2} \sin^2 \theta \, d\theta$$

$$= \frac{3}{2} + \frac{4}{\pi} \cdot \frac{\boxed{\dfrac{2+1}{2}} \; \boxed{\dfrac{0+1}{2}}}{2 \boxed{\dfrac{2+0+2}{2}}} = \frac{3}{2} + \frac{4}{\pi} \cdot \frac{\boxed{\dfrac{3}{2}} \; \boxed{\dfrac{1}{2}}}{2 \boxed{2}} = \frac{3}{2} + \frac{4}{\pi} \cdot \frac{\dfrac{1}{2}\pi}{2} = \frac{3}{2} + 1 = \frac{5}{2}$$

Using the orthogonal property in Eq. (iv), we get

$$c_2 \frac{\pi}{2} = \int_{-1}^{1} \frac{(3x^4 + 2x^3 + x + 2) T_2}{\sqrt{1 - x^2}} \, dx$$

$$= \frac{2}{\pi} \int_{-1}^{1} \frac{(3x^4 + 2x^3 + x + 2)(2x^2 - 1) \, dx}{\sqrt{1 - x^2}}$$

$$= \frac{2}{\pi} \left[6 \int_{-1}^{1} \frac{x^6 \, dx}{\sqrt{1 - x^2}} + 4 \int_{-1}^{1} \frac{x^5 \, dx}{\sqrt{1 - x^2}} + 2 \int_{-1}^{1} \frac{x^3 \, dx}{\sqrt{1 - x^2}} + 4 \int_{-1}^{1} \frac{x^2 \, dx}{\sqrt{1 - x^2}} \right.$$

$$\left. - 3 \int_{-1}^{1} \frac{x^4 \, dx}{\sqrt{1 - x^2}} - 2 \int_{-1}^{1} \frac{x^3 \, dx}{\sqrt{1 - x^2}} - \int_{-1}^{1} \frac{x \, dx}{\sqrt{1 - x^2}} - 2 \int_{-1}^{1} \frac{dx}{\sqrt{1 - x^2}} \right]$$

By using the property of definite integral, we get

$$= \frac{2}{\pi} \left[6 \int_{-1}^{1} \frac{x^6 \, dx}{\sqrt{1 - x^2}} + 0 + 0 + 4 \int_{-1}^{1} \frac{x^2 \, dx}{\sqrt{1 - x^2}} - 3 \int_{-1}^{1} \frac{x^4 \, dx}{\sqrt{1 - x^2}} - 0 - 0 - 2 \int_{-1}^{1} \frac{dx}{\sqrt{1 - x^2}} \right]$$

$$= \frac{45}{12} + 4 \cdot \left(\frac{2}{\pi} \int_{-1}^{1} \frac{x^2 \, dx}{\sqrt{1 - x^2}} \right) - 6 \left(\frac{1}{\pi} \int_{-1}^{1} \frac{x^4 \, dx}{\sqrt{1 - x^2}} \right) - 2 \cdot \left(\frac{2}{\pi} \int_{-1}^{1} \frac{dx}{\sqrt{1 - x^2}} \right)$$

$$= \frac{45}{12} + 4.1 - 6 \cdot \frac{3}{8} - 2.2 = \frac{15}{4} + 4 - \frac{9}{4} - 4 = \frac{15 - 9}{4} = \frac{6}{4} = \frac{3}{2}$$

Hence the required approximation is

$$f(x) = \frac{25}{8} T_0(x) + \frac{5}{2} T_1(x) + \frac{3}{2} T_2(x) = \frac{25}{8} + \frac{5x}{2} + \frac{3}{2}(2x^2 - 1)$$

$$= \frac{25}{8} + \frac{5}{2}x + 3x^2 - \frac{3}{2} = 3x^2 + \frac{5}{2}x + \frac{13}{8}$$

Example 18: Express $2T_0(x) - \frac{1}{4} T_2(x) - \frac{1}{8} T_4(x)$ as polynomials in x.

Solution: We know that $T_0(x) = 1$, $T_2(x) = 2x^2 - 1$, $T_4(x) = 8x^4 - 8x^2 + 1$, so we have

$$2T_0(x) - \frac{1}{4} T_2(x) - \frac{1}{8} T_4(x) = 2 \times 1 - \frac{1}{4}(2x^2 - 1) - \frac{1}{8}(8x^4 - 8x^2 + 1)$$

$$= 2 - \frac{x^2}{2} + \frac{1}{4} - x^4 + x^2 - \frac{1}{8}$$

$$= -x^4 + \frac{1}{2}x^2 + \frac{17}{8}$$

Example 19: Express $1 - x^2 + 2x^4$ as sum of Chebyshev polynomials.

Solution: $1 - x^2 + 2x^4 = T_0(x) - \frac{1}{2}(T_0 + T_2) + \frac{2}{8}(3T_0 + 4T_2 + T_4)$

$$= T_0(x) - \frac{1}{2}T_0(x) - \frac{1}{2}T_2(x) + \frac{3}{4}T_0(x) + T_2(x) + \frac{1}{4}T_4(x)$$

$$= \frac{5}{4}T_0(x) + \frac{1}{2}T_2(x) + \frac{1}{4}T_4(x)$$

6.11 UNIFORM APPROXIMATION

In uniform approximation, we use the uniform norm, i.e. we choose the approximation such that the maximum component of the error vector is minimized. Weierstrass approximation theorem provides the possibility of a uniform approximation, using uniform norm to continuous function on a closed interval. This theorem makes the use of Bernstein polynomials.

$$B_n(f, x) = \sum_{k=0}^{n} f\left(\frac{k}{n}\right) {}^nC_k \, x^k \, (1-x)^{n-k}$$

defined on [0, 1] and it has been shown that

$$\underset{n \to \infty}{\text{Lim }} B_n(f, x) = f(x) \text{ uniformly on } [0, 1]$$

But the convergence of approximation using Bernstein polynomials is very slow. Chebyshev polynomials are the best known polynomials which give a rapid uniform approximation to a continuous function $f(x)$.

6.12 UNIFORM (OR MINIMAX) POLYNOMIAL APPROXIMATION

Consider a continuous function $f(x)$ defined on a closed interval $[a, b]$ be approximated by the polynomial $P_n(x) = c_0 + c_1 x + c_2 x^2 + \dots + c_n x^n$. Then in uniform (or minimax) polynomial approximation, we want to determine the constants $c_0, c_1, c_2, \dots, c_n$ such that the error or deviation

$$E_n(f, c_0, c_1, \dots, c_n) = f(x) - P_n(x)$$

satisfies the minimax principle.

$$\max_{x \in [a,b]} | E_n(f, c_0, c_1, \dots, c_n) | = \min_{x \in [a,b]} | E_n(f, c_0, c_1, \dots, c_n) |$$

First case when $n = 0$. In this case, we aim to approximate the function $f(x)$ by a constant c_0. Consider

$$M = \max_{x \in [a,b]} | (f(x) |, m = \min_{x \in [a,b]} | f(x) |$$

By minimax principle, we require

$$\max_{x \in [a,b]} | f(x) - c_0 | = \min_{x \in [a,b]} | f(x) - c_0 |$$

which gives $M - c_0 = -(m - c_0) = -m + c_0$

or
$$c_0 = \frac{1}{2}(m + M)$$

Also then
$$E_0(f, c_0) = M - \frac{1}{2}(M + m) = \frac{1}{2}(M - m)$$

The graph of the constant minimax approximation has been drawn in Fig. 6.2. From Fig. 6.2, we observe that when the error curve $\varepsilon(x) = f(x) - c_0$ is drawn, the value $\pm E_0(f, c_0)$ is assumed at least twice, once with plus sign and once with minus sign, but always of equal absolute value.

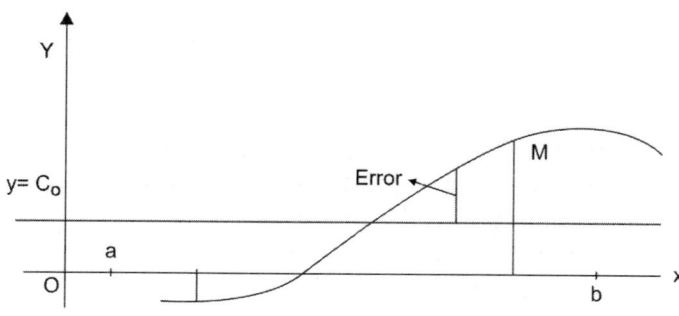

Fig. 6.2

As a generalization of the above concept, we have the following theorem due to Chebyshev.

6.13 CHEBYSHEV EQUIOSCILLATION THEOREM

Let $f(x)$ be continuous on a closed interval $[a, b]$ and $p_n(x)$ be the best uniform approximation according to the minimax principle

$$\max_{x \in [a,b]} | f(x) - p_n(x)| = \min_{x \in [a,b]} | f(x) - p_n(x)| \tag{6.27}$$

We set
$$E_n(f, x) = \max_{x \in [a,b]} | f(x) - p_n(x)| \text{ and } \varepsilon(x) = f(x) - p_n(x) \tag{6.28}$$

Then there are at least $n + 2$ points
$$a = x_0 < x_1 < x_2 \ldots < x_n < x_{n+1} = b$$

where
(i) $\varepsilon(x_i) = \pm E_n$, $i = 0, 1, 2, \ldots, n + 1$
(ii) $\varepsilon(x_i) = -\varepsilon(x_{i+1})$, $i = 0, 1, 2, \ldots, n$

The best uniform approximation is completely and uniquely determined under the above conditions (i) and (ii). It may be observed that the condition (ii) implies $\varepsilon'(x_i) = 0$, $i = 1, 2, \ldots, n$. To clear the above theory we give some examples.

Example 20: Obtain the Chebyshev linear polynomial approximation to the function $f(x) = x^3$ on $[0, 1]$,

Solution: Consider $P_1(x) = c_0 + c_1 x$

The error of approximation ε is given by
$$\varepsilon(x) = x^3 - c_0 - c_1 x$$

We have $\varepsilon'(x) = 3x^2 - c_1$

Assuming the points $x_0 = 0$, $x_1 = \alpha$, $x_2 = 1$ and applying Chebyshev equioscillation theorem, we have

$$\varepsilon(0) = -\varepsilon(\alpha) \text{ or } \varepsilon(0) + \varepsilon(\alpha) = 0$$
$$\varepsilon(\alpha) = -\varepsilon(1) \text{ or } \varepsilon(\alpha) + \varepsilon(1) = 0$$

and
$$\varepsilon'(\alpha) = 0$$

Thus, we have

$$\alpha^3 - \alpha c_1 - 2c_0 = 0$$
$$\alpha^3 - (\alpha + 1) c_1 - 2c_0 + 1 = 0$$

and
$$3\alpha^2 - c_1 = 0$$

These equations give $c_1 = 1$, $\alpha = \dfrac{1}{\sqrt{3}}$ and $c_0 = -\dfrac{1}{3\sqrt{3}}$.

Hence the required Chebyshev linear approximation is given by $P_1(x) = x - \dfrac{1}{3\sqrt{3}}$.

Example 21: Prove that the polynomial of best approximation of degree not exceeding 3 for $|x|$ in the interval $[-1, 1]$ is $x^2 + \dfrac{1}{8}$.

Solution: Consider the polynomial be $P_3(x) = c_0 + c_1 + c_2x^2 + c_3x^3$. Now we want to determine c_0, c_1, c_2, c_3 such that

$$\max_{-1 \le x \le 1} \left\| |x| - P(x) \right\| = \min_{-1 \le x \le 1} \left\| |x| - P(x) \right\|$$

The error of approximation ε is given by

$$\varepsilon(x) = |x| - c_0 - c_1 x - c_2 x^2 - c_3 x^3$$

$$= \begin{cases} x - c_0 - c_1 x - c_2 x^2 - c_3 x^3, & x \ge 0 \\ -x - c_0 - c_1 x - c_2 x^2 - c_3 x^3, & x < 0 \end{cases}$$

Taking the points $x_0 = -1$, $x_1 = -\alpha$, $x_2 = 0$, $x_3 = \alpha$, $x_4 = 1$ and using Chebyshev equioscillation theorem, we get

$$\left. \begin{array}{c} \varepsilon(-1) + \varepsilon(-\alpha) = 0 \\ \varepsilon(-\alpha) + \varepsilon(0) = 0 \\ \varepsilon(0) + \varepsilon(\alpha) = 0 \\ \varepsilon(\alpha) + \varepsilon(1) = 0 \end{array} \right] \tag{i}$$

and
$$\varepsilon'(0) = 0, \ \varepsilon'(-\alpha) = 0, \ \varepsilon'(\alpha) = 0, \tag{ii}$$

From Eq. (ii), we get

$$\varepsilon'(-\alpha) = -1 - c_1 + 2c_2\alpha - 3c_3\alpha^2 = 0$$
$$\varepsilon'(0) = -c_1 = 0$$
$$\varepsilon'(\alpha) = 1 - 3c_3\alpha^2 - 2c_2\alpha - c_1 = 0$$

which gives $c_1 = 0$, $c_3 = 0$ and $c_2 = \dfrac{1}{2\alpha}$.

Remember that $|x|$ is not differentiable at $x = 0$. Its right hand derivative at $x = 0$ is 1 and left hand derivative is -1 and their average is zero. So we have taken $\varepsilon'(0) = 0 - c_1 = -c_1$.

From Eq. (i), we have

$$\varepsilon(-\alpha) + \varepsilon(0) = \varepsilon(0) + \varepsilon(-\alpha) = \alpha - 2c_0 - c_2\alpha^2 = 0$$

and $$\varepsilon(-1) + \varepsilon(-\alpha) = \varepsilon(\alpha) + \varepsilon(1) = 1 + \alpha - 2c_0 - c_2 - c_2\alpha^2 = 0$$

which give $c_2 = 1, C_0 = \dfrac{\alpha - \alpha^2}{2}$.

$$\therefore \quad c_0 = \frac{1}{8}, c_1 = 0, c_2 = 1, c_3 = 0$$

Hence, the best approximation to $|x|$ in the interval $[-1, 1]$ is $x^2 + \dfrac{1}{8}$.

6.14 CHEBYSHEV POLYNOMIAL APPROXIMATION

Let $f(x)$ be a continuous function defined on the interval $[-1, 1]$ and consider $c_0 + c_1x + c_2x^2 + ... + c_nx^n$ be the required minimax (or uniform) polynomial approximation for $f(x)$. Then this problem can be solved with the help of Chebyshev polynomials.

Let $f(x) = \dfrac{a_0}{2} + \displaystyle\sum_{i=1}^{\infty} a_i \cdot T_i(x)$ is the Chebyshev series expansion for $f(x)$. Then the truncated series or the partial sum

$$P_n(x) = \frac{a_0}{2} + \sum_{i=1}^{n} a_i \cdot T_i(x) \tag{6.29}$$

is very nearly the solution to the problem

$$\max_{x\in[-1,1]} \left| f(x) - \sum_{i=0}^{n} c_i x_i \right| = \min_{x\in[-1,1]} \left| f(x) - \sum_{i=0}^{n} c_i x^i \right|$$

i.e. the partial sum of Eq. (6.29) is nearly the best uniform approximation to $f(x)$.

The explanation is given below. We can write

$$f(x) = \frac{a_0}{2} + a_1T_1(x) + a_2T_2(x) + ... + a_nT_n + a_{n+1}T_{n+1}(x) + \text{remainder} \tag{6.30}$$

Neglecting the remainder in Eq. (6.30)

$$f(x) - \left[\frac{a_0}{2} + \sum_{i=1}^{n} a_iT_i(x) \right] = a_{n+1}T_{n+1}(x) \tag{6.31}$$

Since $T_{n+1}(x)$ has $(n + 2)$ equal maxima and minima which alternate in sign, therefore by Chebyshev equioscillation theorem, the polynomial in Eq. (6.29) of degree n is the best uniform approximation to $f(x)$.

Thus to determine the best uniform approximation to a given continuous function defined on $[-1, 1]$, we find it convenient to start with the truncated Chebyshev expansion of the function and then improve upon it by the Chebyshev equioscillation theorem. The polynomial in Eq. (6.29) is called the minimax polynomial. By this process, we can obtain the best lower order approximation, called the minimax approximation to a given polynomial.

Example 22: Find the best lower order approximation to the cubic $2x^3 + 3x^2$ in the interval $[-1, 1]$.

Solution: We have
$$x^3 = \frac{1}{4}[3T_1(x) + T_3(x)]$$

and
$$x^2 = \frac{1}{2}[T_2(x) + T_0(x)]$$

So, we write $2x^3 + 3x^2$ in terms of Chebyshev polynomial as

$$2x^3 + 3x^2 = \frac{2}{4}[3T_1(x) + T_3(x)] + 3x^2$$

$$= 3x^2 + \frac{3}{2}T_1(x) + \frac{1}{2}T_3(x)$$

$$= 3x^2 + \frac{3}{2}x + \frac{1}{2}T_3(x), \qquad \text{since } T_1(x) = x$$

Since $|T_3(x)| \le 1, -1 \le x \le 1$, therefore the polynomial $3x^2 + \frac{3}{2}x$ is the required best lower order approximation to the given cubic with a maximum error $\pm 1/2$ in the range $[-1, 1]$.

Example 23: Find the best uniform approximation of degree 3 or less to x^4 in $[-1, 1]$.

Solution: Write x^4 in terms of Chebyshev polynomials, we get

$$x^4 = \frac{3}{8}T_0 + \frac{1}{2}T_2 + \frac{1}{8}T_4$$

Since T_4 is a polynomial of degree four, therefore, we approximate

$$f(x) = x^4 \text{ by } \frac{3}{8}T_0 + \frac{1}{2}T_2$$

The error of approximation $\varepsilon(x)$ is given by

$$\varepsilon(x) = x^4 - \left(\frac{3}{8}T_0 + \frac{1}{2}T_2\right) = \frac{T_4}{8}$$

Hence the uniform polynomial approximation of degree 3 or less to x^4 is

$$\frac{3}{8}T_0 + \frac{1}{2}T_2 = \frac{3}{8} + \frac{1}{2}(2x^2 - 1) = x^2 - \frac{1}{8}$$

and the maximal error of this approximation on $[-1, 1]$

$$= \max_{x \in [-1,1]} \left|x^4 - \left(\frac{3}{8}T_0 + \frac{1}{2}T_2\right)\right| = \max_{x \in [-1,1]} \frac{1}{8}T_4(x) = \frac{1}{8}$$

since $|T_4(x)| \le 1, -1 \le x \le 1$.

Example 24: The function $P_3(x) = x^3 - 9x^2 - 20x + 5$ is given. Find a second degree polynomial $P_2(x)$ such that $\delta = \max\limits_{x \in [0,4]} |P_3(x) - P_2(x)|$ becomes as small as possible. The value of δ and the value of x for which $|P_3(x) - P_2(x)| = \delta$ should also be given.

Solution: We have $P_3(x) = x^3 - 9x^2 - 20x + 5, 0 \le x \le 4$. We shall first change the interval from [0, 4] to [−1, 1] by using the transformation

$$x = 2(t + 1)$$

So we obtain

$$P_3(t) = 8(t + 1)^3 - 36(t + 1)^2 - 40(t + 1) + 5$$
$$= 8t^3 - 12t^2 - 88t - 63, -1 \le t \le 1.$$

Now expressing each power of t in terms of Chebyshev polynomials, we get

$$P_3(t) = -63T_0 - 88T_1 - 6(T_2 + T_0) + 2(T_3 + 3T_1)$$
$$= -69T_0 - 82T_1 - 6T_2 + 2T_3, \text{ where } -1 \le t \le 1$$

If we truncate this polynomial at T_2, we have

$$\max\limits_{t \in [-1,1]} |P_3(t) - (-69T_0 - 82T_1 - 6T_2)| = \max\limits_{-1 \le t \le 1} |2T_3|$$
$$= 2 = \min\limits_{-1 \le t \le 1} |P_3(t) - (-69T_0 - 82T_1 - 6T_2)|$$

Hence the required approximation is

$$P_2(t) = -69T_0 - 82T_1 - 6T_2$$
$$= -69 - 82t - 6(2t^2 - 1)$$
$$= -12t^2 - 82t - 63$$

for which the maximum absolute error $\delta = 2$ is as small as possible.

Substituting $t = \dfrac{1}{2}(x - 2)$, we obtain

$$P_2(x) = -12\left(\frac{x}{2} - 1\right)^2 - 82\left(\frac{x}{2} - 1\right) - 63 = -3x^2 - 29x + 7$$

We also find that

$$|P_3(x) - P_2(x)| = |x^3 - 6x^2 + 9x - 2| = \delta = 2$$

for $x = 0, 1, 3$ and 4.

6.15 LANCZOS ECONOMIZATION OF POWER SERIES FOR A GENERAL FUNCTION

Firstly we write the given $f(x)$ as a power series in x in the form

$$f(x) = \sum_{i=0}^{\infty} a_i x^i, \quad -1 \le x \le 1 \tag{6.32}$$

Then we change each power of x in Eq. (6.32) in terms of Chebyshev polynomials and we obtain

$$f(x) = \sum_{i=0}^{\infty} c_i T_i(x) \tag{6.33}$$

as the Chebyshev series expansion for $f(x)$ on [−1, 1].

It has been found that for a large number of functions $f(x)$, the series in Eq. (6.33) converges more rapidly than the power series given by Eq. (6.32).

If we truncate series in Eq. (6.33) at $T_n(x)$, then the partial sum

$$P_n(x) = \sum_{i=0}^{n} c_i T_i(x) \tag{6.34}$$

is a good uniform approximation to $f(x)$ in the sense

$$\max_{x \in [-1,1]} |f(x) - P_n(x)| \le |c_{n+1}| + |c_{n+2}| + \ldots \le \varepsilon, \text{ say}$$

For a given ε, it is possible to find the number of terms that should be retained in Eq. (6.34). This process is known as Lanczos economization. Replacing each $T_i(x)$ in Eq. (6.34) by its polynomial form and rearranging the terms, we get the required economized polynomial approximation for $f(x)$.

Example 25: Economize the power series $\sin x \approx x - \dfrac{x^3}{6} + \dfrac{x^5}{120} - \dfrac{x^7}{5040}, -1 \le x \le 1$.

Solution: Since $\dfrac{1}{5040} = 0.000198\ldots$, therefore the truncated series

$$\sin x \approx x - \frac{x^3}{6} + \frac{x^5}{120} \tag{i}$$

will produce a change in the fourth decimal place only.

Write the power series of x in Eq. (i) into Chebyshev polynomials, we get

$$\sin x \approx T_1 - \frac{1}{24}(3T_1 + T_3) + \frac{1}{120 \times 16}(10T_1 + 5T_3 + T_5)$$

or

$$\approx \frac{169}{192} T_1 - \frac{5}{128} T_3 + \frac{1}{1920} T_5 \tag{ii}$$

Since $\dfrac{1}{1920} = 0.00052\ldots$, the truncated series

$$\sin x \approx \frac{169}{192} T_1 - \frac{5}{128} T_3 \tag{iii}$$

will produce a change in the fourth decimal place only. The economized series is therefore given by

$$\sin x \approx \frac{169}{192} T_1 - \frac{5}{128} T_3$$

or

$$\approx \frac{169}{192} x - \frac{5}{128}(4x^3 - 3x)$$

or

$$\approx \frac{383}{384} x - \frac{5}{32} x^3$$

Example 26: Find a uniform polynomial approximation of degree four or less to e^x on $[-1, 1]$ using Lanczos economization with a tolerance of $\varepsilon = 0.02$.

Solution: We know that

$$f(x) = e^x = 1 + x + \frac{x^2}{2} + \frac{x^3}{6} + \frac{x^4}{24} + \frac{x^5}{120} + \dots$$

Since $\dfrac{1}{120} = 0.008\dots$, therefore

$$e^x \simeq 1 + x + \frac{x^2}{2} + \frac{x^3}{6} + \frac{x^4}{24} \tag{i}$$

with a tolerance of $\varepsilon = 0.02$.

Converting each power of x in Eq. (i) in terms of Chebyshev polynomials, we get

$$e^x \simeq T_0 + T_1 + \frac{1}{4}(T_0 + T_2) + \frac{1}{24}(3T_1 + T_3) + \frac{1}{192}(3T_0 + 4T_2 + T_4)$$

or

$$\simeq \frac{81}{64}T_0 + \frac{9}{8}T_1 + \frac{13}{48}T_2 + \frac{1}{24}T_3 + \frac{1}{192}T_4 \tag{ii}$$

we have $\dfrac{1}{192} = 0.005\dots$

$\therefore$ the magnitude of the last term on the right hand side of Eq. (ii) is less than 0.02.

Hence the required economized polynomial approximation for e^x is given by

$$e^x \simeq \frac{81}{64}T_0 + \frac{9}{8}T_1 + \frac{13}{48}T_2 + \frac{1}{24}T_3$$

By changing the Chebyshev polynomials in powers of x, we get

$$e^x \simeq \frac{x^3}{6} + \frac{13}{24}x^2 + x + \frac{191}{192}.$$

Example 27: Suppose we want to approximate the function $f(x) = (3 + x)^{-1}$ on the interval $-1 \le x \le 1$ with a polynomial $P(x)$ such that $\max\limits_{|x| \le 1} | f(x) - P(x)| \le 0.021$.

(a) Show that there does not exist a first degree polynomial satisfying this condition.
(b) Show that there exists a second degree polynomial satisfying this condition.

Solution: We have $\quad f(x) = (3 + x)^{-1} = \dfrac{1}{3}\left(1 + \dfrac{x}{3}\right)^{-1}$

$$= \frac{1}{3} - \frac{1}{9}x + \frac{1}{27}x^2 - \frac{x^3}{81} + \dots, \quad -1 \le x \le 1$$

We approximate $f(x)$ by the second degree polynomial

$$P_2(x) = \frac{1}{3} - \frac{1}{9}x + \frac{x^2}{27}$$

with the error of approximation $\dfrac{1}{81} \simeq 0.0123$, if we approximate $f(x)$ by the first degree

polynomial $P_1(x) = \dfrac{1}{3} - \dfrac{x}{9}$, then the error of approximation is $\dfrac{1}{27} \simeq 0.04$ which is more than the tolerable error 0.021.

Transforming $P_2(x)$ in terms of Chebyshev polynomials, we get

$$P_2(x) = \frac{1}{3} - \frac{1}{9}x + \frac{x^2}{27} = \frac{1}{3}T_0 - \frac{1}{9}T_1 + \frac{1}{54}(T_0 + T_2)$$

$$= \frac{19}{54}T_0 - \frac{1}{9}T_1 + \frac{1}{54}T_2$$

If we truncate $P_2(x)$ at T_1, then maximum error of approximation is $\dfrac{1}{54} = 0.0185$ and the total error becomes $\dfrac{1}{54} + \dfrac{1}{81} = \dfrac{5}{162} \approx 0.03$.

which is again more than the tolerable error 0.021.

Hence there exists no first degree polynomial satisfying the given accuracy.

However, $P_2(x) = \dfrac{1}{3} - \dfrac{1}{9}x + \dfrac{x^2}{27}$ is the second degree polynomial of approximation for $(3 + x)^{-1}$ satisfying the given accuracy.

Example 28: The function f is defined by

$$f(x) = \frac{1}{x}\int_0^x \frac{1 - e^{-t^2}}{t^2}\, dt$$

Approximate f by a polynomial $P(x) = a + bx + cx^2$ such that

$$= \max_{|x| \le 1} |f(x) - P(x)| \le 0.005.$$

Solution: We have $f(x) = \dfrac{1}{x}\int_0^x \left(\dfrac{1 - e^{-t^2}}{t^2}\right) dt$

$$= \frac{1}{x}\int_0^x \frac{1 - \left(1 - t^2 + \dfrac{t^4}{2} - \dfrac{t^6}{6} + \dfrac{t^8}{24} - \dfrac{t^{10}}{120} + \dfrac{t^{12}}{720} - \cdots\right)}{t^2}\, dt$$

$$= \frac{1}{x}\int_0^x \left(1 - \frac{t^2}{2} + \frac{t^4}{6} - \frac{t^6}{24} + \frac{t^8}{120} - \frac{t^{10}}{720} + \cdots\right) dt$$

$$= \frac{1}{x}\left[x - \frac{x^3}{6} + \frac{x^5}{30} - \frac{x^7}{168} + \frac{x^9}{1080} - \frac{x^{11}}{7920} + \cdots\right]$$

$$= 1 - \frac{x^2}{6} + \frac{x^4}{30} - \frac{x^6}{168} + \frac{x^8}{1080} - \frac{x^{10}}{7920} + \cdots \qquad \text{(i)}$$

The tolerable error is 0.005.

Truncating the series in Eq. (i) at x^8, we get

$$P(x) = 1 - \frac{x^2}{6} + \frac{x^4}{30} - \frac{x^6}{168} + \frac{x^8}{1080} \qquad \text{(ii)}$$

Changing the series in Eq. (ii) in terms of Chebyshev polynomials, we get

$$P(x) = T_0 - \frac{1}{12}(T_2 + T_0) + \frac{1}{240}(T_4 + 4T_2 + 3T_0) - \frac{1}{5376}(T_6 + 6T_4 + 15T_2 + 10T_0)$$

$$+ \frac{1}{138240}(T_8 + 8T_6 + 28T_4 + 56T_2 + 35T_0)$$

$$= 0.92755973\,T_0 - 0.06905175\,T_2 + 0.003253\,T_4 - 0.000128\,T_6 + 0.000007\,T_8 \qquad \text{(iii)}$$

Truncating the RHS of Eq. (iii) at T_2, we get

$$P(x) = 0.92755973\,T_0 - 0.06905175\,T_2$$

Changing the Chebyshev polynomials in terms of x, we get

$$P(x) = 0.92755973 - 0.06905175\,(2x^2 - 1)$$

$$= 0.99661148 - 0.13810350\,x^2$$

$$= 0.9966 - 0.1381\,x^2$$

The maximum absolute error in the neglected terms is clearly less than the tolerable error.

Example 29: Find a polynomial $P(x)$ of degree as low as possible such that

$$\max_{|x| \le 1} | e^{x^2} - P(x) | \le 0.05.$$

Solution: we have
$$e^{x^2} = 1 + x^2 + \frac{x^4}{2} + \frac{x^6}{6} + \frac{x^8}{24} + \frac{x^{10}}{120} + \dots$$

$$\simeq 1 + x^2 + \frac{x^4}{2} + \frac{x^6}{6} + \frac{x^8}{24} = P(x)$$

with error in the leading term as $\dfrac{1}{120} \approx 0.0083$.

Converting $P(x)$ in terms of Chebyshev polynomials, we get

$$P(x) = T_0 + \frac{1}{2}(T_0 + T_2) + \frac{1}{16}(3T_0 + 4T_2 + T_4) + \frac{1}{192}(10T_0 + 15T_2 + 6T_4 + T_6)$$

$$+ \frac{1}{3072}(35T_0 + 56T_2 + 28T_4 + 8T_6 + T_8)$$

$$= \frac{1}{3072}[5379\,T_0 + 2600\,T_2 + 316\,T_4 + 24\,T_6 + T_8]$$

We have $\left| \dfrac{1}{3072}(24\,T_6 + T_8) \right| \le 0.0082$

which is less than the required maximal error of approximation 0.05.

Hence we get the required approximation

$$e^{x^2} \approx \frac{1}{3072}(5379\,T_0 + 2600\,T_2 + 316\,T_4)$$

$$= \frac{1}{3072}[5379 + 2600\,(2x^2 - 1) + 316\,(8x^4 - 8x^2 + 1)]$$

$$= \frac{1}{3072}(3095 + 2672\,x^2 + 2528\,x^4)$$

$$= 1.0075 + 0.8698x^2 + 0.8229x^4.$$

EXERCISE 6.2

1. The function f is defined by $f(x) = \int_0^x e^{-t^2/2}\,dt$. Determine the coefficients of a fifth degree polynomial $P_5(x)$ for which $|f(x) - P_5(x)| \le 10^{-4}$ when $|x| \le 1$. (The coefficients should be accurate to within $\pm 2 \times 10^{-5}$).

   [Ans. $0.9997x - 0.1645x^3 + 0.0204x^5$]

2. Approximate $f(x) = \frac{1}{x}\int_0^x \frac{e^t - 1}{t}\,dt$ by a third degree polynomial $P_3(x)$ so that

   $\max_{x \in [-1,\,1]} |f(x) - P_3(x)| \le 0.0003$. $\left[\text{Ans. } \frac{4799}{4800} + \frac{3455}{13824}x + \frac{103}{1800}x^2 + \frac{37}{3456}x^3\right]$

3. Approximate $f(x) = (2x - 1)^3$ by a straight line on the interval $[0, 1]$ so that the maximum norm of the error function is minimized (use Lanczos economization).

   $\left[\text{Ans. } \frac{3(2x - 1)}{4}\right]$

4. Show that the line $\frac{3}{4}(2x - 1)$ is obtained if f is approximated by the method of

   least square with weight function $\frac{1}{\sqrt{x(1 - x)}}$.

5. Find the lowest order polynomial which approximate the function

   $$f(x) = \sum_{r=0}^{4}(-x)^r$$

   in the range $0 \le x \le 1$, with an error less than 0.1. $\left[\text{Ans. } \frac{1}{128}(160x^2 - 168x + 131)\right]$

6. Determine the values of a, b, c and d in the polynomial $P(x) = ax^3 + bx^2 + cx + d$
   which minimizes $\max_{-1 \le x \le 1} |P(x) - |x|\,|$.   [Ans. $a = 0, b = 1, c = 0, d = 1/8$]

7. Determine the polynomial of second degree which is the best approximation in

   maximum norm to $\sqrt{x}$ on the point set $\left(0, \frac{1}{9}, \frac{4}{9}, 1\right)$. $\left[\text{Ans. } \frac{1}{16} + 2x - \frac{9}{8}x^2\right]$

Numerical Differentiation

7.1 INTRODUCTION

In this chapter, we shall study about the numerical differentiation. Actually, the need for numerical differentiation arises from the fact that very often, either $f(x)$ is not explicitly given and only the value of $f(x)$ at certain discrete points are known or $f'(x)$ is difficult to compute analytically

Numerical differentiation methods can be obtained by using any one of the following two techniques:

i. Methods based on interpolation

ii. Methods based on finite difference

7.2 METHODS BASED ON INTERPOLATION

Here we see that how to generate differentiation formulas by differentiating an interpolant. The idea is simple, the first stage is to construct an interpolating polynomial from the data. An approximation of the derivative at any point can be then obtained by a direct differentiation of the interpolant.

By assuming the above procedure and assume that $f(x)$ be a function, the value of $f(x)$ are $f(x_0), f(x_1), ..., f(x_n)$ to the corresponding values of $x = x_0, x_1, ..., x_n$. The Lagrange form of the interpolation polynomial through these points is

$$P_n(x) = \sum_{i=0}^{n} l_i(x) f(x_i) \tag{7.1}$$

where
$$l_i(x) = \frac{(x - x_0)(x - x_1)...(x - x_{i-1})(x - x_{i+1})...(x - x_n)}{(x_i - x_0)(x_i - x_1)...(x_i - x_{i-1})(x_i - x_{i+1})...(x_i - x_n)}$$

The error term in Eq. (7.1) will be

$$E_n(x) = f(x) - P_n(x) = \frac{(x - x_0)(x - x_1)...(x - x_n)}{(n + 1)!} f^{(n+1)}(\xi_n)$$

or
$$f(x) - P_n(x) = \frac{1}{(n + 1)!} f^{(n+1)}(\xi_n) \prod_{i=0}^{n} (x - x_i)$$

where $\xi_n \in [\min(x, x_0, x_1, ..., x_n), \max(x, x_0, x_1, ..., x_n)]$.

Since we are assuming here that the points $x_0, x_1, ..., x_n$ are fixed, we would like to emphasize the dependence of ξ_n on x and hence replace the notation ξ_n by ξ_x. Then

$$f(x) = P_n(x) + E_n = \sum_{i=0}^{n} l_i(x) f(x_i) + \frac{1}{(n+1)!} f^{(n+1)}(\xi_x) w(x) \quad (7.2)$$

where $w(x) = \prod_{i=0}^{n} (x - x_i)$

Differentiating Eq. (7.2)

$$f'(x) = \sum_{i=0}^{n} l_i'(x) f(x_i) + \frac{1}{(n+1)!} f^{(n+1)}(\xi_x) w'(x) + \frac{1}{(n+1)!} \left[\frac{d}{dx} f^{(n+1)}(\xi_x) \right] w(x) \quad (7.3)$$

Let x is one of the interpolating points, i.e. $x \in [x_0, x_1, ..., x_n]$, say x_k, then

$$f'(x_k) = \sum_{i=0}^{n} l_i'(x_k) f(x_i) + \frac{1}{(n+1)!} f^{(n+1)}(\xi_{x_k}) w'(x_k) \quad (7.4)$$

Now

$$w'(x) = \sum_{j=0}^{n} \prod_{\substack{i=0 \\ i \neq j}}^{n} (x - x_i)$$

$$= \sum_{j=0}^{n} [(x - x_0)(x - x_1)...(x - x_{j-1})(x - x_{j+1})...(x - x_n)]$$

Hence, when $w'(x)$ is calculated at point x_k, there is only one term in $w'(x)$ that does not vanish, i.e.

$$w'(x_k) = \prod_{\substack{i=0 \\ i \neq k}}^{n} (x_k - x_i)$$

then the numerical differentiation formula in Eq. (7.4) becomes

$$f'(x_k) = \sum_{i=0}^{n} l_i'(x_k) f(x_i) + \frac{1}{(n+1)!} f^{(n+1)}(\xi_{x_k}) \prod_{\substack{i=0 \\ i \neq k}}^{n} (x_k - x_i) \quad (7.5)$$

The formula in Eq. (7.5) is known as differentiation by interpolation:

When the tabular points are equispaced, we can use Newton's forward or backward difference formula. For $n = 1$ and $k = 0$. This means that we use two interpolation points $(x_0, f(x_0))$ and $(x_1, f(x_1))$, and want to approximate $f'(x_0)$. Then Lagrange interpolation polynomial be

$$f(x) \approx P_1(x) = \frac{x - x_1}{x_0 - x_1} f_0 + \frac{x - x_0}{x_1 - x_0} f_1 \quad (7.6)$$

and

$$f'(x_k) = \frac{f_0}{x_0 - x_1} + + \frac{f_1}{x_1 - x_0} = \frac{f_1 - f_0}{x_1 - x_0}$$

$\therefore$

$$f'(x_k) = \frac{f_1 - f_0}{x_1 - x_0}, k = 0, 1 \quad (7.7)$$

The expression for the error of interpolation can be calculated from second term in the RHS of Eq. (7.5).

$$E_1^{(1)}(x_0) = \frac{1}{2!} f''(\xi)(x_0 - x_1), x_0 < \xi < x_1 \quad (7.8)$$

For $n = 2$ and $k = 0$, we get

$$f(x) \approx P_2(x) = \frac{(x - x_1)(x - x_2)}{(x_0 - x_1)(x_0 - x_2)} f_0 + \frac{(x - x_0)(x - x_2)}{(x_1 - x_0)(x_1 - x_2)} f_1 + \frac{(x - x_0)(x - x_1)}{(x_2 - x_0)(x_2 - x_1)} f_2 \quad (7.9)$$

$$f'(x) = \frac{(2x - x_1 - x_2)}{(x_0 - x_1)(x_0 - x_2)} f_0 + \frac{(2x - x_0 - x_2)}{(x_1 - x_0)(x_1 - x_2)} f_1 + \frac{(2x - x_0 - x_1)}{(x_2 - x_0)(x_2 - x_1)} f_2 \quad (7.10)$$

$$f'(x_0) = \frac{(2x_0 - x_1 - x_2)}{(x_0 - x_1)(x_0 - x_2)} f_0 + \frac{(x_0 - x_2)}{(x_1 - x_0)(x_1 - x_2)} f_1 + \frac{(x_0 - x_1)}{(x_2 - x_0)(x_2 - x_1)} f_2 \quad (7.11)$$

The error in this case is again calculated from the second term in the RHS of Eq. (7.5). So

$$E_2^{(1)}(x_0) = \frac{1}{3!} f'''(\xi)(x_0 - x_1)(x_0 - x_2), \ x_0 < \xi < x_2 \ \left| \begin{array}{l} \because x_1 = x_0 + h \\ x_2 = x_0 + 2h \end{array} \right. \quad (7.12)$$

Again differentiating Eq. (7.10), we get

$$f''(x) = \frac{2f_0}{(x_0 - x_1)(x_0 - x_2)} + \frac{2f_1}{(x_1 - x_0)(x_1 - x_2)} + \frac{2f_2}{(x_2 - x_0)(x_2 - x_1)}$$

$$f''(x_0) = 2\left[\frac{f_0}{(x_0 - x_1)(x_0 - x_2)} + \frac{f_1}{(x_1 - x_0)(x_1 - x_2)} + \frac{f_2}{(x_2 - x_0)(x_2 - x_1)} \right] \quad (7.13)$$

with the error of differentiation

$$E_2^{(2)}(x_0) = \frac{1}{3}(2x_0 - x_1 - x_2) f'''(\xi) + \frac{1}{24}(x_0 - x_1)(x_1 - x_2)\left[f^{iv}(\eta_1) + f^{iv}(\eta_2) \right] \quad (7.14)$$

where $x_0 < \xi, \eta_1, \eta_2 < x_2$.

If the tabular points are equispaced then formula in Eqs (7.7), (7.11) and (7.13) become respectively

$$\left. \begin{array}{l} f'(x_0) = \dfrac{(f_1 - f_0)}{h} \\[3mm] f'(x_0) = \dfrac{-3f_0 + 4f_1 - f_2}{2h} \\[3mm] f''(x_0) = \dfrac{f_0 - 2f_1 + f_2}{h^2} \end{array} \right\} \quad (7.15)$$

with the respective error terms

$$\left. \begin{array}{l} E_1^{(1)}(x_0) = -\dfrac{1}{2}h\, f''(\xi) \text{ of } O(h) \text{ (order of } h\text{)}, \ x_0 < \xi < x_1 \\[3mm] E_2^{(1)}(x_0) = \dfrac{1}{3}h^2\, f'''(\xi) \text{ of } O(h^2) \ x_0 < \xi < x_2 \\[3mm] E_2^{(2)}(x_0) = -h\, f'''(\xi), \ x_0 < \xi < x_2 \end{array} \right\} \quad (7.16)$$

In the same way, we can find the other derivative and respective error.

7.3 METHOD BASED ON FINITE DIFFERENCE

By the definition of shift operator

$$Ef(x) = f(x + h)$$

By using Taylor's series expansion, we get

$$Ef(x) = f(x + h) = f(x) + hf'(x) + \frac{h^2}{2!} f''(x) + \dots$$

$$= f(x) + hDf(x) + \frac{h^2}{2!} D^2 f(x) + \dots$$

$$= \left(1 + hD + \frac{h^2 D^2}{2!} + \dots\right) f(x)$$

$$Ef(x) = e^{hD} f(x) \tag{7.17}$$

where D is the differential operator.

Symbolically, from Eq. (7.17)

$$E = e^{hD} \Rightarrow hD = \log E \tag{7.18}$$

Again from the definition of central difference operator

$$\delta f(x) = f\left(x + \frac{h}{2}\right) - f\left(x - \frac{h}{2}\right) = (E^{1/2} - E^{-1/2}) f(x)$$

or

$$\delta = E^{1/2} - E^{-1/2} = e^{hD/2} - e^{-hD/2}$$

$$= 2 \sin h * \left(\frac{hD}{2}\right) \qquad \text{(here h * is hyperbolic)}$$

Hence

$$hD = 2 \sinh^{-1}(\delta/2) \tag{7.19}$$

Thus, we have from Eq. (7.18)

$$hD = \log E = \log (1 + \Delta) \qquad \because E = 1 + \Delta \tag{7.20}$$

and

$$hD = - \log E^{-1} = - \log (1 - \nabla) \qquad E^{-1} = 1 - \nabla \tag{7.21}$$

So, we can write all the three Eqs (7.20), (7.21) and (7.19) as

$$hD = \begin{cases} \log (1 + \Delta) = \Delta - \dfrac{1}{2}\Delta^2 + \dfrac{1}{3}\Delta^3 - \dots \\[2mm] -\log (1 - \nabla) = \nabla + \dfrac{1}{2}\nabla^2 + \dfrac{1}{3}\nabla^3 + \dots \\[2mm] 2\sinh^{-1}\left(\dfrac{\delta}{2}\right) = \delta - \dfrac{1^2}{2^2 \cdot 3!}\delta^3 + \dots \end{cases} \tag{7.22}$$

Now the general formula is given as

$$h^r D^r = \begin{cases} \Delta^r - \dfrac{1}{2}r\Delta^{r+1} + \dfrac{r(3r + 5)}{24}\Delta^{r+2} - \dots \\[2mm] \nabla^r + \dfrac{1}{2}r\nabla^{r+1} + \dfrac{r(3r + 5)}{24}\nabla^{r+2} + \dots \\[2mm] \mu\delta^r - \dfrac{r + 3}{24}\mu\delta^{r+2} + \dfrac{5r^2 + 52r + 135}{5760}\mu\delta^{r+4} + \dots \quad (r \text{ is odd}) \\[2mm] \delta^r - \dfrac{r}{24}\delta^{r+2} + r\dfrac{(5r + 22)}{5760}\delta^{r+4} - \dots \quad (r \text{ is even}) \end{cases} \tag{7.23}$$

where $\mu = \sqrt{1 + \dfrac{\delta^2}{4}}$ and is used to avoid off-step points in the method.

Putting $r = 1$ and keep only one term in Eq. (7.23), we get

$$hDf(x) = \begin{cases} \Delta f(x) = f(x+h) - f(x) \\ \nabla f(x) = f(x) - f(x-h) \\ \mu\delta\, f(x) = \dfrac{Ef(x) - E^{-1} f(x)}{2} \end{cases}$$

$$\Rightarrow \quad f'(x) = \begin{cases} \dfrac{1}{h}[f(x+h) - f(x)] \\ \dfrac{1}{h}[f(x) - f(x-h)] \\ \dfrac{1}{2h}[f(x+h) - f(x-h)] \end{cases}$$

At point $x = x_k$, we get

$$f'(x_k) = \begin{cases} \dfrac{(f_{k+1} - f_k)}{h} \\ \dfrac{(f_k - f_{k-1})}{h} \\ \dfrac{(f_{k+1} - f_{k-1})}{2h} \end{cases}$$

Similarly for $r = 2$, we have

$$f''(x_k) = \begin{cases} \dfrac{(f_{k+2} - 2f_{k+1} + f_k)}{h^2} \\ \dfrac{(f_k - 2f_{k-1} + f_{k-2})}{h^2} \\ \dfrac{(f_{k+1} - 2f_k + f_{k-1})}{h^2} \end{cases}$$

If we require higher order approximation to the derivative then we use the following formula.

We know the Newton forward formula for interpolation is given as

$$f(x) = f(x_0) + u\Delta f(x_0) + \frac{u(u-1)}{2!}\Delta^2 f(x_0) + \frac{u(u-1)(u-2)}{3!}\Delta^3 f(x_0) + \dots \tag{7.24}$$

where $u = \dfrac{x-a}{h}$

Differentiating Eq. (7.24) with respect to u, we get

$$\frac{df(x)}{du} = \Delta f(x_0) + \frac{(2u-1)}{2}\Delta^2 f(x_0) + \frac{3u^2 - 6u + 2}{6}\Delta^2 f(x_0) + \dots \tag{7.25}$$

But $$\frac{du}{dx} = \frac{1}{h}$$

$$\therefore \quad \frac{df(x)}{dx} = \frac{df(x)}{du} \cdot \frac{du}{dx}$$

$$= \frac{1}{h}\left[\Delta f(x_0) + \left(\frac{2u-1}{2}\right)\Delta^2 f(x_0) + \frac{(3u^2 - 6u + 2)}{6}\Delta^3 f(x_0) + \ldots\right] \quad (7.26)$$

Expression in Eq. (7.26) provides the value of $\frac{dy}{dx}$ at any x which is not tabulated.

If we again differentiating, we get

$$f''(x) = \frac{1}{h^2}\left[\Delta^2 f(x_0) + (u-1)\Delta^3 f(x_0) + \left(\frac{6u^2 - 18u + 11}{12}\right)\Delta^4 f(x_0) + \ldots\right] \quad (7.27)$$

In the same process we could get the formula for first, second derivatives and so on with the help of Newton backward formula for interpolation

$$f(x) = f(x_0 + nh) + u\nabla f(x_0 + nh) + \frac{u(u+1)}{2!}\nabla^2 f(x + nh) +$$

$$\frac{u(u+1)(u+2)}{3!}\nabla^3 f(x_0 + nh) + \ldots \quad (7.28)$$

where $u = \dfrac{x - (x_0 + nh)}{h}$.

Differentiating Eq. (7.28) with respect to u, we get

$$\frac{df(x)}{du} = f'(x)$$

$$= \left[\nabla f(x_0 + nh) + \frac{(2u+1)}{2}\nabla^2 f(x_0 + nh) + \frac{(3u^2 + 6u + 2)}{6}\nabla^3 f(x_0 + nh) + \ldots\right] \quad (7.29)$$

But $\dfrac{du}{dx} = \dfrac{1}{h}$

$$\therefore \quad \frac{df(x)}{dx} = f'(x) = \frac{df(x)}{du} \cdot \frac{du}{dx}$$

$$= \frac{1}{h}\left[\nabla f(x_0 + nh) + \left(\frac{2u+1}{2}\right)\nabla^2 f(x_0 + nh) + \left(\frac{3u^2 + 6u + 2}{6}\right)\nabla^3 f(x_0 + nh) + \ldots\right]$$

Similarly

$$f''(x) = \frac{1}{h^2}\left[\nabla^2 f(x_0 + nh) + (u+1)\nabla^3 f(x_0 + nh) + \left(\frac{6u^2 + 18u + 11}{12}\right)\nabla^4 f(x_0 + nh) + \ldots\right]$$

Example 1: Find $f'(1.3)$ and $f''(1.3)$ from the following table:

x	1.0	1.2	1.4	1.6	1.8	2
f(x)	0.0	0.1280	0.5540	1.2960	2.4320	4.0

Solution: We construct the following divided difference table:

x	$f(x)$	$\Delta f(x)$	$\Delta^2 f(x)$	$\Delta^3 f(x)$	$\Delta^4 f(x)$	$\Delta^5 f(x)$
1.0	0.0					
		0.64				
1.2	0.128		3.725			
		2.13		0.375		
1.4	0.554		3.95		1.5625	
		3.71		1.625		-2.6041
1.6	1.296		4.925		-1.0416	
		5.68		0.7917		
1.8	2.432		5.4			
		7.84				
2.0	4.0					

By using Newton's divided difference formula

$$f(x) = f(x_0) + (x - x_0)\,\Delta f(x_0) + (x - x_0)(x - x_1)\,\Delta^2 f(x_0) + (x - x_0)(x - x_1)(x - x_2)$$
$$\Delta^3 f(x_0) + (x - x_0)(x - x_1)(x - x_2)(x - x_3)\,\Delta^4 f(x_0) + (x - x_0)(x - x_1)$$
$$(x - x_2)(x - x_3)(x - x_4)\,\Delta^5 f(x_0)$$

$$= 0 + (x - 1)\,(0.64) + (x - 1)\,(x - 1.2)\,(3.725) + (x - 1)\,(x - 1.2)\,(x - 1.4)$$
$$(0.375) + (x - 1)\,(x - 1.2)\,(x - 1.4)\,(x - 1.6)\,(1.5625) + (x - 1)\,(x - 1.2)\,(x - 1.4)$$
$$(x - 1.6)\,(x - 1.8)\,(-2.6041)$$

$$= (0.64x - 0.64) + (3.725x^2 - 8.195x + 4.47) + (0.375x^3 - 1.35x^2 + 1.605x - 0.63) +$$
$$(1.5625x^4 - 8.125x^3 + 15.6875x^2 - 13.325x + 4.2) + (-2.6041x^5 + 18.2287x^4 -$$
$$50.51954x^3 + 69.26906x^2 - 46.97380x + 12.59968)$$

$$= -2.6041x^5 + 19.7912x^4 - 58.26954x^3 + 87.33156x^2 - 66.2488x + 19.9997$$

is the required polynomial.

Now $f'(x) = -13.0205x^4 + 79.1648x^3 - 174.80862x^2 + 174.66312x - 66.2488$

$f'(1.3) = -13.0205\,(1.3)^4 + 79.1648\,(1.3)^3 - 174.80862\,(1.3)^2 + 174.66312\,(1.3) - 66.2488$

$\quad = -37.18785 + 173.9251 - 295.42657 + 227.06206 - 66.2488$

$\quad = 2.12394$

Again $f''(x) = -52.082x^3 + 237.4944x^2 - 349.61724x + 174.66312$

$f''(1.3) = -52.082\,(1.3)^3 + 237.4944\,(1.3)^2 - 349.61724\,(1.3) + 174.66312$

$\quad = -114.42415 + 401.365536 - 454.502412 + 174.66312$

$\quad = 7.102094$

Example 2: Using the following data find $f'(6.0)$, error $= O(h)$ and $f''(6.3)$, error $= O(h^2)$.

x	6.0	6.1	6.2	6.3	6.4
$f(x)$	0.1750	-0.1998	-0.2223	-0.2422	-0.2596

Solution: Method of $O(h)$ for $f'(x_0)$ is given by $f'(x_0) = \dfrac{1}{h}[f(x_0 + h) - f(x_0)]$ by method based on finite difference. Hence, with $x_0 = 6.0$, $h = 0.1$, we get

$$f'(6.0) = \frac{1}{0.1}[f(6.1) - f(6.0) = 10[-0.1998 - 0.1750] = -3.748$$

Method of $O(h^2)$ for $f''(x_0)$ is given by

$$f''(x_0) = \frac{1}{h^2}[f(x_0 - h) - 2f(x_0) + f(x_0 + h)]$$

with $x_0 = 6.3$ and $h = 0.1$, we get

$$f''(6.3) = \frac{1}{(0.1)^2}[f(6.3 - 0.1) - 2f(6.3) + f(6.3 + 0.1)]$$

$$= 100[f(6.2) - 2f(6.3) + f(6.4)]$$
$$= 100[-0.2223 - 2 \times (-0.2422) + (-0.2596)]$$
$$= 0.25$$

Example 3: Find $f'(4)$ from the following table:

x	0	1	2	5
$f(x)$	2	5	7	8

Solution: By using Lagrange interpolation formula, we have

$$f(x) = \frac{(x-1)(x-2)(x-5)}{(0-1)(0-2)(0-5)} \cdot (2) + \frac{(x-0)(x-2)(x-5)}{(1-0)(1-2)(1-5)} \cdot (5)$$

$$+ \frac{(x-0)(x-1)(x-5)}{(2-0)(2-1)(2-5)} \cdot (7) + \frac{(x-0)(x-1)(x-2)}{(5-0)(5-1)(5-2)} \cdot (8)$$

$$= -\frac{1}{5}(x-1)(x-2)(x-5) + \frac{5}{4}x(x-2)(x-5)$$

$$- \frac{7}{6}x(x-1)(x-5) + \frac{2}{15}x(x-1)(x-2)$$

$$= \frac{1}{60}x^3 - \frac{11}{20}x^2 + \frac{1}{5}x + 2$$

$$f'(x) = \frac{x^2}{20} - \frac{11}{10}x + \frac{1}{5}$$

$$f'(4) = \frac{16}{20} - \frac{11}{10} \times 4 + \frac{1}{5} = \frac{4}{5} - \frac{22}{5} + \frac{1}{5} = -\frac{17}{5}$$

Example 4: Find $f'(3)$ and $f''(3)$ from the following data:

x	0	2	5	1
$f(x)$	0	8	125	1

Solution: By using Lagrange interpolation formula, we have

$$f(x) = \frac{(x-2)(x-5)(x-1)}{(0-2)(0-5)(0-1)} \cdot (0) + \frac{(x-0)(x-5)(x-1)}{(2-0)(2-5)(2-1)} \cdot (8)$$

$$+ \frac{(x-0)(x-2)(x-1)}{(5-0)(5-2)(5-1)} \cdot (125) + \frac{(x-0)(x-2)(x-5)}{(1-0)(1-2)(1-5)} \cdot (1)$$

$$= -\frac{8}{6}(x^3 - 6x^2 + 5x) + \frac{125}{60}(x^3 - 3x^2 + 2x) + \frac{1}{4}(x^3 - 7x^2 + 10x) = x^3$$

$$f'(x) = 3x^2$$
$$\Rightarrow \qquad f'(3) = 3\,(3)^2 = 27$$
Again $\qquad f''(x) = 6x$
$$\Rightarrow \qquad f''(3) = 6(3) = 18$$

Example 5: Find $f'(1.3)$ and $f''(1.3)$ from the following table:

x	4	5	7	10	11	13
$f(x)$	48	100	294	900	1210	2028

Solution: We construct the following divided difference table:

x	$f(x)$	$\Delta f(x)$	$\Delta^2 f(x)$	$\Delta^3 f(x)$	$\Delta^4 f(x)$	$\Delta^5 f(x)$
4	48					
		52				
5	100		15			
		97		1		
7	294		21		0	
		202		1		0
10	900		27		0	
		310		1		
11	1210		33			
		409				
13	2028					

Hence by using Newton's divided difference formula

$$f(x) = f(x_0) + (x - x_0)\,\Delta f(x_0) + (x - x_0)(x - x_1)\,\Delta^2 f(x_0) + (x - x_0)(x - x_1)(x - x_2)$$
$$\Delta^3 f(x_0) + (x - x_0)(x - x_1)(x - x_2)(x - x_3)\,\Delta^4 f(x_0) + (x - x_0)(x - x_1)$$
$$(x - x_2)(x - x_3)(x - x_4)\,\Delta^5 f(x_0)$$
$$= 48 + (x - 4)\,(52) + (x - 4)\,(x - 5)\,(15) + (x - 4)\,(x - 5)\,(x - 7)\,(1) + 0 + 0$$
$$= 48 + 52\,(x - 4) + 15\,(x^2 - 9x + 20) + (x^3 - 16x^2 + 83x - 140) = x^3 - x^2$$
$$f'(x) = 3x^2 - 2x$$
$$\Rightarrow \qquad f'(8) = 3\,(8)^2 - 2\,(8) = 176$$
Now $f''(x) = 6x - 2$
$$\Rightarrow \quad f''(12) = 6 \times 12 - 2 = 70$$

Example 6: Given the following table, find $y'(5)$ and the maximum value of y:

x	0	2	3	4	7	9
y	4	26	58	112	466	922

Solution: We construct the following Newton's divided difference table:

x	$f(x)$	$\Delta f(x)$	$\Delta^2 f(x)$	$\Delta^3 f(x)$	$\Delta^4 f(x)$	$\Delta^5 f(x)$
0	4					
		11				
2	26		7			
		32		1		
3	58		11		0	
		54		1		0
4	112		16		0	
		118		1		
7	466		22			
		228				
9	922					

By using Newton's divided difference formula

$$y = f(x) = f(x_0) + (x - x_0)\,\Delta f(x_0) + (x - x_0)(x - x_1)\,\Delta^2 f(x_0) + \dots$$

$$= 4 + (x - 0)\,(11) + (x - 0)\,(x - 2)\,(7) + (x - 0)\,(x - 2)\,(x - 3)\,(1) + 0 + 0$$

$$= x^3 + 2x^2 + 3x + 4$$

$$\therefore \qquad y'(x) = 3x^2 + 4x + 3$$

$$\therefore \qquad y'(6) = 3(6)^2 + 4(6) + 3 = 135$$

$y(x)$ is maximum if $y'(x) = 0$

$$\therefore \quad 3x^2 + 4x + 3 = 0$$

But the roots are imaginary. Therefore, there is no extremum value in the range. In fact, it is an increasing curve.

Example 7: The population of a certain town is given below. Find the rate of growth of the population in 1931 and 1971.

Year	x	1931	1941	1951	1961	1971
Population in thousand	y	40.62	60.80	79.95	103.56	132.65

Solution: We construct the difference table:

x	y	Δy	$\Delta^2 y$	$\Delta^3 y$	$\Delta^4 y$
1931	40.62				
		20.18			
1941	60.80		−1.03		
		19.15		5.49	
1951	79.95		4.46		−4.47
		23.61		1.02	
1961	103.56		5.48		
		29.09			
1971	132.65				

We use the same table for backward and forward difference

i. To get $y'(1931)$ we use forward formula

$$y'(x)\,|_{\text{at } x = 1931} = \frac{1}{h}\left[\Delta y_0 - \frac{1}{2}\Delta^2 y_0 + \frac{1}{3}\Delta^2 y_0 - \frac{1}{4}\Delta^4 y_0 + \dots\right]$$

$$= \frac{1}{10}[20.18 - \frac{1}{2}(-1.03) + \frac{1}{3}(5.49) - \frac{1}{4}(-4.47) = 2.36425$$

ii. To get $y'(1971)$ we use the backward formula

$$y'(x)\,|_{\text{at } x = 1971} = \frac{1}{h}\left[\nabla y_n + \frac{1}{2}\nabla^2 y_n + \frac{1}{3}\nabla^3 y_n + \frac{1}{4}\nabla^4 y_n + \dots\right]$$

$$= \frac{1}{10}\left[29.09 + \frac{1}{2}(5.48) + \frac{1}{3}(1.02) + \frac{1}{5}(-4.47)\right] = 3.10525$$

Example 8: The table given below reveals the velocity v of a body during the time 't' specified. Find its acceleration at $t = 1.1$.

t	1.0	1.1	1.2	1.3	1.4
v	43.1	47.7	52.1	56.4	60.8

Solution: v is dependent on time t, i.e. $v = v(t)$. We require acceleration $= \dfrac{dv}{dt}$. Therefore, we have to find $v'(1.1)$

Now we construct a difference table:

t	v	Δv	$\Delta^2 v$	$\Delta^3 v$	$\Delta^4 v$
1.0	43.1				
		4.6			
1.1	47.7		−0.2		
		4.4		0.1	
1.2	52.1		−0.1		0.1
		4.3		0.2	
1.3	56.4		0.1		
		4.4			
1.4	60.8				

By using forward formula

$$\frac{dv}{dt}\bigg|_{t=1.1} = \frac{1}{h}\left[\Delta v_0 - \frac{1}{2}\Delta^2 v_0 + \frac{1}{3}\Delta^3 v_0 - \frac{1}{4}\Delta^4 v_0 + \ldots\right]$$

$$= \frac{1}{0.1}\left[4.4 - \frac{1}{2}(-0.1) + \frac{1}{3}(0.2)\right]$$

$$= 10\,[4.4 + 0.05 + 0.067] = 45.17$$

Example 9: A rod is rotating in a plane. The following table gives the angle θ (in radians) through which the rod has turned for various values of time t (seconds). Calculate the angular velocity and angular acceleration of the rod at $t = 0.6$ seconds.

t	0	0.2	0.4	0.6	0.8	1.0
θ	0	0.12	0.49	1.12	2.02	3.20

Solution: We make the difference table:

t	θ	$\nabla\theta$	$\nabla^2\theta$	$\nabla^3\theta$	$\nabla^4\theta$	$\nabla^5\theta$
0	0					
		0.12				
0.2	0.12		0.25			
		0.37		0.01		
0.4	0.49		0.26		0	
		0.63		0.01		0
0.6	1.12		0.27		0	
		0.90		0.01		
0.8	2.02		0.28			
		1.18				
1.0	3.20					

By using Newton's backward difference formula, if we take 0.6 at $x_0 + nh$, we have
$$\nabla y_n = 0.63,\ \nabla^2 y_n = 0.26,\ \nabla^3 y_n = 0.01$$

$$\frac{d\theta}{dt}\bigg|_{x=0.6} = \frac{1}{h}\left[\nabla y_n + \frac{1}{2}\nabla^2 y_n + \frac{1}{3}\nabla^3 y_n + \ldots\right]$$

$$= \frac{1}{0.2}\left[0.63 + \frac{1}{2}(0.26) + \frac{1}{3}(0.01)\right]$$

$$= 3.8167 \text{ radian/sec.}$$

Also
$$\frac{d^2\theta}{dt^2}\bigg|_{x=0.6} = \frac{1}{h^2}\left[\nabla^2 y_n + \nabla^3 y_n + \frac{11}{12}\nabla^4 y_n + \dots\right]$$

$$= \frac{1}{(0.2)^2}[0.26 + 0.01]$$

$$= 6.75 \text{ radians/sec}^2$$

Example 10: The following table gives the result of an observation. θ is the observed temperature in degree centigrade of a vessel of cooling water, t is the time in minutes from beginning of the observation.

t	1	3	5	7	9
θ	85.3	74.5	67.0	60.5	54.3

Find the approximate rate of cooling at $t = 2.5$ and $t = 7.5$.

Solution: We make the difference table:

t	θ	$\Delta\theta$	$\Delta^2\theta$	$\Delta^3\theta$	$\Delta^4\theta$
1	85.3				
		−10.8			
3	74.5		3.3		
		−7.5		−2.3	
5	67.0		1.0		1.6
		−6.5		−0.7	
7	60.5		0.3		
		−6.2			
9	54.3				

For $t = 2.5$ we follow $u = \dfrac{t - t_0}{h} = \dfrac{2.5 - 1}{2} = \dfrac{1.5}{2} = 0.75$.

Newton's forward difference formula

$$\frac{d\theta}{dt} = \frac{1}{h}\left[\Delta\theta_0 + \left(\frac{2u-1}{2!}\right)\Delta^2\theta_0 + \left(\frac{3u^2 - 6u + 2}{3!}\right)\Delta^3\theta_0 + \left(\frac{3u^3 - 9u^2 + 11u - 3}{12}\right)\Delta^4\theta_0 + \dots\right]$$

$$= \frac{1}{2}\left[-10.8 + \left(\frac{2 \times 0.75 - 1}{2}\right)(3.3) + \left(\frac{3 \times (0.75)^2 - 6 \times 0.75 + 2}{6}\right)(-2.3)\right.$$

$$\left. + \left(\frac{3(0.75)^3 - 9(0.75)^2 + 11(0.75) - 3}{12}\right)(1.6)\right]$$

$$= \frac{1}{2}\left[-10.8 + \frac{0.5 \times 3.3}{2} + 0.3114 + 0.1938\right]$$

$$= -4.7349$$

For $t = 7.5$, we follow $u = \dfrac{t - t_n}{h} = \dfrac{7.5 - 9}{2} = -0.75$

$$\frac{d\theta}{dt} = \frac{1}{h}\left[\nabla\theta_n + \left(\frac{2u+1}{2!}\right)\nabla^2\theta_n + \left(\frac{3u^2 + 6u + 2}{3!}\right)\nabla^3\theta_n + \right.$$

$$\left. + \left(\frac{4u^3 + 18u^2 + 22u + 6}{24}\right)\nabla^4\theta_n + \dots\right]$$

$$= \frac{1}{2}\left[-6.2 + \left(\frac{2 \times -0.75 + 1}{2}\right)(0.3) + \left\{\frac{3(-0.75)^2 + 6(-0.75) + 2}{6}\right\}(-0.7)\right.$$

$$\left. + \left\{\frac{4(-0.75)^3 + 18(-0.75)^2 + 22(-0.75) + 6}{24}\right\}(1.6)\right]$$

$$= \frac{1}{2}[-6.2 + (-0.075) + 0.0948 + (-0.1375)]$$

$$= -2.1059$$

Example 11: Assume that $f(x)$ has a minimum in the interval $x_{n-1} \le x \le x_{n+1}$, where $x_k = x_0 + kh$. Show that the interpolation of $f(x)$ by a polynomial of second degree yields

the approximation $f_n - \frac{1}{8}\left[\frac{(f_{n+1} - f_{n-1})^2}{f_{n+1} - 2f_n + f_{n-1}}\right]$, $f_k = f(x_k)$ for this minimum value of $f(x)$.

Solution: The interpolation polynomial through the points (x_{n-1}, f_{n-1}), (x_n, f_n) and (x_{n+1}, f_{n+1}) is given as

$$f(x) = f(x_{n-1}) + \frac{1}{h}(x - x_{n-1})\,\Delta f_{n-1} + \frac{1}{2!h^2}(x - x_{n-1})(x - x_n)\,\Delta^2 f_{n-1}$$

Since $f(x)$ has a minimum, set $f'(x) = 0$. Therefore

$$f'(x) = \frac{1}{h}\Delta f_{n-1} + \frac{1}{2h^2}(2x - x_{n-1} - x_n)\Delta^2 f_{n-1} = 0$$

which gives $x_{\min} = \frac{1}{2}(x_n + x_{n-1}) - h\frac{\Delta f_{n-1}}{\Delta^2 f_{n-1}}$

Hence, the minimum value of $f(x)$ is

$$f(x_{\min}) = f_{n-1} + \frac{1}{h}\left[\frac{1}{2}(x_n - x_{n-1}) - h\frac{\Delta f_{n-1}}{\Delta^2 f_{n-1}}\right]\Delta f_{n-1}$$

$$+ \frac{1}{2h^2}\left[\frac{1}{2}(x_n - x_{n-1}) - h\frac{\Delta f_{n-1}}{\Delta^2 f_{n-1}}\right]\left[\frac{1}{2}(x_{n-1} - x_n) - h\frac{\Delta f_n}{\Delta^2 f_{n-1}}\right]\Delta^2 f_{n-1}$$

But $x_n - x_{n-1} = h$, we obtain

$$f_{\min} = f_{n-1} + \frac{1}{2}\Delta f_{n-1} - \frac{(\Delta f_{n-1})^2}{2\Delta^2 f_{n-1}} - \frac{1}{8}\Delta^2 f_{n-1}$$

$$= f_n - \Delta f_{n-1} + \frac{1}{8\Delta^2 f_{n-1}}[4\Delta f_{n-1}\Delta^2 f_{n-1} - 4(\Delta f_{n-1})^2 - (\Delta^2 f_{n-1})^2]$$

$$= f_n - \frac{1}{8\Delta^2 f_{n-1}}[(4\Delta f_{n-1} + \Delta^2 f_{n-1})\Delta^2 f_{n-1} + 4(\Delta f_{n-1})^2]$$

Using $\Delta f_{n-1} = f_n - f_{n-1}$, $\Delta^2 f_{n-1} = f_{n+1} - 2f_n + f_{n-1}$ and simplifying, we obtain

$$f_{min} = f_n - \frac{f_{n+1} - 2f_{n-1}f_{n+1} + f_{n-1}^2}{8(f_{n-1} - 2f_n + f_{n-1})} = f_n - \frac{1}{8}\left[\frac{(f_{n+1} - f_{n-1})^2}{f_{n+1} - 2f_n + f_{n-1}}\right]$$

EXERCISE 7.1

1. Find the derivative of $f(x)$ at $x = 0.4$ from the following table:

x	0.1	0.2	0.3	0.4
$f(x)$	1.10517	1.22140	1.34986	1.49182

[Ans. 1.4913

2. Find the first and second derivatives of the function tabulated below at the point $x = 3.0$:

x	3.0	3.2	3.4	3.6	3.8	4.0
y	−14.00	−10.032	−5.296	0.256	0.672	14.00

[Ans. $f'(3) = 18, f''(3) = 18$]

3. Find $f'(5)$ from the following table:

x	0	2	3	4	7	9
$f(x)$	4	26	58	112	466	922

[Ans. $f(5) = 98$]

4. Find the first two derivatives of $(x)^{1/3}$ at $x = 50$ and $x = 56$ given in the table below:

x	50	51	52	53	54	55	56
$y = x^{1/3}$	3.6840	3.7084	3.7325	3.7563	3.7798	3.8030	3.8259

[Ans. 0.02455, −0.0003, 0.02275, −0.0003]

5. Find $f'(2.5)$ from the following table:

x	1.5	1.9	2.5	3.2	4.3	5.9
$f(x)$	3.375	6.059	13.625	29.368	73.907	196.579

[Ans. 16.75]

6. Find $f'(1.2)$ and $f''(1.2)$ from the following table:

x	0	1	2	5
$f(x)$	2	3	12	147

[Ans. 5.72, 9.2]

7. Find $f'(3.5)$ from the following table:

x	1	3	4
$f(x)$	1	27	64

[Ans. 37]

8. The distance covered by an athlete for the 50 metre race is given in the following table:

Time (sec)	0	1	2	3	4	5	6
Distance (metre)	0	2.5	8.5	15.5	24.5	36.5	50

Determine the speed of the athlete at $t = 5$ sec. Correct to two decimal places.

[Ans. 13.13 metre/sec]

9. From the following table, obtain the value of $\dfrac{d^2y}{dx^2}$ at the point $x = 0.96$

x	0.96	0.98	1.00	1.02	1.04
y	0.7825	0.7739	0.7651	0.7563	0.7473

[Ans. −1.925]

10. Find the minimum value of y from the following table

x	0.2	0.3	0.4	0.5	0.6	0.7	
y	0.9182	0.8975	0.8873	0.8862	0.8935	0.9086	[Ans. 0.4623]

7.4 RICHARDSON EXTRAPOLATION METHOD

In order to obtain highly accurate results, we need to use higher order methods which require a large number of function evaluations. As a result, rounding errors may increase. However, it is generally possible to get higher order solutions by combining the computed values found by using a certain lower order method with different step sizes.

To improve the accuracy of the derivative of a function, that is computed by starting with an arbitrarily selected value of h, generally Richardson's extrapolation method is used.

If the error term of the method is known as a power series in h, then by repeating the extrapolation procedure a number of times, we can obtain methods of any arbitrary order. We often take the step sizes as $h, h/2, h/2^2, ...$

From the central difference truncation error, we know that

$$y'(x) = \frac{f(x+h) - f(x-h)}{2h} - \frac{1}{6}h^2\ y'''(x) - \frac{h^4}{120}\ y^{(v)}(x) - ... \quad (7.30)$$

We can write Eq. (7.30) as

$$y'(x) = F(h) + e_2h^2 + e_4h^4 + ... \quad (7.31)$$

or $$y'(x) = F(h) + E_T \quad (7.32)$$

To compute the derivative of a function $y(x)$, E_T is the truncation error given by

$$E_T = e_2h^2 + e_4h^4 + ... \quad (7.33)$$

The basic idea of Richardson's extrapolation is to combine two computed values of $y'(x)$ using the same method but with two different step sizes generally h and $h/2$ to get a higher order method. Thus, we have

$$y'(x) = F(h) + c_1h^2 + c_2h^4 + ... \quad (7.34)$$

and $$y'(x) = F\left(\frac{h}{2}\right) + c_1\frac{h^2}{4} + c_2\frac{h^4}{16} + ... \quad (7.35)$$

where e_i's are constants and $(i = 2, 4, 6, ...)$, independent of h while $F(h)$ and $F(h/2)$ represent approximate values of derivatives. Eliminating c_1 from Eqs (7.35) and (7.34), we get

$$y'(x) = \frac{4F(h/2) - F(h)}{3} + d_1h^4 + O(h^6) \quad (7.36)$$

Let, $$F_1(h/2) = \frac{4F(h/2) - F(h)}{3}$$

then from Eq. (7.36)

$$y'(x) = F_1\left(\frac{h}{2}\right) + d_1h^4 + O(h^6)$$

Thus, we have obtained a fourth order accurate differentiation formula by combining two results which are of second order accurate. By repeating the above concept, we have

$$y'(x) = F_1 \frac{h}{2} + d_1 h^4 + O(h^6) \tag{7.37}$$

or

$$y'(x) = F_1\left(\frac{h}{4}\right) + d_1 \frac{h^4}{16} + O(h^6) \tag{7.38}$$

Eliminating d_1 from Eqs (7.37) and (7.38), we get better approximation as

$$y'(x) = F_2\left(\frac{h}{4}\right) + O(h^6)$$

which is of sixth order accurate, where

$$F_2\left(\frac{h}{4}\right) = \frac{4^2 F_1\left(\frac{h}{2^2}\right) - F_1\left(\frac{h}{2}\right)}{4^2 - 1} \tag{7.39}$$

This extrapolation procedure can be repeated further until the required accuracy is obtained which is called an extrapolation to the limit. The general form of Eq. (7.39) is written as

$$F_j\left(\frac{h}{2^j}\right) = \frac{4^j F_j\left(\frac{h}{2^j}\right) - F_{j-1}\left(\frac{h}{2^{j-1}}\right)}{4^j - 1}, j = 1, 2, 3, \ldots$$

where $F_0(h) = F(h)$

The Extrapolation table is given below:

Order\Step	Second	Fourth	Sixth	Eighth
h	$F(h)$			
$h/2$	$F(h/2)$	$F_1(h)$		
$h/2^2$	$F(h/2^2)$	$F_1(h/2)$	$F_2(h)$	$F_3(h)$
$h/2^3$	$F(h/2^3)$	$F_1(h/2^2)$	$F_2(h/2)$	

The extrapolation procedure can be stopped when $|F_k(h) - F_{k-1}(h/2)| < \epsilon$ where ϵ is the prescribed error tolerance.

Example 12: Using Richardson's extrapolation limit, find $y'(0.05)$ to the function $y = -\frac{1}{x}$ with $h = 0.0128, 0.0064, 0.0032$.

Solution: Let $h = 0.0128$

$$F(h) = \frac{y(x+h) - y(x-h)}{2h} = \frac{-\dfrac{1}{x+h} + \dfrac{1}{x-h}}{2h}$$

$$= \frac{-\left(\dfrac{1}{0.05 + 0.0128}\right) + \left(\dfrac{1}{0.05 - 0.0128}\right)}{2 \times 0.0128}$$

$$= \frac{-15.92357 + 26.88172}{2 \times 0.0128} = 428.0527$$

Now
$$F\left(\frac{h}{2}\right) = \frac{y\left(x + \frac{h}{2}\right) - y\left(x - \frac{h}{2}\right)}{2 \times \frac{h}{2}}, \quad \text{Taking } \frac{h}{2} = 0.0064$$

$$= \frac{-\dfrac{1}{x + \dfrac{h}{2}} + \dfrac{1}{x - \dfrac{h}{2}}}{2 \times \dfrac{h}{2}} = \frac{-\dfrac{1}{0.05 + 0.0064} + \dfrac{1}{0.05 - 0.0064}}{0.0064 \times 2}$$

$$= 406.66276$$

Therefore,
$$F_1\left(\frac{h}{2}\right) = \frac{4F\left(\dfrac{h}{2}\right) - F(h)}{4 - 1} = \frac{(4 \times 406.66276) - 428.0527}{3}$$

$$= 399.53278$$

which is accurate to $O(h^4)$. Halving the step size further, we get

$$F\left(\frac{h}{2^2}\right) = \frac{y\left(x + \dfrac{h}{2^2}\right) - y\left(x - \dfrac{h}{2^2}\right)}{2\left(\dfrac{h}{2^2}\right)} = \frac{-\dfrac{1}{x + \dfrac{h}{2^2}} + \dfrac{1}{x - \dfrac{h}{2^2}}}{2\left(\dfrac{h}{2^2}\right)}$$

Taking
$$\frac{h}{2^2} = 0.0032$$

$$F\left(\frac{h}{2^2}\right) = \frac{-\dfrac{1}{0.05 + 0.0032} + \dfrac{1}{0.05 - 0.0032}}{2 \times 0.0032}$$

$$= 401.64514$$

and
$$F_1\left(\frac{h}{2^2}\right) = \frac{4F\left(\dfrac{h}{2^2}\right) - F\left(\dfrac{h}{2}\right)}{4 - 1} = \frac{4 \times 401.64514 - 406.66276}{3}$$

$$= 399.9726$$

Again
$$F_2\left(\frac{h}{2^2}\right) = \frac{4^2 F_1\left(\dfrac{h}{2^2}\right) - F_1\left(\dfrac{h}{2}\right)}{4^2 - 1} = \frac{4^2 \times 399.9726 - 399.53278}{15}$$

$$= 400.0019$$

The above computation can be given in the table as

h	F	F_1	F_2
0.0128	428.0527		
		399.53278	
0.0064	406.66276		400.0019
		399.9726	
0.0032	401.64514		

Thus, after two steps, $y(0.05) = 400.019$

The exact value is $y'(0.05) = \dfrac{1}{x^2}\Big|_{x=0.05} = 400$

Example 13: By using repeated Richardson's extrapolation find $f'(1.0)$ from the following values:

x	0.6	0.8	0.9	1.0	1.1	1.2	1.4
$f(x)$	0.707178	0.859892	0.925863	0.984007	1.033743	1.074575	1.127986

Applying the approximate formula

$$f'(x_0) = \frac{f(x_0+h)-f(x_0-h)}{2h}$$

with $h = 0.4, 0.2, 0.1$.

Solution: Let $h = 0.4$

$$F(h) = \frac{f(x_0+h)-f(x_0-h)}{2h} = \frac{f(1+0.4)-f(1-0.4)}{2\times0.4}$$

$$= \frac{f(1.4)-f(0.6)}{0.8} = \frac{1.127986-0.707178}{0.8} = 0.526010$$

Now $\qquad F\left(\dfrac{h}{2}\right) = \dfrac{f\left(x_0+\dfrac{h}{2}\right)-f\left(x_0-\dfrac{h}{2}\right)}{2\left(\dfrac{h}{2}\right)}$ $\qquad\Big|$ Taking $\dfrac{h}{2}=0.2$

$$= \frac{f(1+0.2)-f(1-0.2)}{2\times0.2} = \frac{f(1.02)-f(0.8)}{0.4}$$

$$= \frac{1.074575-0.859892}{0.4} = 0.536708$$

Therefore, $\qquad F_1\left(\dfrac{h}{2}\right) = \dfrac{4F\left(\dfrac{h}{2}\right)-F(h)}{4-1} = \dfrac{4\times0.536708-0.526010}{3} = 0.540274$

which is accurate to $O(h^4)$. Halving the step size further, we get

$$F\left(\frac{h}{2^2}\right) = \frac{f\left(x_0+\dfrac{h}{2^2}\right)-f\left(x_0-\dfrac{h}{2^2}\right)}{2\left(\dfrac{h}{2^2}\right)}$$

Taking $\dfrac{h}{2^2}=0.1$

$\therefore \qquad F\left(\dfrac{h}{2^2}\right) = \dfrac{f(1+0.1)-f(1-0.1)}{2\times0.1} = \dfrac{f(1.1)-f(0.9)}{2\times0.1}$

$$= \frac{1.033743-0.925863}{0.2} = 0.539400$$

and
$$F_1\left(\frac{h}{2^2}\right) = \frac{4F\left(\dfrac{h}{2^2}\right) - F\left(\dfrac{h}{2}\right)}{4-1} = \frac{4 \times 0.539400 - 0.536708}{3} = 0.540297$$

Again
$$F_2\left(\frac{h}{2^2}\right) = \frac{4^2 F_1\left(\dfrac{h}{2^2}\right) - F_1\left(\dfrac{h}{2}\right)}{4^2 - 1} = \frac{4^2 \times 0.540297 - 0.540274}{15} = 0.540299$$

The above computation can be given in the table as:

h	$O(h^2)$	$O(h^4)$	$O(h^6)$
0.4	0.526010		
		0.540274	
0.2	0.536708		0.540299
		0.540297	
0.1	0.539400		

Example 14: Compute $f''(0.6)$ from the following table using the formula

$$f''(x_0) = \frac{f(x_0 + h) - 2f(x_0) + f(x_0 - h)}{h^2} + \sum_{i=1}^{\infty} a_i h^{k_i}$$

with $h = 0.4, 0.2, 0.1$ and perform repeated Richardson's extrapolation.

x	$f(x)$
0.2	1.420072
0.4	1.881243
0.5	2.128147
0.6	2.386761
0.7	2.657971
0.8	2.942897
1.0	3.559753

Solution: Using the given formula, we get

$$h = 0.4: f''(0.6) = \frac{f(0.6 + 0.4) - 2f(0.6) + f(0.6 - 0.4)}{(0.4)^2}$$

$$= \frac{f(1.0) - 2f(0.6) + f(0.2)}{0.16}$$

$$= \frac{3.559753 - 2 \times 2.386761 + 1.420072}{0.16} = 1.289394$$

$$h = 0.2: f''(0.6) = \frac{f(0.6 + 0.2) - 2f(0.6) + f(0.6 - 0.2)}{(0.2)^2}$$

$$= \frac{f(0.8) - 2f(0.6) + f(0.4)}{0.04}$$

$$= \frac{2.942897 - 2 \times 2.386761 + 1.881243}{0.04} = 1.265450$$

$$h = 0.1: f''(0.6) = \frac{f(0.6+0.1) - 2f(0.6) + f(0.6-0.1)}{(0.1)^2}$$

$$= \frac{f(0.7) - 2f(0.6) + f(0.5)}{0.01} = \frac{2.657971 - 2 \times 2.386761 + 2.128147}{0.01}$$

$$= 1.259600$$

Applying the Richardson's extrapolation, we get

Now,
$$F_1\left(\frac{h}{2}\right) = \frac{4F\left(\dfrac{h}{2}\right) - F(h)}{4-1} = \frac{4 \times 1.265450 - 1.289394}{3} = 1.257469$$

$$F_1\left(\frac{h}{2^2}\right) = \frac{4F\left(\dfrac{h}{2^2}\right) - F\left(\dfrac{h}{2}\right)}{4-1} = \frac{4 \times 1.259600 - 1.265450}{3} = 1.257650$$

Again
$$F_2\left(\frac{h}{2^2}\right) = \frac{4^2 F_1\left(\dfrac{h}{2^2}\right) - F_1\left(\dfrac{h}{2}\right)}{4^2 - 1} = \frac{16 \times 1.257650 - 1.257469}{15} = 1.257662$$

Extrapolation table:

h	$O(h^2)$	$O(h^4)$	$O(h^6)$
0.4	1.289394		
		1.257469	
0.2	1.265450		1.257662
		1.257650	
0.1	1.259600		

EXERCISE 7.2

1. Given

x	-1	1	2	3	4	5	7
$f(x)$	1	1	16	81	256	625	2401

Using the formula $f'(x_1) = \dfrac{f(x_2) - f(x_0)}{2h}$ and the Richardson extrapolation, find $f'(3)$.　　　　　　　　　　　　　　　　　　　　　　　　　　　[Ans. 108]

2. Consider the function $f(x) = \dfrac{\sin^2\left(\dfrac{\sqrt{x^2+x}}{\cos x - x}\right)}{\sin\left(\dfrac{\sqrt{x}-1}{\sqrt{x^2+1}}\right)}$. Compute $f'(0.25)$ as accurately as possible by using central difference approximation

$$f'(x) = \frac{f(x+h) - f(x-h)}{2h} + O(h^2),$$

with $x = 0.25$ and $h = 0.01$.　　　　　　　　　　　　　　　　　　　[Ans. −9.06669877]

3. Given $f(x) = 5x\,e^{-2x}$. Find the value of $f'(0.35)$ using Richardson extrapolation with central divided difference scheme with $h = 0.25, h = 0.125$. [Ans. 0.7427]

4. Consider $f(x) = xe^x$ with $x_0 = 2.0$ and $h = 0.2$. Use the central difference formula to the first derivative and Richardson extrapolation to give an approximation of order $O(h^4)$. [Ans. 22.1670]

5. Approximate $v(1)$ using the central difference formula and the Richardson extrapolation method.

t in hours	0	1/4	1/2	3/4	1	5/4	3/2	7/4	2
s in miles	0	20	40	50	60	65	70	72	75

[Ans. 0.199]

Numerical Integration

8.1 INTRODUCTION

The process of computing the value of a definite integral $I = \int_a^b f(x)\,dx$ from a set of numerical values of the integrand is called *numerical integration*. If the definite integral under consideration contains the integrand as a function of single variable, the process is called *quadrature*. In numerical integration, we first approximate a polynomial corresponding to given set of values of the integrand and then integration of polynomial is performed.

In this chapter, we shall study the error analysis of *trapezoidal rule*, Simpson's 1/3rd rule, Simpson's 3/8 rule, Boole's rule and Weddle's rule. We also shall study the different types of integration methods such as Gauss–Legendre method, Lobatto integration method etc. and error terms in these methods.

8.2 GENERAL QUADRATURE FORMULA

Consider the integral under consideration is $\int_a^b f(x)\,dx$, where $y = f(x)$.

Let $f(x)$ be given for certain equally distant values of arguments, say $a, a + h, a + 2h,$ ..., $a + nh$. Let the range $(b - a)$ be divided into n equal parts, each of width $h = \dfrac{b - a}{n}$ so that $b = a + nh$.

Let $x_0 = a, x_1 = a + h, x_2 = a + 2h, ..., x_n = a + nh = b$, we have assumed that the $(n + 1)$ ordinates $y_0, y_1, ..., y_n$ are at equal interval.

$\therefore$
$$I = \int_a^b y\,dx = \int_a^{a+nh} f(x)\,dx$$

Now using Newton's forward interpolation formula, we get

$$f(x) = y_0 + u\Delta y_0 + \frac{u(u-1)}{2!}\Delta^2 y_0 + \frac{u(u-1)(u-2)}{3!}\Delta^3 y_0 + ...$$

$$+ \frac{u(u-1)...\{u-(n+1)\}}{n!}\Delta^n y_0$$

where $\quad u = \dfrac{x - x_0}{h} \quad \Rightarrow du = \dfrac{dx}{h}$

$$\therefore \quad I = \int_0^n \left[y_0 + u\Delta y_0 + \frac{u(u-1)}{2!}\Delta^2 y_0 + \frac{u(u-1)(u-2)}{3!}\Delta^3 y_0 + \dots \right] h\, du$$

$$= h\left[ny_0 + \frac{n^2}{2}\Delta y_0 + \frac{\left(\dfrac{n^3}{3}-\dfrac{n^2}{2}\right)}{2!}\Delta^2 y_0 + \frac{\left(\dfrac{n^4}{4}-n^3+n^2\right)}{3!}\Delta^3 y_0 + \dots \right] \quad (8.1)$$

This is the general quadrature formula and is known as Newton–Cote's quadrature formula. We can obtain a number of quadrature formula from this by putting $n = 1, 2, \dots$

8.3 TRAPEZOIDAL RULE

By putting $n = 1$ in Eq. (8.1) means that there are only two paired values and interpolating polynomial is linear. Now

$$\int_{x_0}^{x_1} f(x)\, dx = h\left[y_0 + \frac{1}{2}(y_1 - y_0) \right] = \frac{1}{2}h\,(y_0 + y_1)$$

Similarly, for the next subinterval $x_1 = a + h$ to $x_2 = a + 2h$, we have

$$\int_{x_1}^{x_2} f(x)\, dx = \frac{1}{2}h\,(y_1 + y_2)$$

Continuing this way

$$\int_{x_2}^{x_3} f(x)\, dx = \frac{h}{2}(y_2 + y_3)$$

$$\dots\dots\dots\dots\dots\dots\dots\dots\dots\dots$$
$$\dots\dots\dots\dots\dots\dots\dots\dots\dots\dots$$

$$\int_{x_{n-1}}^{x_n} f(x)\, dx = \frac{h}{2}(y_{n-1} + y_n)$$

Adding all these equations, we get

$$\int_{x_0}^{x_n} f(x)\, dx = \frac{h}{2}[(y_0 + y_n) + 2(y_1 + y_2 + \dots + y_{n-1})]$$

$$= \frac{h}{2}\left[\begin{array}{l} (\text{sum of first and the last ordinates}) \\ + 2(\text{sum of the remaining ordinates}) \end{array} \right]$$

This is known as trapezoidal rule which is applicable only when n is a multiple of 1.
Remarks: Though this method is very simple for calculation purposes of numerical integration, the error in this case is significant. The accuracy of the result can be improved by increasing the number of intervals and decreasing the values of h.

8.3.1 Truncation Error in Trapezoidal Rule

Let $f(x)$ is continuous and possess continuous derivatives in $[x_0, x_n]$. In the neighbourhood of $x = x_0$, we can expand $y = f(x)$ by Taylor's series in powers of $x - x_0$. We get

$$y = y_0 + \frac{(x - x_0)}{1!}y_0' + \frac{(x - x_0)^2}{2!}y_0'' + \dots \quad (8.2)$$

Therefore, $\qquad \displaystyle\int_{x_0}^{x_1} y\, dx = \int_{x_0}^{x_1}\left[y_0 + \frac{(x - x_0)}{1!}y_0' + \frac{(x - x_0)^2}{2!}y_0'' + \dots \right] dx$

$$= \left[y_0 x + \frac{(x - x_0)^2}{2!} y_0' + \frac{(x - x_0)^3}{3!} y_0'' + ... \right]_{x_0}^{x_1}$$

$$= h y_0 + \frac{h^2}{2!} y_0' + \frac{h^3}{3!} y_0'' + ... \tag{8.3}$$

where h is the equal interval length.

Also $\qquad \int_{x_0}^{x_1} y \, dx = \frac{h}{2}(y_0 + y_1) = \text{area of first trapezium} = A_0 \tag{8.4}$

Putting $x = x_1$ in Eq. (8.2)

$$y(x_1) = y_1 = y_0 + \frac{(x_1 - x_0)}{1!} y_0' + \frac{(x_1 - x_0)^2}{2!} y_0'' + ...$$

i.e. $\qquad y_1 = y_0 + \frac{h}{1!} y_0' + \frac{h^2}{2!} y_0'' + ... \tag{8.5}$

From Eqs (8.4) and (8.5), we get

$$A_0 = \frac{h}{2} \left[y_0 + y_0 + \frac{h}{1!} y_0' + \frac{h^2}{2!} y_0'' + ... \right]$$

$$= h y_0 + \frac{h^2}{2} y_0' + \frac{h^3}{2 \times 2!} y_0'' + ... \tag{8.6}$$

Subtracting Eq. (8.6) from Eq. (8.3)

$$\int_{x_0}^{x_1} y \, dx - A_0 = h^3 y_0'' \left[\frac{1}{3!} - \frac{1}{2(2!)} \right] + ... = -\frac{1}{12} h^3 y_0'' + ...$$

Therefore the error in $[x_0, x_1]$ is $-\frac{h^3}{12} y_0''$ (neglecting other terms). Similarly, the error in

$[x_1, x_2]$ is $-\frac{h^3}{12} y_1''$ and error in the interval $[x_{n-1}, x_n]$ is $-\frac{h^3}{12} y_{n-1}''$. Hence the total error is

$$E = -\frac{h^3}{12} (y_0'' + y_1'' + ... + y_{n-1}'')$$

If $y''(\xi)$ be the largest value of $y_0'', y_1'', y_2'', ..., y_{n-1}''$ where $a < \xi < b$.

Hence $\qquad E = -\frac{h^3}{12} [y''(\xi) + y''(\xi) + ... + y''(\xi)] = -\frac{n h^3}{12} y''(\xi))$

$$= -\frac{(b - a)}{12} h^2 y''(\xi) \qquad \left| \begin{array}{l} h = \dfrac{b - a}{n} \Rightarrow (b - a) = nh \end{array} \right.$$

This is the error in trapezoidal rule.

8.4 SIMPSON'S 1/3RD RULE

Putting $n = 2$ in Newton-Cote's quadrature formula, we have

$$\int_{x_0}^{x_2} f(x) \, dx = h \left[2 y_0 + \frac{4}{2} \Delta y_0 + \frac{1}{2} \left(\frac{8}{3} - \frac{4}{2} \right) \Delta^2 y_0 \right] \text{(since other terms vanish)}$$

$$= h\left[2y_0 + 2(y_1 - y_0) + \frac{1}{3}(y_2 - 2y_1 + y_0)\right]$$

$$= \frac{h}{3}[y_2 + 4y_1 + y_0]$$

Similarly, $\int_{x_2}^{x_4} f(x)dx = \frac{h}{3}[y_2 + 4y_3 + y_4]$

$$\cdots\cdots\cdots\cdots \quad \cdots\cdots\cdots\cdots\cdots$$
$$\cdots\cdots\cdots\cdots \quad \cdots\cdots\cdots\cdots\cdots$$

$$\int_{x_{n-2}}^{x_n} f(x)dx = \frac{h}{3}[y_{n-2} + 4y_{n-1} + y_n]$$

Adding all these equations, we get

$$\int_{x_0}^{x_n} f(x)dx = \int_{x_0}^{x_2} f(x)dx + \int_{x_2}^{x_4} f(x)dx + \ldots + \int_{x_{n-2}}^{x_n} f(x)dx$$

$$= \frac{h}{3}[(y_0 + 4y_1 + y_2) + (y_2 + 4y_3 + y_4) + \ldots + (y_{n-2} + 4y_{n-1} + y_n)]$$

$$= \frac{h}{3}[(y_0 + y_n) + 2(y_2 + y_4 + \ldots) + 4(y_1 + y_3 + \ldots)]$$

$$= \frac{h}{3} \text{ [sum of the first and last ordinate + 2 (sum of even ordinates) + 4 (sum of odd ordinates)]}$$

This is called the Simpson's 1/3rd rule which is applicable only when n is a multiple of 2.

8.4.1 Truncation Error in Simpson's 1/3rd Rule

Let $f(x)$ is continuous and possess continuous derivatives in the interval $[x_0, x_n]$. By Taylor expansion of $y = f(x)$ in the neighbourhood of $x = x_0$, we get

$$y = y_0 + \frac{(x - x_0)}{1!} y_0' + \frac{(x - x_0)^2}{2!} y_0'' + \ldots \qquad (8.7)$$

On integrating x_0 to x_2 with respect to x, we get

$$\int_{x_0}^{x_2} y\, dx = \int_{x_0}^{x_2}\left[y_0 + \frac{(x - x_0)}{1!} y_0' + \frac{(x - x_0)^2}{2!} y_0'' + \ldots\right] dx$$

$$= 2hy_0 + 2h^2 y_0' + \frac{8h^3}{3!} y_0'' + \frac{16h^4}{4!} y_0''' + \frac{32h^5}{5!} y_0^{iv} + \ldots \qquad (8.8)$$

Now $\qquad A_1 = \text{area} = \int_{x_0}^{x_2} y\, dx = \frac{h}{3}(y_0 + 4y_1 + y_2) \qquad (8.9)$

where A_1 is the area of the curve in the interval $[x_0, x_2]$. Putting $x = x_1$ in Eq. (8.7), we get

$$y(x_1) = y_1 = y_0 + \frac{(x_1 - x_0)}{1!} y_0' + \frac{(x_1 - x_0)^2}{2!} y_0''' + \ldots = y_0 + \frac{h}{1!} y_0' + \frac{h_2}{2!} y_0'' + \ldots \quad (8.10)$$

Putting $x = x_2 = x_0 + 2h$, in Eq. (8.7), we get

$$y_2 = y_0 + \frac{2h}{1!} y_0' + \frac{4h^2}{2!} y_0'' + \ldots \qquad (8.11)$$

Substituting Eqs (8.10) and (8.11) in Eq. (8.9), we obtain

$$A_1 = 2hy_0 + 2h^2 y_0' + \frac{4h^3}{3} y_0'' + \frac{2h^4}{3} y_0''' + \frac{5h^5}{18} y_0'''' + \dots \qquad (8.12)$$

Now, the error in interval $[x_0, x_2]$ is given by

$$\int_{x_0}^{x_2} y\, dx - A_1 = \left(\frac{4}{15} - \frac{5}{18} \right) h^5 y_0''' + \dots = -\frac{h^5}{90} y_0'''' + \dots \qquad (8.13)$$

Continuing in the same way, we get the error in other subintervals $[x_2, x_4]$, $[x_4, x_6]$, ... etc. Hence the total error E is given by

$$E = -\frac{h^5}{90} [y_0'''' + y_2'''' + y_4'''' + \dots + y_{n-2}''''] \qquad (8.14)$$

Let $y''''(\xi)$ be largest value of the $y_0'''', y_2'''', \dots, y_{n-2}''''$, where $a < \xi < b$.

Hence

$$E = -\frac{h^5}{90} [y''''(\xi) + y''''(\xi) + \dots + y''''(\xi)]$$

$$= -\frac{h^4}{180} y''''(\xi)$$

This is the error in Simpson's 1/3rd rule.

8.5 SIMPSON'S 3/8TH RULE

Putting $n = 3$ in Newton–Cote's quadrature formula, we get

$$\int_{x_0}^{x_3} y\, dx = h \left[3y_0 + \frac{9}{2} \Delta y_0 + \frac{1}{2} \left(\frac{9}{2} \right) \Delta^2 y_0 + \frac{1}{6} \left(\frac{81}{4} - 27 + 9 \right) \Delta^3 y_0 \right]$$

$$= h \left[3y_0 + \frac{9}{2} (y_1 - y_0) + \frac{9}{4} (y_2 - 2y_1 + y_0) + \frac{3}{8} (y_3 - 3y_2 + 3y_1 - y_0) \right]$$

$$= \frac{3h}{8} (y_3 + 3y_2 + 3y_1 + y_0) = \frac{3h}{8} (y_0 + 3y_1 + 3y_2 + y_3)$$

Similarly $\int_{x_3}^{x_6} y\, dx = \frac{3h}{8} (y_3 + 3y_4 + 3y_5 + y_6)$

$$\dots\dots\dots \dots\dots\dots\dots\dots\dots\dots\dots\dots$$
$$\dots\dots\dots \dots\dots\dots\dots\dots\dots$$

$$\int_{x_{n-3}}^{x_n} y\, dx = \frac{3h}{8} (y_{n-3} + 3y_{n-2} + 3y_{n-1} + y_n)$$

Adding the above integrals, we get

$$\int_{x_0}^{x_n} y\, dx = \int_{x_0}^{x_3} y\, dx + \int_{x_0}^{x_6} y\, dx + \dots + \int_{x_{n-3}}^{x_n} y\, dx$$

$$= \frac{3h}{8} [(y_0 + y_n) + 3(y_1 + y_2 + y_4 + y_5 + \dots) + 2(y_3 + y_6 + \dots)]$$

This is known as Simpson's 3/8th rule which is applicable only when n is multiple of 3.

8.5.1 Truncation Error in Simpson's 3/8th Rule

Let $f(x)$ is continuous and possess continuous derivatives in $[x_0, x_n]$. By Taylor expansion of $f(x)$ in the neighbourhood of $x = x_0$, we get

$$y = y_0 + \frac{(x - x_0)}{1!} y_0' + \frac{(x - x_0)^2}{2!} y_0'' + \dots \tag{8.15}$$

$$\int_{x_0}^{x_3} y\,dx = \int_{x_0}^{x_3} \left[y_0 + \frac{(x - x_0)}{1!} y_0' + \frac{(x - x_0)^2}{2!} y_0'' + \dots \right] dx$$

$$= 3h\,y_0 + \frac{9}{2}h^2 y_0' + \frac{9}{2}h^3 y_0'' + \frac{27}{8}h^4 y_0''' + \frac{81}{40}h^5 y_0'''' + \dots \tag{8.16}$$

Now
$$A_1 = \text{area} = \int_{x_0}^{x_3} y\,dx = \frac{3h}{8}[y_0 + 3y_1 + 3y_2 + y_3] \tag{8.17}$$

where A_1 is the area of the curve in the interval $[x_0, x_3]$.

Putting $x = x_1$ in Eq. (8.15), we get

$$y(x_1) = y_1 = y_0 + \frac{(x_1 - x_0)}{1!} y_0' + \frac{(x_1 - x_0)^2}{2!} + y_0'' + \dots = y_0 + \frac{hy_0'}{1!} + \frac{h^2}{2!} y_0'' + \dots \tag{8.18}$$

Putting $x = x_2$ in Eq. (8.15), we get

$$y(x_2) = y_2 = y_0 + \frac{2h}{1!} y_0' + \frac{4h^2}{2!} y_0'' + \dots \tag{8.19}$$

Putting $x = x_3$ in Eq. (8.15), we get

$$y(x_3) = y_3 = y_0 + \frac{3h}{1!} y_0' + \frac{9h^2}{2!} y_0'' + \dots \tag{8.20}$$

Substituting Eqs (8.18), (8.19) and (8.20) in Eq. (8.17)

$$A_1 = \frac{3h}{8}\left[y_0 + 3\left\{ y_0 + \frac{h}{1!} y_0' + \frac{h^2}{2!} y_0'' + \frac{h^3}{3!} y_0''' + \frac{h^4}{4!} y_0'''' + \dots \right\} \right.$$

$$+ 3\left\{ y_0 + \frac{2h}{1!} y_0' + \frac{4h^2}{2!} y_0'' + \frac{8h^3}{3!} y_0''' + \frac{16h^4}{4!} y_0'''' + \dots \right\}$$

$$\left. + \left\{ y_0 + \frac{3h}{1!} y_0' + \frac{9h^2}{2!} y_0'' + \frac{27h^3}{3!} y_0''' + \frac{81h^4}{4!} y_0'''' + \dots \right\} \right]$$

$$= 3hy_0 + \frac{9h^2}{2} y_0' + \frac{9}{2}h^3 y_0'' + \frac{27}{8} y_0'''h^4 + \frac{33}{16}h^5 y_0'''' + \dots \tag{8.21}$$

Now, the error in the interval $[x_0, x_3]$ is given by

$$\int_{x_0}^{x_3} y\,dx - A_1 = \left(\frac{81}{40} - \frac{33}{16} \right) h^5 y_0'''' + \dots + y_{n-3}''''$$

Continuing in the same way we get the error in other subintervals $[x_3, x_6], [x_6, x_9], \dots$ etc. Hence the total error E is given by

$$E = -\frac{3h^5}{80}[y_0'''' + y_3'''' + y_6'''' + \dots + y_{n-3}'''']$$

Let $y''''(\xi)$ be the largest value of the $y_0'''', y_3'''', ..., y_{n-3}''''$ where $a < \xi < b$.

Hence

$$E = -\frac{3}{80} h^5 [y''''(\xi) + y''''(\xi) + ... + y''''(\xi)]$$

$$= -\frac{3}{80} h^5 ny''''(\xi) = -\frac{(b-a)}{80} h^4 y''''(\xi) \quad \text{as} \quad 3nh = b - a$$

This is the error in Simpson's 3/8 rule.

Example 1: Find the approximate value of $I = \int_0^1 \frac{dx}{1+x}$ using (i) Trapezoidal rule, (ii) Simpson's 1/3 rule. Obtain a bound for the error. The exact value of $I = \log 2 = 0.693147$.

Solution: The range of integration is from 0 to 1. Dividing the interval into 8 parts each of width $h = \frac{1-0}{8} = \frac{1}{8}$, the values of $\frac{1}{1+x}$ is given below. Let $y = f(x) = \frac{1}{1+x}$

x	0	1/8	2/8	3/8	4/8	5/8	6/8	7/8	1
y	1	8/9	4/5	8/11	2/3	8/13	4/7	8/15	1/2
	y_0	y_1	y_2	y_3	y_4	y_5	y_6	y_7	y_8

By Trapezoidal rule

$$\int_0^1 \frac{dx}{1+x} = \frac{h}{2}[(y_0 + y_8) + 2(y_1 + y_2 + y_3 + y_4 + y_5 + y_6 + y_7)]$$

$$= \frac{\frac{1}{8}}{2}\left[\left(1 + \frac{1}{2}\right) + 2\left(\frac{8}{9} + \frac{4}{5} + \frac{8}{11} + \frac{2}{3} + \frac{8}{13} + \frac{4}{7} + \frac{8}{15}\right)\right]$$

$$= 0.694122$$

Error $= 0.694122 - 0.693147 = 0.000975$

The error in the trapezoidal rule

$$|E| \leq \left|\frac{(b-a)h^2}{12}\right| \max_{0 \leq x \leq 1} |f''(\xi)|$$

$$\leq \left|\frac{(1-0)\left(\frac{1}{8}\right)^2}{12}\right| \max_{0 \leq x \leq 1} \left|\frac{2}{(1+x)^3}\right|$$

$$= 0.001302 \times 2 = 0.002604$$

ii. Now, using Simpson's 1/3 rule

$$\int_0^1 \frac{dx}{1+x} = \frac{h}{3}[(y_0 + y_8) + 4(y_1 + y_3 + y_5 + y_7) + 2(y_2 + y_4 + y_6)]$$

$$= \frac{1}{8 \times 3}\left[\left(1 + \frac{1}{2}\right) + 4\left(\frac{8}{9} + \frac{8}{11} + \frac{8}{13} + \frac{8}{15}\right) + 2\left(\frac{4}{5} + \frac{2}{3} + \frac{4}{7}\right)\right]$$

$$= 0.693154$$

Error $= 0.693154 - 0.693147 = 0.000007$

The error in the Simpson's 1/3rd rule

$$|E| \le \left|\frac{(b-a)}{180} h^4\right| \max_{0 \le x \le 1} |f''''(\xi)|$$

$$\le \left|\frac{(1-0)}{180}\left(\frac{1}{8}\right)^4\right| \max_{0 \le x \le 1} \left|\frac{24}{(1+x)^5}\right|$$

$$\le \frac{1}{180 \times 4096} \times 24 = 0.0000326$$

Example 2: Evaluate $\int_0^1 e^x dx$ by Simpson's 1/3rd rule correct to five decimal places, by proper choice of h.

Solution: Here, interval length $= b - a = 1$.

Let $y = e^x \Rightarrow y''' = e^x$

$$\text{Error} = |E| \le \left|\frac{(b-a)}{180} h^4\right| \max_{0 \le x \le 1} |y''''(\xi)|$$

$$\le \left|\frac{(1-0)}{180}(h)^4\right| \max_{0 \le x \le 1} |e^x|$$

$$\le \frac{1}{180} h^4 e$$

We require $(E) < 10^{-6}$

$$\frac{h^4 e}{180} < 10^{-6} \quad \Rightarrow \quad h < \left(\frac{180 \times 10^{-6}}{e}\right)^{1/4} = 0.148$$

Hence, we take $h = 0.1$ to have the accuracy required.

$$\therefore \quad \int_0^1 e^x dx = \frac{0.1}{3}[(1+e) + 2(e^{0.2} + e^{0.4} + e^{0.6} + e^{0.8}) + 4(e^{0.1} + e^{0.3} + e^{0.5} + e^{0.7} + e^{0.9})]$$

$$= 1.718283$$

By actual integration, $\int_0^1 e^x dx = (e^x)_0^1 = e - 1 = 1.71828183$ correct to five decimal places, the answer is 1.71828.

Example 3: Compute the error in the evaluation of $\int_4^{5.2} \log x \, dx$ by Simpson's 1/3rd rule.

Solution: Let $y = f(x) = \log x$, then

$$f''''(x) = -\frac{6}{x^4}$$

Let the points are taken at width of 0.2 i.e. $h = 0.2$.

$$\text{Error} = |E| \le \left|\frac{(b-a)}{180} h^4\right| \max_{4 \le x \le 5.2} |f''''(\xi)|$$

$$\le \left|\left(\frac{5.2-4}{180}\right)(0.2)^4\right| \max_{4 \le x \le 5.2} \left|-\frac{6}{x^4}\right|$$

since $f''''(\xi)$ has maximum value at $x = 4$

Therefore $\qquad |E| \le \left|\left(\dfrac{5.2 - 4}{180}\right)(0.2)^4\right| \left|\left(\dfrac{-6}{4^4}\right)\right| = 0.00000025$

Example 4: If trapezoidal rule is to be used to compute $\int_0^1 e^{-x^2}\,dx$ with an error at most $\dfrac{1}{2} \times 10^{-4}$, how many points should be used?

Solution: We have error in trapezoidal rule

$$|E| \le \left|\dfrac{(b - a)}{12}h^2\right| \max_{a \le x \le 6} |f''(\xi)|$$

Here $\qquad f(x) = e^{-x^2}$

$\qquad\qquad f'(x) = -2xe^{-x^2}$

and $\qquad f''(x) = -2[xe^{-x^2} \cdot (-2x) + e^{-x^2}] = 4x^2 e^{-x^2} - 2e^{-x^2}$

$\qquad\qquad\qquad = (4x^2 - 2)e^{-x^2}$

Now $\qquad |f''(0)| = |-2| = 2$

$\qquad\qquad |f''(1)| = |2e^{-1}| = 0.7358$

Therefore $|f''(0)| \le 2$ in the interval $[0, 1]$ and the error in absolute value will be not greater than $\dfrac{1}{6}h^2$. So, to have an error of at most $\dfrac{1}{2} \times 10^{-4}$, we require

$$\dfrac{1}{6}h^2 \le \dfrac{1}{2} \times 10^{-4}$$

i.e. $\qquad\qquad h \le 0.01732$

and also as $\qquad h = \dfrac{1 - 0}{n} = \dfrac{1}{n}$

So, we require $\qquad \dfrac{1}{n} \le 0.01732 \quad \text{or} \quad n \le 58$

i.e. at least 58 points are required for the desired accuracy of given integration.

Example 5: Find $I = \int_1^5 x^4\,dx$ by using Simpson's 3/8th rule. Also find an error bound for this estimation. The exact value is 624.8. Determine the error. Is it within the bound?

Solution: Let $y = f(x) = x^4 \Rightarrow f''''(x) = 24$.

Here $b - a = 5 - 1 = 4$. We divide the interval $[5, 1]$ into six parts each of width $h = \dfrac{4}{6} = \dfrac{2}{3}$.

The values of y are given below:

x	1	5/3	7/3	3	11/3	13/3	5
y	1	625/81	2401/81	27	14641/81	28561/81	625
	y_0	y_1	y_2	y_3	y_4	y_5	y_6

By using Simpson's 3/8 rule

$$I = \int_1^5 x^4 dx = \frac{3h}{8}[(y_0 + y_6) + 3(y_1 + y_2 + y_4 + y_5) + 2(y_3)]$$

$$= \frac{3 \times 2}{8 \times 3}\left[(1 + 625) + 3\left(\frac{625}{81} + \frac{2401}{81} + \frac{14641}{81} + \frac{28561}{81}\right) + 2 \times 27\right]$$

$$= \frac{1}{4}[626 + 3 \times 570.716 + 54] = 598.037$$

Error $= 598.037 - 624.8 = -26.763$

The error bound in Simpson's 3/8 rule

$$|E| \leq \left|\frac{(b-a)}{80} h^4\right| \max_{1 \leq x \leq 5} |f''''(\xi)|$$

$$\leq \left|\frac{(5-1)}{80} \times \left(\frac{2}{3}\right)^4\right| \max_{1 \leq x \leq 5} |24|$$

$$\leq \left|\frac{4}{80} \times \frac{16}{27}\right| \times 24 = 0.7111$$

∴ The error $26.763 > 0.7111$. So the error is not within bound.

Example 6: Solve the Example 5 by using trapezoidal rule.

Solution: Here we take $n = 4$. So, $h = \dfrac{b-a}{n} = \dfrac{5-1}{4} = 1$. The values of y are given below

x	1	2	3	4	5
y	1	16	81	256	625
	y_0	y_1	y_2	y_3	y_4

By using trapezoidal rule, we get

$$I = \int_1^5 x^4 dx = \frac{h}{2}[(y_0 + y_4) + 2(y_1 + y_2 + y_3)]$$

$$= \frac{1}{2}[(1 + 625) + 2(16 + 81 + 256)] = 666$$

$|\text{Error}| = 666 - 624.80 = 41.2$

The error bound in trapezoidal rule

$$|E| \leq \left|\frac{(b-a)}{12} h^2\right| \max_{1 \leq x \leq 5} |f''(\xi)|$$

$$\leq \left|\frac{(5-1)}{12} \times 1^2\right| \max_{1 \leq x \leq 5} |12x^2|$$

$$\leq \left|\frac{4}{12} \times 1\right| \times 12 \times 25 = 100$$

So, the error $41.2 < 100$. Hence the error is within bound.

Example 7: Evaluate $I = \int_0^2 \dfrac{dx}{1+x^2}$ using Simpson's 1/3rd rule. How large should n be chosen in order to ensure that $|E| \leq 5 \times 10^{-6}$.

Solution: The range of integration is from 0 to 2. Dividing the interval into 6 parts each of width $h = \dfrac{2-0}{6} = \dfrac{1}{3}$, the values of $\dfrac{1}{1+x^2}$ is given below.

Let $y = f(x) = \dfrac{1}{1+x^2}$

x	0	1/3	2/3	1	4/3	5/3	2
$f(x)$	1	9/10	9/13	1/2	9/25	9/34	1/5

By Simpson's 1/3rd rule

$$\int_0^2 \frac{dx}{1+x^2} = \frac{h}{3}[(y_0 + y_6) + 4(y_1 + y_3 + y_5) + 2(y_2 + y_4)]$$

$$= \frac{1}{3 \times 3}\left[\left(1 + \frac{1}{5}\right) + 4\left(\frac{9}{10} + \frac{1}{2} + \frac{9}{34}\right) + 2\left(\frac{9}{13} + \frac{9}{25}\right)\right]$$

$$= \frac{1}{9}\left[\frac{6}{5} + 4(1.6647) + 2(1.0523)\right] = 1.1070$$

Also

$$f''''(x) = 24\left[\frac{5x^4 - 10x^2 + 1}{(1+x^2)^5}\right]$$

$$\max_{0 \leq x \leq 2} |f''''(x)| = f''''(0) = 24$$

Then

$$|E| \leq \left|\frac{(b-a)}{180} h^4\right| \max_{0 \leq x \leq 2} |f''''(\xi)|$$

$$\leq \left|\frac{(2-0)}{180} h^4\right| \max_{0 \leq x \leq 2} \left|24\left[\frac{5x^4 - 10x^2 + 1}{(1+x^2)^5}\right]\right|$$

$$\leq \frac{48h^4}{180} = \frac{4h^4}{15}$$

Then $|E| \leq 5 \times 10^{-6}$ is true if

$$\frac{4h^4}{15} \leq 5 \times 10^{-6}$$

$\Rightarrow$
$$h \leq 0.0658$$
$$n \geq 30.39$$

Therefore, selection of $n \geq 31$ will give the desired error bound.

Example 8: A train is moving at the speed of 30 m/sec. Suddenly brakes are applied. The speed of the train per second after t seconds is given by

Time (t)	0	5	10	15	20	25	30	35	40	45
Speed (v)	30	24	19	16	13	11	10	8	7	5

Apply Simpson's 3/8th rule to determine the distance moved by the train in 45 seconds.

Solution: If s meter is the distance covered in t seconds, then

$$\frac{ds}{dt} = v$$

$$\Rightarrow \qquad [s]_{t=0}^{t=45} = \int_0^{45} v\, dt$$

Since the number of subintervals is 9 (multiple of 3), hence by using Simpson's 3/8th rule

$$\int_0^{45} v\, dt = \frac{3h}{8}[(v_0 + v_9) + 3(v_1 + v_2 + v_4 + v_5 + v_7 + v_8) + 2(v_3 + v_6)]$$

$$= \frac{15}{8}[(30 + 5) + 3(24 + 19 + 13 + 11 + 8 + 7) + 2(16 + 10)]$$

$$= 624.375 \text{ metres}$$

Hence the distance moved by the train in 45 seconds is 624.375 meter.

Example 9: The speedo meter of a car, t seconds after its start is shown in the following table:

t	0	12	24	36	48	60	72	84	96	108	120
v	0	3.60	10.08	18.9	21.6	18.54	10.26	5.4	4.5	5.4	9.0

Using Simpson's 1/3rd rule, find the distance travelled by the car in 2 minutes.

Solution: If s metres is the distance covered in t seconds, then

$$\frac{ds}{dt} = v$$

$$\Rightarrow \qquad [s]_{t=0}^{t=120} = \int_0^{120} v\, dt$$

Since the number of sub-interval is 10 (multiple of 2), hence, by using Simpson's 1/3rd rule

$$\int_0^{120} v\, dt = \frac{h}{3}[(v_0 + v_{10}) + 4(v_1 + v_3 + v_5 + v_7 + v_9) + 2(v_2 + v_4 + v_6 + v_8)]$$

$$= \frac{12}{3}[(0 + 9) + 4(3.6 + 18.9 + 18.54 + 5.4 + 5.4) + 2(10.08 + 21.6 + 10.26 + 4.5)]$$

$$= 1236.96 \text{ metres}$$

Hence, the distance travelled by car in 2 minutes is 1236.96 metres.

Example 10: Evaluate $\int_0^1 \dfrac{dx}{1+x^2}$ using (a) Trapezoidal rule, (b) Simpson's 3/8 rule.

Solution: Here $y = f(x) = \dfrac{1}{1+x^2}$. We divide the range [0, 1] in 6 equal parts so that

$$h = \frac{b-a}{n} = \frac{1-0}{6} = \frac{1}{6}$$

The values of $f(x) = \dfrac{1}{1+x^2}$ are as tabulated below:

x	1	1/6	1/3	1/2	2/3	5/6	1
$f(x)$	1	36/37	9/10	4/5	9/13	36/61	1/2
	y_0	y_1	y_2	y_3	y_4	y_5	y_6

(a) Using trapezoidal rule

$$\int_0^1 \frac{dx}{1+x^2} = \frac{h}{2}[(y_0 + y_6) + 2(y_1 + y_2 + y_3 + y_4 + y_5)]$$

$$= \frac{1}{12}\left[\left(1+\frac{1}{2}\right) + 2\left(\frac{36}{37} + \frac{9}{10} + \frac{4}{5} + \frac{9}{13} + \frac{36}{61}\right)\right]$$

$$= 0.784241.$$

(b) Using Simpson's 1/3rd rule

$$\int_0^1 \frac{dx}{1+x^2} = \frac{3h}{8}[(y_0 + y_6) + 2y_3 + 3(y_1 + y_2 + y_4 + y_5)]$$

$$= \left(\frac{3}{8}\right)\left(\frac{1}{6}\right)\left[\left(1+\frac{1}{2}\right) + 2\left(\frac{4}{5}\right) + 3\left(\frac{36}{37} + \frac{9}{10} + \frac{9}{13} + \frac{36}{61}\right)\right]$$

$$= 0.785396$$

Example 11: A curve is drawn to pass through the points given by the following table:

x	1	1.5	2	2.5	3	3.5	4
y	2	2.4	2.7	2.8	3	2.6	2.1

Estimate the area bounded by the curve, the x-axis and the lines $x = 1$, $x = 4$.

Solution: By Simpson's 1/3rd rule, we have required area

$$= \frac{h}{3}[(y_0 + y_6) + 4(y_1 + y_3 + y_5) + 2(y_2 + y_4)]$$

$$= \frac{0.5}{3}[(2 + 2.1) + 4(2.4 + 2.8 + 2.6) + 2(2.7 + 3)]$$

$$= 7.783 \text{ units of area}$$

EXERCISE 8.1

1. $I = \int_{-1}^4 x^3 dx$

   i. Calculate the exact value of I.

   ii. Approximate I using the trapezoidal rule. Find an error bound for approximation. Calculate the error. Is it within bound?   [Ans. 63.75, 67.5, 3.75, yes]

2. Evaluate $I = \int_0^2 \dfrac{dx}{1+x^2}$ by using the trapezoidal method. How large should n be choosen in order to ensure that $|E| \le 5 \times 10^{-6}$. [Ans. $n \ge 517$]

3. Find a value for n that guarantees an error of no more than 10^{-5} in the approximation by the Simpson's rule of $\int_2^4 x^{1/2}\, dx$. [Ans. $n = 7$]

4. Find the minimum number of subintervals required to approximate $\int_0^4 x^2\, dx$ with an error less than 0.00001. [Ans. $|E| = 0, n = 2$]

5. Determine the value of n so that the trapezoidal rule will approximate the value of $\int_0^1 (1+x^2)^{1/2}\, dx$ with an error that is less than 0.01. [Ans. $n = 3$]

6. Determine a value of n so that the Simpson's rule will approximate the value of $\int_0^1 \cos x^2\, dx$ with an error that is less than 0.001. [Ans. $n \ge 2.2 \Rightarrow n = 3$]

7. Let $I = \int_1^3 \dfrac{1}{x}\, dx$. Find an error bound by using trapezoidal rule. [Ans. 0.0533]

8. Let $I = \int_0^1 e^{-x^2}\, dx$. Estimate I using the following rule with the indicated number n of subintervals. Find an error bound in the estimation in each case:
   i. The trapezoidal rule, $n = 10$
   ii. Simpson's rule, $n = 4$
   [Ans. i. $I = 0.74621079$, $|E| = 0.0016666$; ii. $I = 0.74685537$, $|E| = 0.000434028$]

9. A function f is unknown but experiment has determined its following values:

x	0.0	0.2	0.4	0.6	0.8	1.0	1.2	1.4	1.6
$f(x)$	0.52	0.57	0.63	0.77	0.79	0.80	0.98	1.83	2.07

Find the best Simpson's rule approximation for $\int_{0.0}^{1.6} f(x)\, dx$ using data from the table. Assuming $|f'''''(x)| \le 3$ on $[0.0, 1.6]$, compute a bound for the magnitude of the error in the approximation. [Ans. 1.55133, $|E| = 0.00768$]

10. Evaluate $\int_{0.5}^{0.7} x^{1/2}\, e^{-x}\, dx$ approximately by using a suitable formula.
[Ans. $I = 0.0848271$]

11. Use Simpson's rule dividing the range into ten equal parts, to show that
$\int_0^1 \dfrac{\log(1+x^3)}{1+x^2} = 0.1730.$

12. Evaluate $\int_4^{5.2} \log_e x\, dx$ by:
   i. Simpson's 1/3rd rule
   ii. Simpson's 3/8th rule [Ans. i. 1.82784726, ii. 1.82784707]

13. Compute the value of the definite integral $\int_{0.2}^{1.4} (\sin x - \log_e x + e^x)\, dx$ by:
   i. the trapezoidal rule
   ii. Simpson's 1/3rd rule
   iii. Simpson's 3/8th rule [Ans. i. 4.05617, ii. 4.05106, iii. 4.05116]

8.6 BOOLE'S RULE

Putting $n = 4$ in Newton-Cote's quadrature formula, we have

$$\int_{x_0}^{x_4} ydx = h\left[4y_0 + \frac{16}{2}\Delta y_0 + \left(\frac{\frac{64}{3} - 8}{2}\right)\Delta^2 y_0 + \frac{(64 - 64 + 16)}{6}\Delta^3 y_0\right.$$

$$\left. + \frac{\left(\frac{4^5}{5} - \frac{3 \times 4^4}{2} + \frac{11}{3} \times 4^3 - 3 \times 4^2\right)}{24}\Delta^4 y_0\right] \quad \text{(Since other terms vanish)}$$

$$= h\left[4y_0 + 8\Delta y_0 + \frac{20}{3}\Delta^2 y_0 + \frac{8}{3}\Delta^3 y_0 + \frac{28}{90}\Delta^4 y_0\right]$$

$$= h\left[4y_0 + 8(y_1 - y_0) + \frac{20}{3}(y_2 - 2y_1 + y_0) + \frac{8}{3}(y_3 - 3y_2 + 3y_1 - y_0)\right.$$

$$\left. + \frac{28}{90}(y_4 - 4y_3 + 6y_2 - 4y_1 + y_0)\right]$$

$$= h\left[\frac{14}{45}y_0 + \frac{64}{45}y_1 + \frac{24}{45}y_2 + \frac{64}{45}y_3 + \frac{14}{45}y_4\right]$$

$$\int_{x_0}^{x_4} ydx = \frac{2h}{45}(7y_0 + 32y_1 + 12y_2 + 32y_3 + 7y_4)$$

Similarly, $\int_{x_0}^{x_8} ydx = \frac{2h}{45}(7y_4 + 32y_5 + 12y_6 + 32y_7 + 7y_8)$

....

....

$$\int_{x_{n-4}}^{x_n} ydx = \frac{2h}{45}(7y_{n-4} + 32y_{n-3} + 12y_{n-2} + 32y_{n-1} + 7y_n)$$

Adding the above integrals, we get

$$\int_{x_0}^{x_n} ydx = \frac{2h}{45}[7y_0 + 32y_1 + 12y_2 + 32y_3 + 14y_4 + 32y_5 + 12y_6 + 32y_7 + 14y_8 + ...]$$

This is known as Boole's rule which is applicable only when n is multiple of 4.

8.6.1 Truncation Error in Boole's Rule

Let $f(x)$ is continuous and possess continuous derivative in $[x_0, x_n]$. By Taylor's expansion of $f(x)$ in the neighbour of $x = x_0$, we get

$$y = y_0 + \frac{(x - x_0)}{1!}y_0' + \frac{(x - x_0)^2}{2!}y_0'' + ... \quad (8.22)$$

Using it, we follow

$$\int_{x_0}^{x_4} ydx = \int_{x_0}^{x_4}\left[y_0 + \frac{(x - x_0)}{1!}y_0' + \frac{(x - x_0)^2}{2!}y_0'' + ...\right]dx$$

$$= 4hy_0 + 8h^2 y_0' + \frac{32}{3} h^3 y_0'' + \frac{32}{3} h^4 y_0'''$$

$$+ \frac{128}{15} h^5 y_0'''' + \frac{256}{45} h^6 y_0''''' + \frac{1024}{315} h^7 y_0'''''' + \dots \qquad (8.23)$$

Now $A_1 = \text{area} = \int_{x_0}^{x_4} y \, dx = \frac{2h}{45} (7y_0 + 32y_1 + 12y_2 + 32y_3 + 7y_4)$ \qquad (8.24)

where A_1 is the area of the curve in the interval $[x_0, x_4]$.

Putting $x = x_1$ in Eq. (8.22), we get

$$y(x_1) = y_1 = y_0 + \frac{(x_1 - x_0)}{1!} y_0' + \frac{(x_1 - x_0)^2}{2!} y_0'' + \dots = y_0 + hy_0' + \frac{h^2}{2!} y_0'' + \dots \qquad (8.25)$$

Putting $x = x_2$ in Eq. (8.22), we get

$$y(x_2) = y_2 = y_0 + \frac{(x_2 - x_0)}{1!} y_0' + \frac{(x_2 - x_0)^2}{2!} y_0'' + \dots = y_0 + \frac{2h}{1!} y_0' + \frac{4h^2}{2!} y_0'' + \dots \quad (8.26)$$

Putting $x = x_3$ in Eq. (8.22), we get

$$y(x_3) = y_3 = y_0 + \frac{(x_3 - x_0)}{1!} y_0' + \frac{(x_3 - x_0)^2}{2!} y_0'' + \dots = y_0 + \frac{3h}{1!} y_0' + \frac{9h^2}{2!} y_0'' + \dots \quad (8.27)$$

Putting $x = x_4$ in Eq. (8.22), we get

$$y(x_4) = y_4 = y_0 + \frac{(x_4 - x_0)}{1!} y_0' + \frac{(x_4 - x_0)^2}{2!} y_0'' + \dots = y_0 + \frac{4h}{1!} y_0' + \frac{16h^2}{2!} y_0'' + \dots \quad (8.28)$$

Substituting Eqs (8.25)–(8.28) in Eq. (8.24), we get

$$A_1 = \frac{2h}{45} \left[7y_0 + 32 \left\{ y_0 + \frac{hy_0'}{1!} + \frac{h^2}{2!} y_0'' + \frac{h^3}{3!} y_0''' + \frac{h^4}{4!} y_0'''' + \dots \right\} \right.$$

$$+ 12 \left\{ y_0 + \frac{2h}{1!} y_0' + \frac{4h^2}{2!} y_0'' + \frac{8h^3}{3!} y_0''' + \frac{16h^4}{4!} y_0'''' + \dots \right\}$$

$$+ 32 \left\{ y_0 + \frac{3h}{1!} y_0' + \frac{9h^2}{2!} y_0'' + \frac{27h^3}{3!} y_0''' + \frac{81h^4}{4!} y_0'''' + \dots \right\}$$

$$\left. + 7 \left\{ y_0 + \frac{4h}{1!} y_0' + \frac{16h^2}{2!} y_0'' + \frac{64h^3}{3!} y_0''' + \frac{256h^4}{4!} y_0'''' + \dots \right\} \right]$$

$$A_1 = 4hy_0 + 8h^2 y_0' + \frac{32}{3} h^3 y_0'' + \frac{32}{3} h^4 y_0''' + \frac{128}{15} h^5 y_0''''$$

$$+ \frac{256}{45} h^6 y_0''''' + \frac{264}{81} h^7 y_0'''''' + \dots \qquad (8.29)$$

Now, the error in interval $[x_0, x_4]$ is given by

$$\int_{x_0}^{x_4} y \, dx - A_1 = \left(\frac{1024}{315} - \frac{264}{81} \right) h^7 y_0'''''' + \dots$$

$$= \frac{-8}{945} h^7 y_0'''''' + \dots$$

Continuing in the same way, we get the error in the other subinterval $[x_4, x_8]$, $[x_8, x_{12}]$, ... etc, Hence the total error E is given by

$$E = -\frac{8}{945} h^7 [y_0'''''' + y_4'''''' + y_8'''''' + ... + y_{n-4}'''''']$$

Let $y''''''(\xi)$ be the largest value of the $y_0'''''', y_4'''''', ..., y_{n-4}''''''$ where $a < \xi < b$.

Hence
$$E = -\frac{8}{945} h^7 [y''''''(\xi) + y''''''(\xi) + ... + y''''''(\xi)] = -\frac{8}{945} nh^7 y''''''(\xi)$$

$$= -\frac{(b-a)2h^6}{945} y''''''(\xi) \qquad |\, 4nh = b-a$$

This is the error in Boole's rule.

8.7 WEDDLE'S RULE

Putting $n = 6$ in Newton-Cote's quadrature formula, we have

$$\int_{x_0}^{x_6} ydx = h\left[6y_0 + \frac{36}{2}\Delta y_0 + \frac{\left(\dfrac{216}{3} - \dfrac{36}{2}\right)}{2!}\Delta^2 y_0 \right.$$

$$+ \frac{1}{3!}\left(\frac{6^4}{4} - 6^3 + 6^2\right)\Delta^3 y_0 + \frac{1}{4!}\left(\frac{6^5}{5} - \frac{3\times 6^4}{2} + \frac{11\times 6^3}{3} - 3\times 6^2\right)\Delta^4 y_0$$

$$+ \frac{1}{120}\left(\frac{6^6}{6} - 2\times 6^5 + \frac{35}{4}\times 6^4 - \frac{50}{3}\times 6^3 + 12\times 6^2\right)\Delta^5 y_0$$

$$\left. + \frac{1}{720}\left(\frac{6^7}{7} - \frac{5\times 6^6}{2} + 17\times 6^5 - \frac{225}{4}\times 6^4 + \frac{274}{3}\times 6^3 - 60\times 6^2\right)\Delta^6 y_0 \right]$$

$$= h\left[6y_0 + 18\Delta y_0 + \frac{54}{2}\Delta^2 y_0 + 24\Delta^3 y_0 + \frac{246}{20}\Delta^4 y_0 \right.$$

$$\left. + \frac{66}{20}\Delta^5 y_0 + \frac{246}{840}\Delta^6 y_0 \right]$$

$$= h\left[6y_0 + 18(y_1 - y_0) + 27(y_2 - 2y_1 + y_0) + 24(y_3 - 3y_2 + 3y_1 - y_0) \right.$$

$$+ \frac{123}{10}(y_4 - 4y_3 + 6y_2 - 4y_1 + y_0) + \frac{33}{10}(y_5 - 5y_4 + 10y_3$$

$$\left. -10y_2 + 5y_1 - y_0) + \frac{41}{140}(y_6 - 6y_5 + 15y_4 - 20y_3 + 15y_2 - 6y_1 + y_0) \right]$$

$$= \frac{3h}{10}[y_0 + 5y_1 + y_2 + 6y_3 + y_4 + 5y_5 + y_6] \qquad \left| \therefore \frac{41}{42} \approx 1 \right.$$

Similarly,

$$\int_{x_6}^{x_{12}} ydx = \frac{3h}{10}[y_6 + 5y_7 + y_8 + 6y_9 + y_{10} + 5y_{11} + y_{12}]$$

....

....

$$\int_{x_{n-6}}^{x_n} y\,dx = \frac{3h}{10}[y_{n-6} + 5y_{n-5} + y_{n-4} + 6y_{n-3} + y_{n-2} + 5y_{n-1} + y_n]$$

Adding the above integrals, we get

$$\int_{x_0}^{x_n} y\,dx = \frac{3h}{10}[y_0 + 5y_1 + y_2 + 6y_3 + y_4 + 5y_5 + 2y_6 + 5y_7 + y_8 + 6y_9 + y_{10} + 5y_{11} + 2y_{12} + \dots]$$

This is known as Weddle's rule which is applicable only when n is multiple of 6.

Example 12: The velocity v of a particle at distance s from a point on its path is given by the table below:

s in metre	0	10	20	30	40	50	60
v in m/sec	47	58	64	65	61	52	38

Estimate the time taken to travel 60 meter by using Weddle's rule.

Solution: We know $\dfrac{ds}{dt} = v$

$\therefore$ $\qquad dt = \dfrac{ds}{v}$

$$t = \int \frac{ds}{v}$$

To get t, we have to integrate $1/v$ from 0 to 60.

$\therefore$
$$t = \int_0^{60} \frac{ds}{v}$$

$$= \frac{3h}{10}[y_0 + 5y_1 + y_2 + 6y_3 + y_4 + 5y_5 + 2y_6]$$

$$= \frac{3 \times 10}{10}[0.02128 + 5 \times 0.01724 + 0.01562$$

$$+ 6 \times 0.01538 + 0.01639 + 5 \times 0.01923 + 0.02632]$$

$$= 3 \times 0.35424 = 1.06272$$

Example 13: Evaluate $\int_4^{5.2} \log x\,dx$ by Weddle's rule.

Solution: Let $y = \log x$, we divide the whole range (5.2, 4) into 12 equal parts.

$\therefore$
$$h = \frac{b-a}{12} = \frac{5.2-4}{12} = 0.1$$

x	log x
4.0	1.38629436
4.1	1.41098697
4.2	1.43508452
4.3	1.45861502
4.4	1.48160454
4.5	1.50407740
4.6	1.52605630
4.7	1.54756251

4.8	1.56861592
4.9	1.58923520
5.0	1.60943791
5.1	1.62924050
5.2	1.64865863

By Weddle's rule, we have

$$\int_4^{5.2} \log x\,dx = \frac{3h}{10}[y_0 + 5y_1 + y_2 + 6y_3 + y_4 + 5y_5 + 2y_6 + 5y_7 + y_8 + 6y_9 + y_{10} + 5y_{11} + y_{12}]$$

$$= \frac{3 \times 0.1}{10}[1.38629436 + 5 \times 1.41098697 + 1.43508452 + 6 \times 1.45861502$$

$$+ 1.48160454 + 5 \times 1.50407740 + 2 \times 1.52605630 + 5 \times 1.54756251$$

$$+ 1.56861592 + 6 \times 1.58923520 + 1.60943791 + 5 \times 1.62924050 + 1.64865863]$$

$$= 1.82784740$$

Example 14: Find the distance between two stations from the following data consisting of the speed $v(t)$ of an electric train at various times t after leaving one station until it stops at the next station. Apply Boole's rule.

v (miles/hr)	0	13	33	39.5	40	40	36	15	0
t (min)	0	0.5	1	1.5	2	2.5	3	3.5	4

Solution: If s miles is the distance covered in t seconds, then

$$\frac{ds}{dt} = v \implies ds = vdt$$

Therefore

$$[s]_{t=0}^{t=4} = \int_0^4 vdt$$

By using Boole's rule

$$\int_0^4 vdt = \frac{2h}{45}[7y_0 + 32y_1 + 12y_2 + 32y_3 + 14y_4 + 32y_5 + 12y_6 + 32y_7 + 7y_8]$$

We have

$$h = 0.5 \implies h = \frac{0.5}{60} = \frac{1}{120}$$

Therefore, $s = \int_0^4 vdt = \frac{2 \times 0.5}{60 \times 45}[7 \times 0 + 32 \times 13 + 12 \times 33 + 32 \times 39.5 + 14 \times 40$

$$+ 32 \times 40 + 12 \times 36 + 32 \times 15 + 7 \times 0]$$

$$= 1.788 \text{ miles}$$

Example 15: Find the value of integral $\int_1^2 \frac{1}{x} dx$ by using Boole's rule with strips. Determine the error by direct integration.

Solution: Let $y = \frac{1}{x}$. We divide the whole range of integration in 8 equal parts, so that

$$h = \frac{b-a}{8} = \frac{2-1}{8} = \frac{1}{8} = 0.125$$

The table is as below:

x	1	1.125	1.250	1.375	1.5	1.625	1.75	1.875	2.0
y	1	0.889	0.8	0.727	0.667	0.615	0.571	0.533	0.5

Using Boole's rule, we have

$$\int_1^2 \frac{1}{x} dx = \frac{2h}{45}[7y_0 + 32y_1 + 12y_2 + 32y_3 + 14y_4 + 32y_5 + 12y_6 + 32y_7 + 7y_8]$$

$$= \frac{2 \times 0.125}{45}[7 \times 1 + 32 \times 0.889 + 12 \times 0.8 + 32 \times 0.727 + 14 \times 0.667$$

$$+ 32 \times 0.615 + 12 \times 0.571 + 32 \times 0.533 + 7 \times 0.5]$$

$$= 0.69299$$

Actual value of integral $= [\log x]_1^2 = 0.69315$

Amount of error $= 0.69299 - 0.69315 = -0.00016$.

EXERCISE 8.2

1. Find by Weddle's rule, the value of $I = \int_{0.4}^{1.6} \frac{x}{\sin hx} dx$ by taking 12 sub-intervals.

[Ans. 1.0101996]

2. Evaluate $\int_{0.2}^{1.4} (\sin x - \log_e x + e^x) dx$ approximately using Weddle's rule correct to 4 decimals.

[Ans. 4.051]

3. Evaluate $\int_0^4 \frac{dx}{1 + x^2}$ using Boole's rule taking $h = 0.5$. Compare the results with the actual value and indicate the error.

4. Use Boole's rule to compute the integral $\int_0^{\pi/2} \sqrt{\sin x} \, dx$.

[Ans. 1.1812387]

8.8 NUMERICAL INTEGRATION

We approximate the integral

$$I = \int_a^b w(x) f(x) dx \tag{8.30}$$

by a linear combination of the values of $f(x)$ in the form

$$I = \int_a^b w(x) f(x) dx = \sum_{i=0}^n \lambda_i f(x_i) \tag{8.31}$$

where x_i ($i = 0, 1, 2, ..., n$) are called the nodes which are distributed in the interval $[a, b]$ and λ_i are $n + 1$ unknown coefficients, called weights of the integration method or the quadrature rule in Eq. (8.31). $w(x) > 0$ is called the weight function. The error of integration is given by

$$R_n = \int_a^b w(x) f(x) dx - \sum_{i=0}^n \lambda_i f(x_i) \tag{8.32}$$

An integration method of the form in Eq. (8.31) is said to be of order p, if it produces exact values ($R_n \equiv 0$), when $f(x)$ is a polynomial of degree $\leq p$.

Since in Eq. (8.31), we have $(2n + 2)$ constants [i.e. we have $(n + 1)$ nodes x_i's and $(n + 1)$ weights λ_i's], the method can be made exact for polynomials of degree $\leq (2n + 1)$. Thus the method in Eq. (8.31) can be maximum order of $(2n + 1)$. The order will be reduced if some of the nodes are known in advance.

For a method of order m, we have

$$\int_a^b w(x) x^k \, dx = \sum_{i=0}^n \lambda_i x_i^k = 0, \, k = 0, 1, ..., m \tag{8.33}$$

which determine the weights λ_i's and nodes x_i's. The error of integration is obtained by

$$R_n = \frac{C}{(m+1)!} f^{(m+1)}(\xi), \quad a < \xi < b \tag{8.34}$$

where

$$C = \int_a^b w(x) x^{m+1} dx - \sum_{i=0}^n \lambda_i x_i^{m+1} \tag{8.35}$$

8.9 NEWTON–COTE'S INTEGRATION METHOD

When $w(x) = 1$ and the nodes x_i's are uniformly distributed in the interval $[a, b]$ with $x_0 = a$ and $x_n = b$ and the spacing $h = \dfrac{b-a}{n}$. Since the nodes x_i's, $x_i = x_0 + ih$, $i = 0, 1, 2, ...,$ n are known, we have only to determine the weights λ_i's, $i = 0, 1, 2, ..., n$. These methods are known as Newton–Cotes integration method of order n. When both the end points of the interval of integration are used as nodes in the methods, the methods are called closed type methods, otherwise, it is called open type methods.

So, if we put $w(x) = 1$ in Eq. (8.30), we get

$$I = \int_{x_0}^{x_0 + nh} f(x) dx \tag{8.36}$$

We replace $f(x)$ by y_x in Lagrange's interpolating polynomial, given by

$$y_n = \frac{(x - x_1)(x - x_2)...(x - x_n)}{(x_0 - x_1)(x_0 - x_2)...(x_0 - x_n)} y_0 + \frac{(x - x_0)(x - x_2)...(x - x_n)}{(x_1 - x_0)(x_1 - x_2)...(x_1 - x_n)} y_1$$

$$+ ... + \frac{(x - x_0)(x - x_1)...(x - x_{i-1})(x - x_{i+1})...(x - x_n)}{(x_i - x_0)(x_i - x_1)...(x_i - x_{i-1})(x_i - x_{i+1})...(x_i - x_n)} y_i$$

$$+ ... + \frac{(x - x_0)(x - x_1)...(x - x_{n-1})}{(x_n - x_0)(x_n - x_1)...(x_n - x_{n-1})} y_n \quad . \tag{8.37}$$

where $x_i = x_0 + ih$, $i = 0, 1, 2, ..., n$

Let $x = x_0 + ph$ \hfill (8.38)

Hence

$$y_n = \sum_{i=1}^n \frac{(x - x_0)(x - x_1)(x - x_2)...(x - x_{i-1})(x - x_{i+1})...(x - x_n)}{(x_i - x_0)(x_i - x_1)(x_i - x_2)...(x_i - x_{i-1})(x_i - x_{i+1})...(x_i - x_n)} y_i$$

$$= \sum_{i=0}^n \frac{ph(p-1)h(p-2)h...(p-i+1)h(p-i-1)h...(p-n)h}{(ih)(i-1)h...(1h)(-1 \cdot h)(-2h)...(i-n)h} y_i$$

$$= \sum_{i=0}^{n} \frac{p(p-1)(p-2)\dots(p-i+1)(p-i-1)\dots(p-n)}{[i(i-1)(i-2)\dots 1](-1)^{n-i}[1, 2, \dots (n-i)]} y_i$$

or $$y_n = \sum_{i=0}^{n} \frac{(-1)^{n-1} p(p-1)(p-2)\dots(p-n)}{(i)!(n-i)!((p-i))} y_i \qquad (8.39)$$

It may be noted here that for equally spaced data, Lagrange's formula is the same as Newton–Gregory forward difference formula.

Hence from Eq. (8.36), we have

$$I \approx \int_{x_0}^{x_0+nh} y_n \, dx$$

$$\approx \int_{x_0}^{x_0+nh} \sum_{i=0}^{n} \frac{(-1)^{n-i} y_i \, p(p-1)(p-2)\dots(p-n)}{(i)!(n-i)!(p-i)} \, dx$$

$$\approx h \sum_{i=0}^{n} \int_{0}^{n} \frac{(-1)^{n-i}}{(i)!(n-i)!} y_i \, \frac{p(p-1)(p-2)\dots(p-n)}{p-i} \, dp \qquad (8.40)$$

since $x = x_0 + ph \Rightarrow dx = hdp$.

If we take $$I \approx h \sum_{i=0}^{n} \lambda_i y_i \qquad (8.41)$$

then Eq. (8.40) gives, on comparison

$$\lambda_i = \frac{(-1)^{n-i}}{(i)!(n-i)!} \int_{0}^{n} \frac{p(p-1)(p-2)\dots(p-n)}{p-i} \, dp \qquad (8.42)$$

λ_i is called the Cotes's number and Eq. (8.41) gives Newton–Cotes's quadrature formula for equally spaced data.

For different values of n, Eq. (8.42) gives different values of λ_i where $i = 0, 1, 2, \dots, n$.

8.9.1 Particular Cases

1. If $n = 1$, then $i = 0, 1$

Hence, by Eq. (8.42) $$\lambda_0 = \frac{(-1)^{1-0}}{0!(1-1)!} \int_{0}^{1} \frac{p(p-1)}{p-0} \, dp = (-1) \int_{0}^{1} (p-1) \, dp$$

$$= (-1)\left(\frac{p^2}{2} - p\right)_{0}^{1} = (-1)\left[\frac{1}{2} - 1\right] = \frac{1}{2}$$

$$\Rightarrow \qquad \lambda_0 = \frac{1}{2}$$

$$\lambda_1 = \int_{0}^{1} \frac{p(p-1)}{p-1} \, dp = \frac{1}{2}$$

2. If $n = 2$, then $i = 0, 1, 2$

Hence, by Eq. (8.42) $$\lambda_0 = \frac{(-1)^{2-0}}{0!(2!)} \int_{0}^{2} \frac{p(p-1)(p-2)}{p-0} \, dp = \frac{1}{3}$$

$$\lambda_1 = \frac{(-1)^{2-1}}{1!(2-1)!} \int_0^2 \frac{p(p-1)(p-2)}{p-1}\, dp = \frac{4}{3}$$

$$\lambda_2 = \frac{(-1)^{2-2}}{2!(2-2)!} \int_0^2 \frac{p(p-1)(p-2)}{p-2}\, dp = \frac{1}{3}$$

Similarly, we can find Cote's numbers for $n = 3, 4, \ldots$

8.9.2 Closed Type Methods

From Eq. (8.31), we know that

$$I = \int_a^b w(x)\, f(x)\, dx = \sum_{i=0}^n \lambda_i\, f(x_i)$$

If we take $w(x) = 1$ and $n = 1$, we obtain trapezoidal rule

$$\int_a^b f(x)\, dx = \frac{h}{2}[f(a) + f(b)] \tag{8.43}$$

where $h = \dfrac{b-a}{n} = \dfrac{b-a}{1} = b - a$

The error term is given as

$$R_1 = -\frac{h^3}{12}\, f''(\xi),\ a < \xi < b \tag{8.44}$$

For $n = 2$ in Eq. (8.31), we get the Simpson's rule

$$\int_a^b f(x)\, dx = \frac{h}{3}\left[f(a) + 4f\left(\frac{a+b}{2}\right) + f(b) \right] \tag{8.45}$$

where $h = \dfrac{b-a}{2}$. The error term for $m = 2$ is given by

$$R_2 = \frac{C}{(2+1)!}\, f^{(2+1)}(\xi) = \frac{C}{3!}\, f'''(\xi),\quad a < \xi < b$$

where

$$C = \int_a^b x^{2+1}\, dx - \frac{(b-a)}{2 \times (3)}\left[a^3 + 4\left(\frac{a+b}{2}\right)^3 + b^3 \right]\ \ |\because f(x) = x^3$$

$$= \int_a^b x^3\, dx - \frac{(b-a)}{6}\left[a^3 + 4\left(\frac{a+b}{2}\right)^3 + b^3 \right] = 0$$

and hence the method is exact for polynomial of degree 3 also. The error term for $m = 3$ is given by

$$R_2 = \frac{c}{(3+1)!}\, f^{(3+1)}(\xi) = \frac{C}{4!}\, f''''(\xi),\quad a < \xi < b$$

where

$$C = \int_a^b x^{3+1}\, dx - \frac{b-a}{6}\left[a^4 + 4\left(\frac{a+b}{2}\right)^4 + b^4 \right] = \frac{(b-a)^5}{120}$$

Hence, the error approximation is given by

$$R_2 = -\frac{(b-a)^5}{2880} f''''(\xi) = -\frac{h^5}{90} f''''(\xi), \, a < \xi, b \qquad \left| h = \frac{b-a}{2} \right.$$

For $n = 3$ in Eq. (8.31), we get the Simpson's 3/8 rule

$$\int_a^b f(x)dx = \frac{3h}{8} [f(a) + 3f(a+h) + 3f(a+2h) + f(b)]$$

The error term is given by

$$R_3 = -\frac{3}{80} h^5 f''''(\xi), \quad a < \xi < b$$

and hence the above method is also third order method.

8.10 GAUSSIAN INTEGRATION METHOD

When we determined both the nodes and the weights in the following integration method, then the methods are called Gaussian integration methods. The formula is as

$$I = \int_a^b w(x) f(x) dx = \sum_{i=0}^{n} \lambda_i f(x_i) \qquad (8.46)$$

where x_i's are nodes and λ_i's are called the weights of the integration method.

If the nodes x_i's in Eq. (8.46) are choosen as zeros of an orthogonal polynomial, orthogonal with respect to the weight function $w(x)$ on the interval $[a, b]$, then the method in Eq. (8.46) has order $2n + 1$ and all the weights $\lambda_i > 0$.

Here we are giving the proof of Eq. (8.46).

Proof: Let $f(x)$ be a polynomial of degree less than or equal to $2n + 1$. Let $p_n(x)$ be the Lagrange interpolating polynomial of degree less than or equal to n, interpolating the data (x_i, f_i), $i = 0, 1, 2, ..., n$, we have

$$p_n(x) = \sum_{i=0}^{n} l_i(x) f(x_i) \qquad (8.47)$$

where $\qquad l_i(x) = \frac{\pi(x)}{(x - x_i) \pi'(x_i)}$

The polynomial $[f(x) - p_n(x)]$ has roots at $x_0, x_1, ..., x_n$. Hence, it can be written as

$$f(x) - p_n(x) = q_{n+1}(x) r_n(x) \qquad (8.48)$$

where $r_n(x)$ is a polynomial of degree at most n and $q_{n+1}(x_i) = 0$, $i = 0, 1, 2, ..., n$.

Now integrating the Eq. (8.48) from a to b, we get

$$\int_a^b [f(x) - p_n(x)] w(x)dx = \int_a^b w(x) q_{n+1}(x) r_n(x)dx$$

or $\qquad \int_a^b w(x) f(x) dx = \int_a^b w(x) p_n(x)dx + \int_a^b w(x) q_{n+1}(x) r_n(x)dx$

The second integral on the right hand side is zero, if $p_{n+1}(x)$ is an orthogonal polynomial, orthogonal with respect to the weight function $w(x)$, to all polynomial of degree less than or equal to n.

So, we have

$$\int_a^b w(x)\,f(x)\,dx = \int_a^b w(x)\,p_n(x)\,dx$$

$$= \sum_{i=0}^{n} f(x_i) \int_a^b l_i(x)\,w(x)\,dx = \sum_{i=0}^{n} \lambda_i\,f(x_i)$$

where

$$\lambda_i = \int_a^b w(x)\,l_i(x)\,dx$$

Thus

$$\int_a^b w(x)\,f(x)\,dx = \sum_{i=0}^{n} \lambda_i\,f(x_i)$$

This proves that formula in Eq. (8.46) is exact for polynomial of degree $\leq 2n + 1$.

Note: Any finite interval $[a, b]$ can be transformed to $[-1, 1]$ by using the transformation

$$x = \frac{(b-a)}{2}t + \frac{(a+b)}{2}$$

We consider the integral in the form

$$\int_{-1}^{1} w(x)\,f(x)\,dx = \sum_{i=0}^{n} \lambda_i\,f(x_i) \tag{8.49}$$

8.11 GAUSS–LEGENDRE INTEGRATION METHOD

Consider the integral in the form

$$\int_{-1}^{1} w(x)\,f(x)\,dx = \sum_{i=0}^{\infty} \lambda_i\,f(x_i) \tag{8.50}$$

The nodes x_i's are the zeros of the Legendre polynomials. We know that the Legendre polynomial is

$$P_{n+1}(x) = \frac{1}{2^{n+1}\,(n+1)!}\,\frac{d^{n+1}}{dx^{n+1}}(x^2 - 1)^{n+1} \tag{8.51}$$

From Eq. (8.51), we get

$$P_0(x) = 1,\ P_1(x) = x,\ P_2(x) = \frac{1}{2}(3x^2 - 1),\ P_3(x) = \frac{1}{2}(5x^3 - 3x) \text{ etc.}$$

The Legendre polynomials are orthogonal on interval $[-1, 1]$ with respect to the weight function $w(x) = 1$. The methods in Eq. (8.50) are of order $2n + 1$ and are called Gauss–Legendre integration methods. When $w(x) = 1$, then the formula in Eq. (8.50) becomes

$$\int_{-1}^{1} f(x)\,dx = \sum_{i=0}^{n} \lambda_i\,f(x_i) \tag{8.52}$$

Here all the nodes x_i's and weights λ_i's are unknown.

  i. **Two point formula when $n = 1$:** From Eq. (8.52)

$$\int_{-1}^{1} f(x)\,dx = \lambda_0 f(x_0) + \lambda_1 f(x_1) \tag{8.53}$$

Here we have four unknowns, λ_0, λ_1, x_0, x_1. Making the method exact for polynomial of degree $2n + 1 = (2 \times 1 + 1) = 3$. So, it is sufficient to make the method exact for $f(x) = 1, x, x^2, x^3$. Thus, we get

$$f(x) = 1 \quad \Rightarrow \quad \int_{-1}^{1} dx = \lambda_0 + \lambda_1 \text{ since } f(x) = f(x_0) = f(x_1) = 1$$

or $\qquad\qquad 2 = \lambda_0 + \lambda_1.$ \hfill (8.54)

Similarly, $\qquad f(x) = x \qquad \Rightarrow 0 = \lambda_0 x_0 + \lambda_1 x_1$ \hfill (8.55)

$$= x^2 \qquad \Rightarrow \frac{2}{3} = \lambda_0 x_0^2 + \lambda_1 x_1^2$$ \hfill (8.56)

$$= x^3 \qquad \Rightarrow 0 = \lambda_0 x_0^3 + \lambda_1 x_1^3$$ \hfill (8.57)

Solving for λ_0, λ_1, x_0 and x_1, we get

$$\lambda_0 = \lambda_1 = 1, x_0 = -x_1 = \frac{1}{\sqrt{3}}$$

Hence the two points Gauss–Legendre method is given by

$$\int_{-1}^{1} f(x)dx = f\left(-\frac{1}{\sqrt{3}}\right) + f\left(\frac{1}{\sqrt{3}}\right)$$ \hfill (8.58)

The error term $\qquad R_1 = \dfrac{C}{(3+1)!} f^{(3+1)}(\xi); \quad -1 < \xi < 1$

$$= \frac{C}{4!} f''''(\xi)$$

and the error constant

$$C = \int_{-1}^{1} x^{3+1} dx - \left[f\left(-\frac{1}{\sqrt{3}}\right) + f\left(\frac{1}{\sqrt{3}}\right)\right]$$

$$= \int_{-1}^{1} x^4 dx - \left(\frac{1}{9} + \frac{1}{9}\right) \text{ since } f(x) = x^4, f\left(-\frac{1}{\sqrt{3}}\right) = \frac{1}{9}, f\left(\frac{1}{\sqrt{3}}\right) = \frac{1}{9}$$

$$= \frac{8}{45}$$

Therefore, $R_1 = \dfrac{1}{135} f''''(\xi); \quad -1 < \xi < 1$

ii. **Three point formula when $n = 2$:** We have

$$\int_{-1}^{1} f(x)dx = \lambda_0 f(x_0) + \lambda_1 f(x_1) + \lambda_2 f(x_2)$$ \hfill (8.59)

The method has 6 unknowns, λ_0, λ_1, λ_2, x_0, x_1, x_2 and it can be made exact for polynomial of degree up to 5 ($2n + 1 = 2 \times 2 + 1 = 5$). So, it is sufficient to make the method exact for $f(x) = 1, x, x^2, x^3, x^4, x^5$. Thus, we get

$$\lambda_0 + \lambda_1 + \lambda_2 = 2$$ \hfill (8.60)

$$\lambda_0 x_0 + \lambda_1 x_1 + \lambda_2 x_2 = 0$$ \hfill (8.61)

$$\lambda_0 x_0^2 + \lambda_1 x_1^2 + \lambda_2 x_2^2 = \frac{2}{3}$$ \hfill (8.62)

$$\lambda_0 x_0^3 + \lambda_1 x_1^3 + \lambda_2 x_2^3 = 0 \tag{8.63}$$

$$\lambda_0 x_0^4 + \lambda_1 x_1^4 + \lambda_2 x_2^4 = \frac{2}{5} \tag{8.64}$$

$$\lambda_0 x_0^5 + \lambda_1 x_1^5 + \lambda_2 x_2^5 = 0 \tag{8.65}$$

On solving the Eqs (8.60)–(8.65), we get

$$x_0 = \pm\sqrt{\frac{3}{5}}, \; x_1 = 0, \; x_2 = \mp\sqrt{\frac{3}{5}}, \; \lambda_0 = \frac{5}{9}, \; \lambda_1 = \frac{8}{9}, \; \lambda_2 = \frac{5}{9}$$

Hence the three point Gauss–Legendre method is given by

$$\int_{-1}^{1} f(x)\,dx = \frac{5}{9} f\left(-\sqrt{\frac{3}{5}}\right) + \frac{8}{9} f(0) + \frac{5}{9} f\left(\sqrt{\frac{3}{5}}\right)$$

The error constant $C = \int_{-1}^{1} x^6\,dx - \frac{1}{9}\left[5f\left(-\sqrt{\frac{3}{5}}\right) + 8f(0) + 5f\left(\sqrt{\frac{3}{5}}\right)\right]$

$$= \left(\frac{x^7}{7}\right)_{-1}^{1} - \frac{1}{9}\left[5\left(-\sqrt{\frac{3}{5}}\right)^6 + 8\times 0 + 5\left(\sqrt{\frac{3}{5}}\right)^6\right]$$

$$= \frac{2}{7} - \frac{1}{9}\left[5\times\frac{27}{125} + \frac{5\times 27}{125}\right] = \frac{8}{175}$$

and the error term is

$$R_2 = \frac{C}{6!} f''''''(\xi) = \frac{1}{15750} f''''''(\xi), \quad -1 < \xi < 1$$

We are giving the following table for nodes and the corresponding weights of the method for $n \leq 4$.

Nodes and weights for the Gauss–Legendre integration methods		
n	nodes (x_i)	Weight (λ_i)
1	± 0.5773502692	1.0000000000
2	0.0000000000	0.8888888889
	± 0.7745966692	0.5555555556
3	± 0.3399810436	0.6521451549
	± 0.8611363116	0.3478548451
4	0.0000000000	0.5688888889
	± 0.5384693101	0.4786286705
	± 0.9061798459	0.2369268851

8.12 LOBATTO INTEGRATION METHOD

Here we take $w(x) = 1$ and the two end points -1 and 1 are always taken as nodes. The remaining $(n-1)$ nodes and the $(n+1)$ weights are to be determined. The integration methods of the form

$$\int_{-1}^{1} f(x)\,dx = \lambda_0 f(-1) + \sum_{i=1}^{n-1} \lambda_i f(x_i) + \lambda_n f(1) \tag{8.66}$$

are called the Lobatto integration methods and are of order $2n - 1$. Hence this method can be made exact for polynomial of degree upto $2n - 1$.

i. **Three point formula when $n = 2$:** For $n = 2$, we have

$$\int_{-1}^{1} f(x)dx = \lambda_0 f(-1) + \lambda_1 f(x_1) + \lambda_2 f(1) \tag{8.67}$$

Making the formula exact for $f(x) = 1, x, x^2, x^3$, we get

$$f(x) = 1 \Rightarrow \lambda_0 + \lambda_1 + \lambda_2 = 2 \tag{8.68}$$

$$= x \Rightarrow -\lambda_0 + \lambda_1 x_1 + \lambda_2 = 0 \tag{8.69}$$

$$= x^2 \Rightarrow \lambda_0 + \lambda_1 x_1^2 + \lambda_2 = \frac{2}{3} \tag{8.70}$$

$$= x^3 \Rightarrow -\lambda_0 + \lambda_1 x_1^3 + \lambda_2 = 0 \tag{8.71}$$

Subtracting Eq. (8.69) from Eq. (8.71), we have

$$\lambda_1 x_1 (x_1^2 - 1) = 0 \Rightarrow x_1 = 0 \quad | \text{ Since } x_1 \neq \pm 1$$

From Eq. (8.69) and (8.70), we get

$$\lambda_0 = \lambda_2 = \frac{1}{3}$$

From Eq. (8.68), we get

$$\lambda_1 = \frac{4}{3}$$

$\therefore$ Lobatto method is given by

$$\int_{-1}^{1} f(x)dx = \frac{1}{3}[f(-1) + 4f(0) + f(1)] \tag{8.72}$$

and the error constant is given by

$$C = \int_{-1}^{1} x^4 \, dx - \frac{1}{3}[f(-1) + 4f(0) + f(1)]$$

since $f(x) = x^4$ gives $f(-1) = 1, f(1) = 1, f(0) = 0$

The error term is

$$R_2 = \frac{C}{4!} f''''(\xi) = -\frac{1}{90} f''''(\xi), \quad -1 < \xi < 1$$

ii. **Four point formula when $n = 3$:** For $n = 3$, we have

$$\int_{-1}^{1} f(x)dx = \lambda_0 f(-1) + \lambda_1 f(x_1) + \lambda_2 f(x_2) + \lambda_3 f(1) \tag{8.73}$$

Here we have 6 unknowns, $\lambda_0, \lambda_1, \lambda_2, \lambda_3, x_1, x_2$. Making the method exact for polynomial of degree upto $(2n - 1)$ i.e. $2 \times 3 - 1 = 5$. So, it is sufficient to make the method exact for $f(x) = 1, x, x^2, x^3, x^4, x^5$. Thus

for $f(x) = 1$, we have $\qquad \lambda_0 + \lambda_1 + \lambda_2 + \lambda_3 = 2 \tag{8.74}$

for $f(x) = x$, we have $\qquad -\lambda_0 + \lambda_1 x_1 + \lambda_2 x_2 + \lambda_3 = 2 \tag{8.75}$

for $f(x) = x^2$, we have $\qquad \lambda_0 + \lambda_1 x_1^2 + \lambda_2 x_2^2 + \lambda_3 = \frac{2}{3} \tag{8.76}$

for $f(x) = x^3$, we have $-\lambda_0 + \lambda_1 x_1^3 + \lambda_2 x_2^3 + \lambda_3 = 0$ (8.57)

for $f(x) = x^4$, we have $\lambda_0 + \lambda_1 x_1^4 + \lambda_2 x_2^4 + \lambda_3 = \dfrac{2}{5}$ (8.78)

for $f(x) = x^5$, we have $\lambda_0 + \lambda_1 x_1^5 + \lambda_2 x_2^5 + \lambda_3 = 0$ (8.79)

On solving Eqs (8.74)–(8.79), we get

$$\lambda_0 = \frac{1}{6}, \lambda_1 = \frac{5}{6}, \lambda_2 = \frac{5}{6}, \lambda_3 = \frac{1}{6}, x_1 = -x_2 = \frac{1}{\sqrt{5}}$$

$\therefore$ Lobatto method for $n = 3$ is given by

$$\int_{-1}^{1} f(x)\,dx = \frac{1}{6}\left[f(-1) + 5f\left(-\frac{1}{\sqrt{5}}\right) + 5f\left(\frac{1}{\sqrt{5}}\right) + f(1) \right]$$

and the error constant is

$$C = \int_{-1}^{1} x^6\,dx - \frac{1}{6}\left[f(-1) + 5f\left(-\frac{1}{\sqrt{5}}\right) + 5f\left(\frac{1}{\sqrt{5}}\right) + f(1) \right]$$

$$= \frac{2}{7} - \frac{1}{6}\left[1 + 5\left(\frac{1}{125}\right) + 5\left(\frac{1}{125}\right) + 1\right] = -\frac{32}{525}$$

Therefore, the error term is

$$R_3 = \frac{C}{6!}\, f''''''(\xi) = -\frac{32}{525(6!)}\, f''''''(\xi); \quad -1 < \xi < 1$$

We are giving the following table for nodes and the corresponding weights of the method for $n \le 5$.

\multicolumn{3}{c}{**Nodes and weights for the Lobatto integration methods**}
n	nodes (x_i)	Weight (λ_i)
2	± 1.00000000 0.00000000	0.33333333 1.33333333
3	± 1.00000000 ± 0.44721360	0.16666667 0.83333333
4	± 1.00000000 ± 0.65465367 0.00000000	0.10000000 0.54444444 0.71111111
5	± 1.00000000 ± 0.76505532 ± 0.28523152	0.06666667 0.37847496 0.55485837

8.13 RADAU INTEGRATION METHOD

Consider $w(x) = 1$ and the lower limit -1 is fixed as a node. The remaining n nodes and $n + 1$ weights are to be determined. The integration methods of the form

$$\int_{-1}^{1} f(x)\,dx = \lambda_0\, f(-1) + \sum_{i=1}^{n} \lambda_i\, f(x_i)$$ (8.80)

are called Radau integration methods and are of $2n$ degree.

i. **Two point formula when $n = 1$:** For $n = 1$, we have

$$\int_{-1}^{1} f(x)dx = \lambda_0 f(-1) + \lambda_1 f(x_1) \tag{8.81}$$

We observe that the method has three unknowns and it can be made exact for polynomial of degree upto 2 ($2n = 2 \times 1 = 2$).

For $f(x) = 1, x, x^2$, we get

$$f(x) = 1 \quad \Rightarrow \lambda_0 + \lambda_1 = 2 \tag{8.82}$$

$$= x \quad \Rightarrow -\lambda_0 + \lambda_1 x_1 = 0 \tag{8.83}$$

$$= x^2 \quad \Rightarrow \lambda_0 + \lambda_1 x_1^2 = \frac{2}{3} \tag{8.84}$$

Adding the Eqs (8.82) and (8.83), we get

$$\lambda_1(1 + x_1) = 2$$

Adding Eqs (8.83) and (8.84), we get

$$\lambda_1 x_1(1 + x_1) = \frac{2}{3}$$

From the above

$$\frac{\lambda_1 x_1(1 + x_1)}{\lambda_1(1 + x_1)} = \frac{\frac{2}{3}}{2} = \frac{1}{3} \Rightarrow x_1 = \frac{1}{3}$$

Again, $\quad \lambda_1\left(1 + \frac{1}{3}\right) = 2 \Rightarrow \lambda_1 = \frac{3}{2}$

From Eq. (8.82), we get

$$\lambda_0 = \frac{1}{2}$$

$\therefore$ The Radau methods is given by

$$\int_{-1}^{1} f(x)\,dx = \frac{1}{2}f(-1) + \frac{3}{2}f\left(\frac{1}{3}\right)$$

$$= \frac{1}{2}\left[f(-1) + 3f\left(\frac{1}{3}\right)\right] \tag{8.85}$$

Now the error constant is given by

$$C = \int_{-1}^{1} x^3 dx - \frac{1}{2}f(-1) + 3f\left(\frac{1}{3}\right)$$

$$= 0 - \frac{1}{2}\left[-1 + 3\left(\frac{1}{27}\right)\right] = \frac{4}{9}$$

Therefore, the error term is

$$R_1 = \frac{C}{3!}f'''(\xi) = \frac{2}{27}f'''(\xi), \quad -1 < \xi < 1$$

ii. **Three point formula when $n = 2$:** For $n = 2$, we have from Eq. (8.80)

$$\int_{-1}^{1} f(x)dx = \lambda_0 f(-1) + \lambda_1 f(x_1) + \lambda_2 f(x_2) \tag{8.86}$$

The method has five unknowns i.e. $\lambda_0, \lambda_1, \lambda_2, x_1$ and x_2 and it can be made exact for polynomial of degree up to $2 \times 2 = 4$.

For $f(x) = 1, x, x^2, x^3, x^4$ we get

$$f(x) = 1 \quad \Rightarrow \lambda_0 + \lambda_1 + \lambda_2 = 2 \tag{8.87}$$

$$f(x) = x \quad \Rightarrow -\lambda_0 + \lambda_1 x_1 + \lambda_2 x_2 = 0 \tag{8.88}$$

$$f(x) = x^2 \quad \Rightarrow \lambda_0 + \lambda_1 x_1^2 + \lambda_2 x_2^2 = \frac{2}{3} \tag{8.89}$$

$$f(x) = x^3 \quad \Rightarrow -\lambda_0 + \lambda_1 x_1^3 + \lambda_2 x_2^3 = 0 \tag{8.90}$$

$$f(x) = x^4 \quad \Rightarrow \lambda_0 + \lambda_1 x_1^4 + \lambda_2 x_2^4 = \frac{2}{5} \tag{8.91}$$

On solving Eqs (8.87)–(8.91), we get

$$\lambda_0 = \frac{2}{9}, \lambda_1 = \frac{16 + \sqrt{6}}{18}, \lambda_2 = \frac{16 - \sqrt{6}}{18}$$

$$x_1 = \frac{1 - \sqrt{6}}{5}, x_2 = \frac{1 + \sqrt{6}}{5}$$

∴ The Radau integration method is given by

$$\int_{-1}^{1} f(x)\,dx = \frac{2}{9} f(-1) + \left(\frac{16 + \sqrt{6}}{18} \right) f\left(\frac{1 - \sqrt{6}}{5} \right) + \frac{16 - \sqrt{6}}{18} f\left(\frac{1 + \sqrt{6}}{5} \right)$$

The error constant is given by

$$C = \int_{-1}^{1} x^5\,dx - \left[\frac{2}{9} f(-1) + \left(\frac{16 + \sqrt{6}}{18} \right) f\left(\frac{1 - \sqrt{6}}{5} \right) + \left(\frac{16 - \sqrt{6}}{18} \right) f\left(\frac{1 + \sqrt{6}}{5} \right) \right]$$

$$= \frac{8}{75}$$

The error term is given by

$$R_2 = \frac{C}{5!} f''''(\xi) = \frac{8}{75(5!)} f''''(\xi), \quad -1 < \xi < 1$$

The nodes and the corresponding weights for the above method are given in table below:

Nodes and weights for Radau integration methods		
n	*nodes (x_i)*	*Weight (λ_i)*
1	−1.0000000	0.5000000
	0.3333333	1.5000000
2	−1.0000000	0.2222222
	−0.2898979	1.0249717
	0.6898979	0.7528061
3	−1.0000000	0.1250000
	−0.5753189	0.6576886
	0.1810663	0.7763870
	0.8228241	0.4409244

Example 16: Determine a, b and c such that the formula

$$\int_0^h f(x)\,dx = h\left[af(0) + bf\left(\frac{h}{3}\right) + cf(h)\right]$$

is exact for polynomial of as high degree as possible and determine the order of the truncation error.

Solution: Making the method exact for polynomial of degree up to 2, we obtain

$$f(x) = 1 \quad \Rightarrow \quad h = h(a+b+c), \quad \text{or } a+b+c = 1$$

$$= x \quad \Rightarrow \quad \frac{h^2}{2} = h\left(\frac{bh}{3} + ch\right), \quad \text{or } \frac{1}{3}b + c = \frac{1}{2}$$

$$= x^2 \quad \Rightarrow \quad \frac{h^3}{3} = h\left(\frac{bh^2}{9} + ch^2\right) \quad \text{or } \frac{b}{9} + c = \frac{1}{3}$$

Solving the above equations, we get $a = 0$, $b = \dfrac{3}{4}$, $c = \dfrac{1}{4}$. Hence, the required formula is

$$\int_0^h f(x)\,dx = h\left[0 + \frac{3}{4}f\left(\frac{h}{3}\right) + \frac{1}{4}f(h)\right]$$

$$= \frac{h}{4}\left[3f\left(\frac{h}{3}\right) + f(h)\right]$$

The truncation error of the formula is given by

$$\text{TE} = \frac{c}{3!}f'''(\xi), \quad 0 < \xi < h$$

where $\quad C = \displaystyle\int_0^h x^3\,dx - \frac{h}{4}\left[3\,\frac{h^3}{27} + h^3\right]$ since $f(x) = x^3$ gives $f\left(\dfrac{h}{3}\right) = \dfrac{h^3}{27}$, $f(h) = h^3$

$$= \frac{h^4}{4} - \frac{30\,h^4}{4 \times 27} = -\frac{h^4}{36}$$

Hence, we have

$$\text{TE} = -\frac{h^4}{216}f'''(\xi) = O(h^4)$$

Example 17: Determine the weights and nodes in the quadrature formula

$$\int_{-1}^1 f(x)\,dx = \sum_{i=1}^4 A_i\, f(x_i)$$

with $x_1 = -1$ and $x_4 = 1$, so that the formula becomes exact for polynomial of highest possible degree.

Solution: Making the method

$$\int_{-1}^1 f(x)\,dx = A_1 f(-1) + A_2 f(x_2) + A_3 f(x_3) + A_4 f(1) \qquad \text{(i)}$$

exact for $f(x) = 1, x, x^2, x^3, x^4, x^5$, we have the equations

$$f(x) = 1 \quad \Rightarrow \quad A_1 + A_2 + A_3 + A_4 = 2 \tag{ii}$$

$$= x \quad \Rightarrow \quad -A_1 + A_2 x_2 + A_3 x_3 + A_4 = 0 \tag{iii}$$

$$= x^2 \Rightarrow \quad A_1 + A_2 x_2^2 + A_3 x_3^2 + A_4 = \frac{2}{3} \tag{iv}$$

$$= x^3 \Rightarrow \quad -A_1 + A_2 x_2^3 + A_3 x_3^2 + A_4 = 0 \tag{v}$$

$$= x^4 \Rightarrow \quad A_1 + A_2 x_2^4 + A_3 x_3^4 + A_4 = \frac{2}{5} \tag{vi}$$

$$= x^5 \Rightarrow \quad -A_1 + A_2 x_2^5 + A_3 x_3^5 + A_4 = 0 \tag{vii}$$

Subtracting Eq. (ii) from Eq. (iv), Eq. (v) from Eq. (iii), Eq. (vi) from Eq. (iv) and Eq. (vii) from Eq. (v), we get

$$\frac{4}{3} = A_2(1 - x_2^2) + A_3(1 - x_3^2)$$

$$0 = A_2 x_2(1 - x_2^2) + A_3 x_3(1 - x_3^2)$$

$$\frac{4}{15} = A_2 x_2^2(1 - x_2^2) + A_3 x_3^2(1 - x_3^2)$$

$$0 = A_2 x_2^3(1 - x_2^2) + A_3 x_3^3(1 - x_3^2)$$

Removing A_3 from the above equations, we get

$$\frac{4}{3} x_3 = A_2(1 - x_2^2)(x_3 - x_2)$$

$$-\frac{4}{15} = A_2 x_2(1 - x_2^2)(x_3 - x_2)$$

$$\frac{4}{15} x_3 = A_2 x_2^2(1 - x_2^2)(x_3 - x_2)$$

which gives $x_2 x_3 = -\dfrac{1}{5}, x_2 = -x_3 = \dfrac{1}{\sqrt{5}}$ and $A_1 = A_4 = \dfrac{1}{6}, A_2 = A_3 = \dfrac{5}{6}.$

The error term of method is given by

$$\text{TE} = \frac{C}{6!} f''''''(\xi), \quad -1 < \xi < 1$$

where

$$C = \int_{-1}^{1} x^6 dx - [A_1 + A_2 x_2^6 + A_3 x_3^6 + A_4] = \frac{2}{7} - \frac{26}{75}$$

$$= -\frac{32}{525}$$

Hence, we have $\quad \text{TE} = -\dfrac{32}{(6!)525} f''''''(\xi)$

Example 18: Consider the quadrature formula

$$\int_a^b f(x)\, dx = \sum_{i=0}^{n} w_i f(x_i)$$

where $w_i > 0$ and the rule is exact for $f(x) = 1$. If $f(x_i)$ have the error at most by $(0.5)\,10^{-k}$, show that the error in the quadrature rule is not greater than $10^{-k}\left(\dfrac{b-a}{2}\right)$.

Solution: We have $w_i > 0$. Since the quadrature rule is exact for $f(x) = 1$, we have

$$\sum_{i=0}^{n} w_i = b - a$$

We also have 　　$| \text{Error} | = \left| \sum_{i=0}^{n} w_i [f(x_i) - f^*(x_i)] \right|$

$$\leq (0.5)(10^{-k}) \sum_{i=0}^{n} w_i = \frac{1}{2}(b-a) 10^{-k}$$

Example 19: Find a quadrature formula

$$\int_0^1 \frac{f(x)\,dx}{\sqrt{x(1-x)}} = \alpha_1 f(0) + \alpha_2 f\left(\frac{1}{2}\right) + \alpha_3 f(1)$$

which is exact for polynomials of highest possible degree. Then use the formula on

$\int_0^1 \frac{dx}{\sqrt{x - x^3}}$ and compare with the exact value.

Solution: Making the method exact for polynomials of degree upto 2, i.e. we take $f(x) = 1, x, x^2$, then

$$\text{for } f(x) = 1 \quad \Rightarrow I_1 = \int_0^1 \frac{dx}{\sqrt{x(1-x)}} = \alpha_1 + \alpha_2 + \alpha_3$$

$$\text{for } f(x) = x \quad \Rightarrow I_2 = \int_0^1 \frac{x\,dx}{\sqrt{x(1-x)}} = \frac{1}{2}\alpha_2 + \alpha_3$$

$$\text{for } f(x) = x^2 \quad \Rightarrow I_3 = \int_0^1 \frac{x^2 dx}{\sqrt{x(1-x)}} = \frac{1}{4}\alpha_2 + \alpha_3$$

where 　　$I_1 = \int_0^1 \frac{dx}{\sqrt{x(1-x)}} = 2\int_0^1 \frac{dx}{\sqrt{1-(2x-1)^2}}$

Let $2x - 1 = t$, then $2dx = dt$, and $x = 0 \Rightarrow t = -1$, $x = 1 \Rightarrow t = 1$.

Now 　　$I_1 = 2\int_{-1}^1 \frac{\dfrac{dt}{2}}{\sqrt{1-t^2}} = \int_{-1}^1 \frac{dt}{\sqrt{1-t^2}} = [\sin^{-1} t]_{-1}^1 = \pi$

$$I_2 = \int_0^1 \frac{x\,dx}{\sqrt{x(1-x)}} = 2\int_0^1 \frac{x\,dx}{\sqrt{1-(2x-1)^2}} = \frac{1}{2}\int_{-1}^1 \frac{(t+1)\,dt}{\sqrt{1-t^2}}$$

$$= \frac{1}{2}\int_{-1}^1 \frac{t\,dt}{\sqrt{1-t^2}} + \frac{1}{2}\int_{-1}^1 \frac{dt}{\sqrt{1-t^2}}$$

$$= \frac{\pi}{2} \quad \text{[First integrand will be zero because } \int_{-1}^1 f(x)\,dx = 0$$
$$\text{if } f(x) \text{ is odd function]}$$

$$I_3 = \int_0^1 \frac{x^2 dx}{\sqrt{x(1-x)}} = 2\int_0^1 \frac{x^2 dx}{\sqrt{1-(2x-1)^2}}$$

$$= \frac{1}{4}\int_{-1}^{1} \frac{(t+1)^2}{\sqrt{1-t^2}}\,dt = \frac{1}{4}\int_{-1}^{1} \frac{t^2\,dt}{\sqrt{1-t^2}} + \frac{1}{2}\int_{-1}^{1} \frac{t\,dt}{\sqrt{1-t^2}} + \frac{1}{4}\int_{-1}^{1} \frac{dt}{\sqrt{1-t^2}}$$

$$= \frac{1}{4}\times 2\int_{0}^{1} \frac{t^2\,dt}{\sqrt{1-t^2}} + 0 + \frac{1}{4}\pi$$

$$= \frac{1}{2}\int_{0}^{1} \frac{t^2\,dt}{\sqrt{1-t^2}} + \pi_4 = A_1 + \frac{\pi}{4}$$

where
$$A_1 = \frac{1}{2}\int_{0}^{1} \frac{t^2\,dt}{\sqrt{1-t^2}}$$

Let $t = \sin\theta$, then $dt = \cos\theta\,d\theta$, we have

$$A_1 = \frac{1}{2}\int_{0}^{\frac{\pi}{2}} \frac{\sin^2\theta\cos\theta\,d\theta}{\sqrt{1-\sin^2\theta}} = \frac{1}{2}\int_{0}^{\frac{\pi}{2}} \sin^2\theta\,d\theta$$

$$= \frac{1}{2}\int_{0}^{\frac{\pi}{2}} \sin^2\theta\cos^0\theta\,d\theta = \frac{1}{2}\cdot\frac{\left\lfloor\frac{2+1}{2}\right.\left\lfloor\frac{0+1}{2}\right.}{2\left\lfloor\frac{2+0+2}{2}\right.}$$

$$= \frac{1}{2}\frac{\left\lfloor\frac{3}{2}\right.\left\lfloor\frac{1}{2}\right.}{2\lfloor\overline{2}}=\frac{1}{2}\frac{\frac{1}{2}\sqrt{\pi}\cdot\sqrt{\pi}}{2.1}$$

$$= \frac{\pi}{8}$$

Hence
$$I_3 = \frac{\pi}{8} + \frac{\pi}{4} = \frac{3\pi}{8}$$

Hence, we have the equation

$$\alpha_1 + \alpha_2 + \alpha_3 = \pi$$

$$\frac{\alpha_2}{2} + \alpha_3 = \frac{\pi}{2}$$

$$\frac{\alpha_2}{4} + \alpha_3 = \frac{3\pi}{8}$$

On solving these equations, we get $\alpha_1 = \frac{\pi}{4}, \alpha_2 = \frac{\pi}{2}, \alpha_3 = \frac{\pi}{4}$.

The quadratic formula is given by

$$\int_{0}^{1} \frac{f(x)\,dx}{\sqrt{x(1-x)}} = \frac{\pi}{4}\left[f(0) + 2f\left(\frac{1}{2}\right) + f(1)\right]$$

We now use this formula to evaluate

$$I = \int_{0}^{1} \frac{dx}{\sqrt{x-x^3}} = \int_{0}^{1} \frac{dx}{\sqrt{1+x}\,\sqrt{x(1-x)}} = \int_{0}^{1} \frac{f(x)\,dx}{\sqrt{x(1-x)}}$$

where $$f(x) = \frac{1}{\sqrt{1+x}}$$

We obtain
$$I = \int_0^1 \frac{dx}{\sqrt{x-x^3}} = \int_0^1 \frac{f(x)dx}{\sqrt{x(1-x)}} = \frac{\pi}{4}\left[1 + 2\sqrt{\frac{2}{3}} + \frac{1}{\sqrt{2}}\right] = 2.62331$$

The exact value is $I = 2.62205755$.

Example 20: Find the value of integral $I = \int_2^3 \frac{\cos 2x}{1+\sin x}\,dx$ using Gauss–Legendre two and three point integration rules.

Solution: We have $$I = \int_2^3 \frac{\cos 2x}{1+\sin x}\,dx$$

By transforming the limits -1 to 1, we set

$$x = \left(\frac{b-a}{2}\right)t + \frac{a+b}{2}$$

$$= \left(\frac{3-2}{2}\right)t + \frac{3+2}{2} = \frac{t+5}{2}$$

$$\therefore \quad I = \int_{-1}^1 \frac{\cos 2\left(\dfrac{t+5}{2}\right)}{1+\sin\left(\dfrac{t+5}{2}\right)}\frac{dt}{2} = \frac{1}{2}\int_{-1}^1 \frac{\cos(t+5)\,dt}{1+\sin\left(\dfrac{t+5}{2}\right)}$$

Using Gauss–Legendre two point formula

$$\int_{-1}^1 f(x)dx = f\left(\frac{1}{\sqrt{3}}\right) + f\left(-\frac{1}{\sqrt{3}}\right)$$

$$\int_{-1}^1 f(t)dt = f\left(\frac{1}{\sqrt{3}}\right) + f\left(-\frac{1}{\sqrt{3}}\right)$$

where
$$f(t) = \frac{1}{2}\left[\frac{\cos(t+5)}{1+\sin\left(\dfrac{t+5}{2}\right)}\right]$$

$$f\left(\frac{1}{\sqrt{3}}\right) = \frac{1}{2}\left[\frac{\cos\left(\dfrac{1}{\sqrt{3}}+5\right)}{1+\sin\left(\dfrac{\dfrac{1}{\sqrt{3}}+5}{2}\right)}\right] = \frac{1}{2}(0.56558356)$$

$$f\left(-\frac{1}{\sqrt{3}}\right) = \frac{1}{2}\left[\frac{\cos\left(-\dfrac{1}{\sqrt{3}}+5\right)}{1+\sin\left(\dfrac{-\dfrac{1}{\sqrt{3}}+5}{2}\right)}\right] = \frac{1}{2}(-0.15856672)$$

$$\therefore \qquad I = \int_{-1}^{1} f(t)\,dt = \frac{1}{2}[0.56558356 - 0.15856672] = 0.20350842$$

Using Gauss–Legendre three point formula

$$\int_{-1}^{1} f(x)\,dx = \frac{1}{9}\left[5f\left(-\sqrt{\frac{3}{5}}\right) + 8f(0) + 5f\sqrt{\frac{3}{5}}\right]$$

We obtain
$$I = \frac{1}{18}[-1.26018516 + 1.41966658 + 3.48936887] = 0.20271391$$

Example 21: Integrate by Gaussian quadrature $(n = 3)$ $\int_{1}^{2} \frac{dx}{1+x^3}$.

Solution: To transform the limit -1 to 1, we set

$$x = \left(\frac{b-a}{2}\right)t + \left(\frac{a+b}{2}\right) = \left(\frac{2-1}{2}\right)t + \left(\frac{2+1}{2}\right) = \frac{t}{2} + \frac{3}{2}$$

$$= \frac{t+3}{2} \quad \Rightarrow \quad dx = \frac{dt}{2}$$

$$\therefore \qquad I = \int_{1}^{2} \frac{dx}{1+x^3} = \int_{-1}^{1} \frac{\dfrac{dt}{2}}{1 + \left(\dfrac{t+3}{2}\right)^3}$$

$$= \frac{1}{2}\int_{-1}^{1} \frac{dt}{1 + \dfrac{(t+3)^3}{8}} = \int_{-1}^{1} f(t)\,dt$$

where
$$f(t) = \frac{1}{2}\left[\frac{1}{1 + \dfrac{(t+3)^3}{8}}\right]$$

Using the Gauss–Legendre four point formula

$$\int_{-1}^{1} f(x)\,dx = 0.652145\,[f(0.339981) + f(-0.339981)]$$
$$+ 0.347855\,[f(0.861136) + f(-0.861136)]$$

We obtain $I = \dfrac{1}{2}[0.652145(0.176760 + 0.298268) + 0.347855 \times (0.122020 + 0.449824)]$

$$= 0.254353.$$

Example 22: Obtain an approximate value of $I = \int_{-1}^{1}(1-x^2)^{1/2}\cos x\,dx$ by using Gauss–Legendre integration method for $n = 2, 3$.

Solution: We have Gauss–Legendre three point formula as

$$\int_{-1}^{1} f(x)\,dx = \frac{5}{9}f\left(-\sqrt{\frac{3}{5}}\right) + \frac{8}{5}f(0) + \frac{5}{9}f\left(\sqrt{\frac{3}{5}}\right)$$

We have
$$f(x) = (1-x^2)^{1/2}\cos x$$

$$\therefore \qquad f\left(-\sqrt{\frac{3}{5}}\right) = \left(1 - \frac{3}{5}\right)^{1/2} \cos\left(-\sqrt{\frac{3}{5}}\right) = 0.452017988$$

$$f(0) = 1$$

$$f\left(\sqrt{\frac{3}{5}}\right) = \left(1 - \frac{3}{5}\right)^{1/2} \cos\left(\sqrt{\frac{3}{5}}\right) = 0.452017988$$

$$I = \int_{-1}^{1} f(x)\,dx = \frac{5}{9} \times 0.452017988 + \frac{8 \times 1}{9} + \frac{5}{9} \times 0.452017988 = 1.391131$$

We have Gauss–Legendre four point formula

$$\int_{-1}^{1} f(x)\,dx = 0.652145\,[f(0.339981) + f(-0.339981)$$
$$+ 0.347855\,[f(0.861136) + f(-0.861136)]$$

We obtain $I = 0.652145[\sqrt{1 - (0.339981)^2}\,\cos(0.339981) + \sqrt{1 - (-0.339981)^2}$

$$\cos(-0.339981)] + 0.347855\,[\sqrt{1 - (0.861136)^2}\,\cos(0.861136$$

$$+ \sqrt{1 - (-0.861136)^2}\,\cos(-0.861136)]$$

$$= 2 \times 0.652145\,[\sqrt{1 - (0.339981)^2}\,\cos(0.339981)]$$

$$+ 2 \times 0.347855\,[\sqrt{1 - (0.861136)^2}\,\cos(0.861136)]$$

$$= 1.3868837$$

Example 23: Evaluate $I = \int_{0}^{4} t e^{2t}\,dt$ by Gauss–Legendre formula for $n = 1, 2$.

Solution: By transform the limit -1 to 1 we set

$$t = \frac{(b - a)}{2} x + \frac{a + b}{2}$$

$$= \left(\frac{4 - 0}{2}\right) x + \left(\frac{4 + 0}{2}\right) = 2x + 2 = 2(x + 1)$$

Using Gauss–Legendre two point formula

$$\int_{-1}^{1} f(t)\,dt = f\left(-\frac{1}{\sqrt{3}}\right) + f\left(\frac{1}{\sqrt{3}}\right)$$

$$I = \int_{-1}^{1} 2(x + 1)\,e^{2(2x + 2)}\,2dx = 4\int_{-1}^{1} (x + 1)\,e^{4x + 4}\,dx$$

$$= 4\left(-\frac{1}{\sqrt{3}} + 1\right)e^{-\frac{4}{\sqrt{3}} + 4} + 4\left(\frac{1}{\sqrt{3}} + 1\right)e^{\frac{4}{\sqrt{3}} + 4}$$

$$= 9.16765732 + 3468.376279 = 3477.543936$$

Now using Gauss-Legendre three point formula

$$I = \int_{-1}^{1} f(t)\,dt = \frac{5}{9} f(-\sqrt{0.6}) + \frac{8}{9} f(0) + \frac{5}{9} f(\sqrt{0.6})$$

$$= \frac{5}{9} \times 4(-\sqrt{0.6} + 1)\, e^{4 - 4 \times \sqrt{0.6}} + \frac{8}{9} \times 4e^4 + \frac{5}{9} \times 4(1 + \sqrt{0.6}) \times e^{4 + 4\sqrt{0.6}}$$

$$= 1.233995303 + 194.1267557 + 4771.745938 = 4967.106689$$

Example 24: Evaluate $I = \dfrac{1}{\sqrt{2\pi}} \displaystyle\int_{0}^{1.64} e^{-\frac{x^2}{2}}\, dx$ by Gauss–Legendre formula for $n = 1, 2$.

Solution: By transform the limit -1 to 1, we set

$$x = \left(\frac{b-a}{2} \right) t + \left(\frac{a+b}{2} \right)$$

$$= \left(\frac{1.64 - 0}{2} \right) t + \left(\frac{1.64 + 0}{2} \right) = 0.82(1 + t)$$

$$dx = 0.82\, dt$$

Since

$$I = \frac{1}{\sqrt{2\pi}} \int_{0}^{1.64} e^{-\frac{x^2}{2}}\, dx = \frac{0.82}{\sqrt{2\pi}} \int_{-1}^{1} e^{-\frac{1}{2}[0.82(1+t)]^2}\, dt$$

$$= \frac{0.82}{\sqrt{2\pi}} \int_{-1}^{1} f(t)\, dt$$

where

$$f(t) = e^{-\frac{1}{2}[0.82(1+t)]^2}$$

i. **Two point formula ($n = 1$):**

$$I = \frac{0.82}{\sqrt{2\pi}} \int_{-1}^{1} f(t)\, dt = \frac{0.82}{\sqrt{2\pi}} \left[f\left(-\frac{1}{\sqrt{3}} \right) + f\left(\frac{1}{\sqrt{3}} \right) \right]$$

$$= \frac{0.82}{\sqrt{2\pi}} \left[e^{-\frac{1}{2}\left\{ 0.82\left(1 - \frac{1}{\sqrt{3}} \right) \right\}^2} + e^{-\frac{1}{2}\left\{ 0.82\left(1 + \frac{1}{\sqrt{3}} \right) \right\}^2} \right]$$

$$= 0.32713267\,(0.94171147 + 0.43323413)$$

$$= 0.44978962$$

ii. **Three point formula ($n = 2$):**

$$I = \frac{0.82}{\sqrt{2\pi}} \int_{-1}^{1} f(t)\, dt = \frac{0.82}{\sqrt{2\pi}} \left[\frac{5}{9} f\left(-\sqrt{0.6} \right) + \frac{8}{9} f(0) + \frac{5}{9} f\left(\sqrt{0.6} \right) \right]$$

$$= \frac{0.82}{\sqrt{2\pi}} \left\{ \frac{5}{9} e^{-\frac{1}{2}[0.82(1 - \sqrt{0.6})]^2} + \frac{8}{9} e^{-\frac{1}{2}[0.82(1 - 0)]^2} + \frac{5}{9} e^{-\frac{1}{2}[0.82(1 + \sqrt{0.6})]^2} \right\}$$

$$= 0.32713267\,(0.54614659 + 0.63509351 + 0.19271450)$$

$$= 0.44946544$$

Example 25: The Radau quadrature formula is given by

$$\int_{-1}^{1} f(x)\,dx = B_1\, f(-1) + \sum_{K=1}^{n} H_K\, f(x_K) + R$$

Determine x_k, H_k and R for $n = 1$.

Solution: Making the method $\int_{-1}^{1} f(x)\,dx = B_1\, f(-1) + H_1 f(x) + R$ exact for $f(x) = 1$, x and x^2, we obtain the equations

$$f(x) = 1 \;\Rightarrow\; B_1 + H_1 = 2$$
$$= x \;\Rightarrow\; -B_1 + H_1 x_1 = 0$$
$$= x^2 \;\Rightarrow\; B_1 + H_1 x_1^2 = \frac{2}{3}$$

On solving the above equations, we get $x_1 = \dfrac{1}{3}$, $H_1 = \dfrac{3}{2}$, $B_1 = \dfrac{1}{2}$.

Hence, we obtain the method

$$\int_{-1}^{1} f(x)\,dx = \frac{1}{2} f(-1) + \frac{3}{2} f\left(\frac{1}{3}\right)$$

The error term is given by

$$R = \frac{C}{3!} f'''(\xi), \quad -1 < \xi < 1$$

where

$$C = \int_{-1}^{1} x^3 dx - [-B_1 + H_1 x_1^3] = \frac{4}{9}$$

Hence, we have

$$R = \frac{4}{9(3!)} f'''(\xi) = \frac{2}{27} f'''(\xi), \quad -1 < \xi < 1$$

Example 26: The Lobatto quadrature formula is given by

$$\int_{-1}^{1} f(x)\,dx = B_1 f(-1) + B_2 f(1) \sum_{K=1}^{n-1} H_K f(x_K) + R$$

Determine x_K, H_K and R for $n = 3$.

Solution: Making the method

$$\int_{-1}^{1} f(x)\,dx = B_1 f(-1) + B_2 f(1) + H_1 f(x_1) + H_2 f(x_2) + R \tag{i}$$

exact for $f(x) = 1$, x, x^2, x^3, x^4, x^5, we obtain the equations

$$f(x) = 1 \;\Rightarrow\; B_1 + B_2 + H_1 + H_2 = 2 \tag{ii}$$
$$= x \;\Rightarrow\; -B_1 + B_2 + H_1 x_1 + H_2 x_2 = 0 \tag{iii}$$
$$= x^2 \;\Rightarrow\; B_1 + B_2 + H_1 x_1^2 + H_2 x_2^2 = \frac{2}{3} \tag{iv}$$
$$= x^3 \;\Rightarrow\; -B_1 + B_2 + H_1 x_1^3 + H_2 x_2^3 = 0 \tag{v}$$
$$= x^4 \;\Rightarrow\; B_1 + B_2 + H_1 x_1^4 + H_2 x_2^4 = \frac{2}{5} \tag{vi}$$
$$= x^5 \;\Rightarrow\; -B_1 + B_2 + H_1 x_1^5 + H_2 x_2^5 = 0 \tag{vii}$$

or $\quad H_1(1-x_1^2) + H_2(1-x_2^2) = \dfrac{4}{3}\quad$ [On solving Eqs (ii and iv)

$H_1(1-x_1^2)x_1 + H_2(1-x_2^2)x_2 = 0\quad$ [On solving Eqs (iii and v)

$H_1(1-x_1^2)x_1^2 + H_2(1-x_2^2)x_2^2 = \dfrac{4}{15}\quad$ [On solving Eqs (iv and vi)

$H_1(1-x_1^2)x_1^3 + H_2(1-x_2^2)x_2^3 = 0\quad$ [On solving Eqs (v and vii)

or $\quad H_1(1-x_1^2)(x_2-x_1) = \dfrac{4}{3}x_2$

$H_1(1-x_1^2)(x_2-x_1)x_1 = -\dfrac{4}{15}$

$H_1(1-x_1^2)(x_2-x_1)x_1^2 = \dfrac{4}{15}x_2$

Solving the above equations, we get

$$x_1 x_2 = -\frac{1}{5} \text{ and } x_1 = -x_2$$

The solution is obtained as $x_1 = \dfrac{1}{\sqrt{5}}, x_2 = -\dfrac{1}{\sqrt{5}}; H_1 = H_2 = \dfrac{5}{6}; B_1 = B_2 = \dfrac{1}{6}.$

Hence the method is given by

$$\int_{-1}^{1} f(x)\,dx = \frac{1}{6}\left[f(-1) + f(1) + \frac{5}{6}\left\{ f\left(\frac{1}{\sqrt{5}}\right) + f\left(-\frac{1}{\sqrt{5}}\right) \right\} \right]$$

The error term is

$$R = \frac{C}{6!} f''''''(\xi), \quad -1 < \xi < 1$$

where

$$C = \int_{-1}^{1} x^6\,dx - [B_1 + B_2 + H_1 x_1^6 + H_2 x_2^6]$$

$$= \frac{2}{7} - \left[\frac{1}{6} + \frac{1}{6} + \frac{5}{6}\left(\frac{1}{\sqrt{5}}\right)^6 + \frac{5}{6}\left(-\frac{1}{\sqrt{5}}\right)^6 \right]$$

$$= \frac{2}{7} - \left[\frac{2}{6} + \frac{5}{6}\cdot\frac{1}{5^3} + \frac{5}{6}\cdot\frac{1}{5^3} \right] = \frac{2}{7} - \left(\frac{1}{3} + \frac{1}{75}\right)$$

$$= -\frac{32}{525}$$

Hence, we have $\qquad R = -\dfrac{32}{(6!)\,525} f''''''(\xi), \quad -1 < \xi < 1$

Example 27: Evaluate $I = \int_0^1 \dfrac{dx}{2x^2 + 2x + 1}$ using the Lobatto three point and Radau three point formula, compare with the exact solution.

Solution: Using the transformation we convert the limit 0 to 1 into -1 to 1

$$x = \frac{(b-a)}{2}t + \frac{a+b}{2}$$

$$= \frac{(1-0)}{2}t + \frac{(1+0)}{2} = \frac{t+1}{2}$$

$\Rightarrow$ $$dx = \frac{dt}{2}$$

Therefore $$I = \int_{-1}^{1} \frac{\dfrac{dt}{2}}{2\left(\dfrac{t+1}{2}\right)^2 + 2\left(\dfrac{t+1}{2}\right) + 1} = \int_{-1}^{1} \frac{\dfrac{dt}{2}}{\left(\dfrac{t+1}{2}\right)^2 + (t+1) + 1}$$

$$= \int_{-1}^{1} \frac{dt}{(t+1)^2 + 2(t+1) + 2} = \int_{-1}^{1} \frac{dt}{(t+1)^2 + 2t + 4} = \int_{-1}^{1} f(t)\,dt$$

where $$f(t) = \frac{1}{(t+1)^2 + 2t + 4}$$

By using Lobatto three point formula $(n = 2)$

$$I = \frac{1}{3}[f(-1) + 4f(0) + f(1)]$$

$$= \frac{1}{3}\left[\frac{1}{-2+4} + 4 \cdot \frac{1}{1^2+4} + \frac{1}{2^2 + 2\times1 + 4}\right]$$

$$= \frac{1}{3}\left[\frac{1}{2} + \frac{4}{5} + \frac{1}{10}\right] = 0.46667$$

By using Radau three point formula

$$I = \frac{2}{9}f(-1) + \frac{16+\sqrt{6}}{18}f\left(\frac{1-\sqrt{6}}{5}\right) + \frac{16-\sqrt{6}}{18}f\left(\frac{1+\sqrt{6}}{5}\right)$$

$$= \frac{2}{9} \times \frac{1}{2} + \left(\frac{16+\sqrt{6}}{18}\right)\frac{1}{\left(\dfrac{1-\sqrt{6}}{5}+1\right)^2 + 2\times\left(\dfrac{1-\sqrt{6}}{5}\right) + 4}$$

$$+ \left(\frac{16-\sqrt{6}}{18}\right) \times \frac{1}{\left(\dfrac{1+\sqrt{6}}{5}+1\right)^2 + 2\times\left(\dfrac{1+\sqrt{6}}{5}\right) + 4}$$

$$= 0.11111 + 0.26118 + 0.09141 = 0.4637$$

The exact solution is $$I = \tan^{-1}3 - \frac{\pi}{4} = 0.46365$$

Example 28: Obtain the approximate value of $I = \int_{-1}^{1} e^{-x^2}\cos x\,dx$ using

i. Radau integration method for $n = 2, 3$

ii. Lobatto integration method for $n = 2, 3$.

Solution: We have
$$I = \int_{-1}^{1} e^{-x^2} \cos x \, dx = \int_{-1}^{1} f(x) \, dx$$

where
$$f(x) = e^{-x^2} \cos x$$

Using Radau three point formula ($n = 2$)

$$I = \int_{-1}^{1} e^{-x^2} \cos x \, dx = \int_{-1}^{1} f(x) \, dx$$

$$= \frac{2}{9} f(-1) + \frac{16 + \sqrt{6}}{18} f\left(\frac{1 - \sqrt{6}}{5}\right) + \left(\frac{16 - \sqrt{6}}{18}\right) f\left(\frac{1 + \sqrt{6}}{5}\right)$$

$$= 0.081738 + \left(\frac{16 + \sqrt{6}}{18}\right) \times (-0.289898) + \left(\frac{16 - \sqrt{6}}{18}\right) \times 0.689898$$

$$= 1.307951$$

Using Radau four point formula ($n = 3$)

$$I = \int_{-1}^{1} f(x) \, dx = 0.125000 \, f(-1) + 0.657689 \, f(-0.575319)$$

$$+ 0.776387 \, f(0.181066) + 0.440924 \, f(0.822824)$$

$$= 1.312610$$

Using Lobatto three point formula ($n = 2$)

$$I = \int_{-1}^{1} f(x) \, dx = \frac{1}{3} [f(-1) + 4f(0) + f(1)]$$

$$= \frac{1}{3} [e^{-1} \cos(-1) + 4 \cos 0 + e^{-1} \cos 1]$$

$$= \frac{1}{3} [2e^{-1} \cos 1 + 4] = \frac{1}{3} [4 + 2 \times 0.19876611]$$

$$= \frac{4.397532221}{3} = 1.465844$$

Using Lobatto four point formula ($n = 3$)

$$I = \int_{-1}^{1} f(x) \, dx = 0.166667 [f(-1) + f(1)] + 0.833333 [f(0.447214) + f(-0.447214)]$$

$$= 0.166667 [e^{-1} \cos(-1) + e^{-1} \cos 1] + 0.833333$$

$$[e^{-(0.447214)^2} \cos(-0.447214) + e^{-(-0.447214)^2} \cos(0.447214)]$$

$$= 1.296610$$

Example 29: Evaluate the integral $I = \int_{0}^{2} \frac{dx}{3 + 4x}$ by using

   i. Radau integration method for $n = 1, 2$

   ii. Lobatto integration method for $n = 2, 3$.

Solution: The use of Radau and Lobatto integration method, the interval $[0, 2]$ is to be reduced in $[-1, 1]$.

For transformation $x = \left(\dfrac{a+b}{2}\right) + \left(\dfrac{b-a}{2}\right)t$

$$= \left(\dfrac{2+0}{2}\right) + \left(\dfrac{2-0}{2}\right)t$$

or $\qquad = 1 + t$

$\Rightarrow \qquad dx = dt$

$\therefore \qquad I = \displaystyle\int_0^2 \dfrac{dx}{3+4x} = \int_{-1}^1 \dfrac{dt}{3+4(1+t)} = \int_{-1}^1 \dfrac{dt}{4t+7} = \int_{-1}^1 f(t)\,dt$

where $\qquad f(t) = \dfrac{1}{4t+7}$

i. Using Radau two point formula $(n = 1)$

$$I = \int_{-1}^1 f(t)\,dt = \frac{1}{2}\left[f(-1) + 3f\left(\frac{1}{3}\right)\right]$$

$$= \frac{1}{2}\left[\frac{1}{-4+7} + 3 \times \frac{1}{\frac{4}{3}+7}\right] = \frac{1}{2}\left[\frac{1}{3} + \frac{9}{25}\right]$$

$$= 0.34667$$

Using Radau three point formula $(n = 2)$

$$I = \int_{-1}^1 f(t)\,dt = \frac{2}{9}f(-1) + \frac{16+\sqrt{6}}{18}f\left(\frac{1-\sqrt{6}}{5}\right) + \left(\frac{16-\sqrt{6}}{18}\right)f\left(\frac{1+\sqrt{6}}{5}\right)$$

$$= \frac{2}{9}\left(\frac{1}{-4+7}\right) + \frac{16+\sqrt{6}}{18}\left[\frac{1}{4\left(\frac{1-\sqrt{6}}{5}\right)+7}\right] + \frac{16-\sqrt{6}}{18}\left[\frac{1}{4\left(\frac{1+\sqrt{6}}{5}\right)+7}\right]$$

$$= 0.074074 + 1.0249716\,(0.17122091) + 0.752806\,(0.102463)$$

$$= 0.32670$$

ii. Using Lobatto three point formula $(n = 2)$

$$I = \frac{1}{3}[f(-1) + 4f(0) + f(1)] = \frac{1}{3}\left[\frac{1}{-4+7} + \frac{4}{7} + \frac{1}{4+7}\right]$$

$$= \frac{1}{3}\left[\frac{1}{3} + \frac{4}{7} + \frac{1}{11}\right] = 0.33189$$

Using Lobatto four point formula $(n = 3)$

$$I = \int_{-1}^1 f(t)\,dt = 0.166667[f(-1) + f(1)] + 0.833333\,[f(0.447214) + f(-0.447214]$$

$$= 0.166667\left[\frac{1}{-4+7} + \frac{1}{4+7}\right] + 0.833333 \times$$

$$\left[\frac{1}{4 \times 0.447214 + 7} + \frac{1}{4 \times -0.447214 + 7}\right]$$

$$= 0.0707072 + 0.094817 + 0.159914$$

$$= 0.325438$$

Example 30: Evaluate the integral $I = \int_0^1 \dfrac{dx}{1+x}$ using Lobatto three point formula and Radau two point integration method.

Solution: Transform the limit [0, 1] into [−1, 1] by using the transformation

$$x = \left(\frac{b-a}{2}\right)t + \frac{a+b}{2} = \left(\frac{1-0}{2}\right)t + \frac{1+0}{2} = \frac{t+1}{2}$$

$$\Rightarrow \qquad dx = \frac{dt}{2}$$

$$\therefore \qquad I = \int_0^1 \frac{dx}{1+x} = \int_{-1}^1 \frac{\dfrac{dt}{2}}{1+\dfrac{t+1}{2}} = \int_{-1}^1 \frac{dt}{t+3} = \int_{-1}^1 f(t)\,dt$$

where $\qquad f(t) = \dfrac{1}{t+3}$

Using Lobatto three point formula is given as

$$I = \frac{1}{3}[f(-1) + 4f(0) + f(1)] = \frac{1}{3}\left[\frac{1}{-1+3} + \frac{4}{0+3} + \frac{1}{1+3}\right]$$

$$= \frac{1}{3}\left[\frac{1}{2} + \frac{4}{3} + \frac{1}{4}\right] = 0.69444$$

Using Radau two point formula

$$I = \frac{1}{2}\left[f(-1) + 3f\left(\frac{1}{3}\right)\right] = \frac{1}{2}\left[\frac{1}{-1+3} + 3\left(\frac{1}{\dfrac{1}{3}+3}\right)\right]$$

$$= \frac{1}{2}\left[\frac{1}{2} + \frac{9}{10}\right] = 0.7$$

EXERCISE 8.3

1. Evaluate the integral $I = \int_0^1 \dfrac{dx}{1+x}$ using Gauss–Legendre three point formula.

[Ans. 0.693122]

2. Evaluate integral $I = \int_1^2 \dfrac{2x\,dx}{1+x^4}$ using Gauss–Legendre one point, two point and three point quadrature rule. Compare with the exact solution.

[Ans. 0.4948, 0.5434, 0.5406]

3. Evaluate the integrals $I = \int_0^2 \dfrac{dx}{3+4x}$ by Gauss–Legendre two point and three point formula.

[Ans. 0.320610, 0.324390]

4. Evaluate $\int_0^2 \dfrac{dx}{x^2+2x+10}$ by Gauss–Legendre two point and three point formula.

[Ans. 0.154639, 0.154548]

5. Determine x_i and A_i in the quadrature formula below so that σ, the order of approximation will be as high as possible

$$\int_{-1}^{1}(2x^2+1)\,f(x)\,dx = A_1 f(x_1) + A_2 f(x_2) + A_3 f(x_3) + R$$

What is the value of σ? Answer with four significant digits.　　　　[Ans. $\sigma = 5$]

6. Determine the coefficients in the formula

$$\int_{0}^{2h} x^{-\frac{1}{2}} f(x)\,dx = (2h)^{\frac{1}{2}}[A_0 f(0) + A_1 f(h) + A_2 f(2h)] + R$$

and calculate the remainder R, when $f'''(x)$ is constant.

$$\left[\text{Ans. } A_0 = \frac{12}{15},\ A_1 = \frac{16}{15},\ A_2 = \frac{2}{15},\ R = \frac{8\sqrt{2}}{315}\,h^{\frac{7}{2}}\,f'''(\xi) \right]$$

7. Evaluate the integral $I = \dfrac{dx}{x^2+1}$ in the interval $[0, 1]$ using the Lobatto and Radau three point formula.　　　　[Ans. 0.78333, 0.78584]

8. Evaluate the integral $\displaystyle\int_{0}^{1}\dfrac{dx}{2x^2+2x+3}$ using Lobatto and Radau three point formula.　　　　[Ans. 0.22751, 0.227906]

9. Apply Gauss–Legendre two point formula to evaluate i. $\displaystyle\int_{0}^{1}\dfrac{dx}{1+x^2}$ ii. $\displaystyle\int_{-1}^{1}\dfrac{dx}{1+x^2}$.

[Ans. i. 0.75, ii. 1.5]

10. Using Gaussian three point formula evaluate $\displaystyle\int_{-1}^{1}(3x^2+5x^4)dx$.　　　　[Ans. 4]

11. Evaluate $\displaystyle\int_{-2}^{2} e^{-\frac{x}{2}}\,dx$ by Gauss–Legendre two point formula.　　　　[Ans. 4.6854]

12. Evaluate $\displaystyle\int_{2}^{3}\dfrac{dt}{1+t}$ by Gauss–Legendre two point and three point formula.

[Ans. 0.28188, 0.28767]

13. By using Lobatto and Radau three point formula, evaluate $\displaystyle\int_{0}^{1} e^x dx$.

14. By using Lobatto integration formula ($n = 2$) and Radau integration formula for n = 1, 2, evaluate i. $\displaystyle\int_{2}^{4}(2x^4+2)dx$ ii. $\displaystyle\int_{5}^{12}\dfrac{dx}{x}$.

UNIT IV

Chapter 9

Solution of Simultaneous Linear Algebraic Equations

9.1 INTRODUCTION

The system of linear equations generally arises in the field of science and engineering. In lower classes, linear equations can be solved by the use of Cramer's rule or by matrix methods. These methods become tedious and time consuming when the number of unknowns in the system is large. After the availability of computers, numerical methods can be used which are appropriate for computer operation. The numerical methods of solution of system of linear can be classified into two types namely, i. direct ii. iterative.

9.2 DIRECT METHODS

These methods are used to solve the problem by a finite sequence of operations. In the absence of rounding errors, direct methods would deliver an exact solution. In these methods, the amount of computation involved can be specified in advance. They are independent of the desired accuracy.

9.3 ITERATIVE METHODS

These methods provide a sequence of approximation that converges when the number of steps tend to infinity. Here we will study only direct method.

9.4 GAUSS ELIMINATION METHOD

This method is based on the elimination of the unknowns by transforming the given system into an equivalent system with upper triangular coefficient matrix using elementary row operations that could be solved by back substitution.

Consider the n linear equation in n unknown

$$
\left.
\begin{aligned}
a_{11}x_1 + a_{12}x_2 + \ldots + a_{1n}x_n &= b_1 \\
a_{21}x_1 + a_{22}x_2 + \ldots + a_{2n}x_n &= b_2 \\
a_{31}x_1 + a_{32}x_2 + \ldots + a_{3n}x_n &= b_3 \\
\cdots\cdots\cdots\cdots\cdots\cdots\cdots\cdots\cdots\cdots\cdots \\
\cdots\cdots\cdots\cdots\cdots\cdots\cdots\cdots\cdots\cdots\cdots \\
a_{n1}x_1 + a_{n2}x_2 + \ldots + a_{nn}x_n &= b_n
\end{aligned}
\right\}
\qquad (9.1)
$$

where a_{ij} and b_i are known constants and x_i's are unknowns.

237

We know that this system can be written in matrix form as

$$AX = B \tag{9.2}$$

where

$$A = \begin{bmatrix} a_{11} & a_{12} & \cdots & a_{1n} \\ a_{21} & a_{22} & \cdots & a_{2n} \\ a_{31} & a_{32} & \cdots & a_{3n} \\ \cdots & \cdots & \cdots & \cdots \\ \cdots & \cdots & \cdots & \cdots \\ a_{n1} & a_{n2} & \cdots & a_{nn} \end{bmatrix}, \quad X = \begin{bmatrix} x_1 \\ x_2 \\ \cdots \\ \cdots \\ \cdots \\ x_n \end{bmatrix}$$

and

$$B = \begin{bmatrix} b_1 \\ b_2 \\ \cdots \\ \cdots \\ b_n \end{bmatrix} \tag{9.3}$$

In Eq. (9.3), A is called the coefficient matrix of Eq. (9.1). If matrix B is also included in A, we have the matrix

$$[A:B] = \begin{bmatrix} a_{11} & a_{12} & \cdots & a_{1n} & b_1 \\ a_{21} & a_{22} & \cdots & a_{2n} & b_2 \\ \cdots & \cdots & \cdots & \cdots & \cdots \\ \cdots & \cdots & \cdots & \cdots & \cdots \\ a_{n1} & a_{n2} & \cdots & a_{nn} & b_n \end{bmatrix} \tag{9.4}$$

which is called the augmented matrix of Eq. (9.1). Now we reduce the augmented matrix $[A:B]$ into upper triangular matrix using row transformation only. To obtain this, we multiply the first row of Eq. (9.4) (if $a_{11} \neq 0$) by $-\dfrac{a_{i1}}{a_{11}}$ and add to the ith row of $[A:B]$, where $i = 2, 3, ..., n$. By doing this, all elements in the first column of $[A:B]$ are made to zero except a_{11}. Now Eq. (9.4) is of the form

$$\begin{bmatrix} a_{11} & a_{12} & \cdots & a_{1n} & b_1 \\ 0 & b_{22} & \cdots & b_{2n} & c_2 \\ \cdots & \cdots & \cdots & \cdots & \cdots \\ \cdots & \cdots & \cdots & \cdots & \cdots \\ 0 & b_{n2} & \cdots & b_{nn} & c_n \end{bmatrix} \tag{9.5}$$

Now take the pivot b_{22}. Considering b_{22} as the pivot, i.e. $b_{22} \neq 0$, we will construct all elements below b_{22} in the second column of Eq. (9.5) as zero. To obtain this, we multiply second row of Eq. (9.5) by $-\dfrac{b_{i2}}{b_{22}}$ and add to the corresponding elements of the ith row $(i = 3, 4, ..., n)$. Now all elements below b_{22} are reduced to zero. Now Eq. (9.5) reduces to

$$\left[\begin{array}{ccccc|c} a_{11} & a_{12} & a_{13} & \cdots & a_{1n} & b_1 \\ 0 & b_{22} & b_{23} & \cdots & b_{2n} & c_2 \\ 0 & 0 & c_{33} & \cdots & c_{3n} & d_3 \\ \cdots & \cdots & \cdots & \cdots & \cdots & \cdots \\ \cdots & \cdots & \cdots & \cdots & \cdots & \cdots \\ 0 & 0 & c_{n3} & \cdots & c_{nn} & d_n \end{array}\right] \qquad (9.6)$$

Now taking c_{33} as the pivot, using elementary row operation, we make all elements below c_{33} as zeros if we repeat this process till the nth equation, all elements below the leading diagonal elements of A are made to zero.

Hence, after all these operations, we get the augmented matrix of the form

$$\left[\begin{array}{cccccc} a_{11} & a_{12} & a_{13} & \cdots & a_{1n} & b_1 \\ 0 & b_{22} & b_{23} & \cdots & b_{2n} & c_2 \\ 0 & 0 & c_{33} & \cdots & c_{3n} & d_3 \\ \cdots & \cdots & \cdots & \cdots & \cdots & \cdots \\ \cdots & \cdots & \cdots & \cdots & \cdots & \cdots \\ 0 & 0 & 0 & \cdots & \alpha_{nn} & M_n \end{array}\right] \qquad (9.7)$$

This process is called triangularization and reduced matrix is called upper triangular matrix.

From Eq. (9.7), we get the new systems of equation of the type

$$\left.\begin{array}{r} a_{11}x_1 + a_{12}x_2 + a_{13}x_3 + \ldots + a_{1n}x_n = b_1 \\ b_{22}x_2 + b_{23}x_3 + \ldots + b_{2n}x_n = c_2 \\ c_{33}x_3 + \ldots + c_{3n}x_n = d_3 \\ \cdots\cdots\cdots\cdots\cdots\cdots\cdots\cdots\cdots \\ \alpha_{nn}x_n = M_n \end{array}\right\} \qquad (9.8)$$

From these equations $x_1, x_2, x_3, \ldots, x_n$ can be obtained by back substitution.

The diagonal elements of the coefficient matrix corresponding to the system of linear equations are called pivot elements.

Clearly, the above computations will fail, if any one of the pivots $a_{11}, a_{22}, \ldots, a_{nn}$ vanishes. In this case, method can be modified by rearranging the equations, i.e. the rows, so that the pivot is non zero. This can be easily effected on a computer and is called partial pivoting.

If partial pivoting is not possible, then matrix A is called singular, i.e. the Eq. (9.1) have no solution. In fact, the number of non zero elements in the principal diagonal of Eq. (9.7) is the rank of the matrix A.

Example 1: Solve the system by Gauss-elimination method $2x + 3y - z = 5$; $4x + 4y - 3z = 3$ and $2x - 3y + 2z = 2$.

Solution: Given system is equivalent to

$$\underset{A}{\left[\begin{array}{ccc} 2 & 3 & -1 \\ 4 & 4 & -3 \\ 2 & -3 & 2 \end{array}\right]} \underset{X}{\left[\begin{array}{c} x \\ y \\ z \end{array}\right]} = \underset{B}{\left[\begin{array}{c} 5 \\ 3 \\ 2 \end{array}\right]}$$

Here augmented matrix is

$$[A:B] = \begin{bmatrix} 2 & 3 & -1 & | & 5 \\ 4 & 4 & -3 & | & 3 \\ 2 & -3 & 2 & | & 2 \end{bmatrix}$$

Taking $a_{11} = 2$ as the pivot, reduce all the elements below that to zero, we get

$$\sim \begin{bmatrix} 2 & 3 & -1 & | & 5 \\ 0 & -2 & -1 & | & -7 \\ 0 & -6 & 3 & | & -3 \end{bmatrix} \quad \begin{array}{l} R_2 + (-2)\,R_1 \quad \text{i.e.}\ R_{21}\,(-2) \\ R_3 + (-1)\,R_1 \quad \text{i.e.}\ R_{31}\,(-1) \end{array}$$

Taking $a_{22} = -2$ as the pivot, reduce all the element below that to zero, we get

$$\sim \begin{bmatrix} 2 & 3 & -1 & | & 5 \\ 0 & -2 & -1 & | & -7 \\ 0 & 0 & 6 & | & 18 \end{bmatrix} \quad R_{32}\,(-3)$$

From this, we get

$$2x + 3y - z = 5$$
$$-2y - z = -7$$
$$6z = 18$$

Solving above equations using back substitution, we get $z = 3$, $y = 2$, $x = 1$.

Example 2: Solve the equations $10x - y + 2z = 4$; $x + 10y - z = 3$; $2x + 3y + 20z = 7$ using the Gauss-elimination method.

Solution: The given system is diagonally dominant and therefore no pivoting is necessary.

Given system is equivalent to

$$\underbrace{\begin{bmatrix} 10 & -1 & 2 \\ 1 & 10 & -1 \\ 2 & 3 & 20 \end{bmatrix}}_{A} \underbrace{\begin{bmatrix} x \\ y \\ z \end{bmatrix}}_{X} = \underbrace{\begin{bmatrix} 4 \\ 3 \\ 7 \end{bmatrix}}_{B}$$

Hence, the augmented matrix is

$$[A:B] = \begin{bmatrix} 10 & -1 & 2 & | & 4 \\ 1 & 10 & -1 & | & 3 \\ 2 & 3 & 20 & | & 7 \end{bmatrix}$$

$$\sim \begin{bmatrix} 10 & -1 & 2 & | & 4 \\ 0 & \dfrac{101}{10} & -\dfrac{6}{5} & | & \dfrac{13}{5} \\ 0 & \dfrac{16}{5} & \dfrac{98}{5} & | & \dfrac{31}{5} \end{bmatrix} \quad R_{21}\left(-\dfrac{1}{10}\right),\ R_{31}\left(-\dfrac{1}{5}\right)$$

$$\sim \begin{bmatrix} 10 & -1 & 2 & \bigm| & 4 \\ 0 & \dfrac{101}{10} & -\dfrac{6}{5} & \bigm| & \dfrac{13}{5} \\ 0 & 0 & \dfrac{20180}{1010} & \bigm| & \dfrac{5430}{1010} \end{bmatrix} \quad R_{32}\left(-\dfrac{32}{101}\right)$$

The equivalent system is

$$10x - y + 2z = 4$$

$$\frac{101}{10}y - \frac{6}{5}z = \frac{13}{5}$$

$$\frac{20180}{1010}z = \frac{5430}{1010}$$

Using back substitution, we get $x_3 = 0.269$, $x_2 = 0.289$, $x_1 = 0.375$.

Example 3: Solve the following system of equation by Gauss-elimination method:

$$x_1 + 2x_2 + 3x_3 + 4x_4 = 10$$
$$7x_1 + 10x_2 + 5x_3 + 2x_4 = 40$$
$$13x_1 + 6x_2 + 2x_3 - 3x_4 = 34$$
$$11x_1 + 14x_2 + 8x_3 - x_4 = 65$$

Solution: The system is equivalent to

$$\begin{bmatrix} 1 & 2 & 3 & 4 \\ 7 & 10 & 5 & 2 \\ 13 & 6 & 2 & -3 \\ 11 & 14 & 8 & -1 \end{bmatrix} \begin{bmatrix} x_1 \\ x_2 \\ x_3 \\ x_4 \end{bmatrix} = \begin{bmatrix} 10 \\ 40 \\ 34 \\ 64 \end{bmatrix}$$

$$\qquad A \qquad\qquad X \ = \ B$$

Hence the augmented matrix is

$$[A\!:\!B] = \begin{bmatrix} 1 & 2 & 3 & 4 & \bigm| & 10 \\ 7 & 10 & 5 & 2 & \bigm| & 40 \\ 13 & 6 & 2 & -3 & \bigm| & 34 \\ 11 & 14 & 8 & -1 & \bigm| & 64 \end{bmatrix}$$

$$\sim \begin{bmatrix} 1 & 2 & 3 & 4 & \bigm| & 10 \\ 0 & -4 & -16 & -26 & \bigm| & -30 \\ 0 & -20 & -37 & -55 & \bigm| & -96 \\ 0 & -8 & -25 & -45 & \bigm| & -46 \end{bmatrix} \quad R_{21}(-7),\ R_{31}(-13),\ R_{41}(-11)$$

$$\sim \begin{bmatrix} 1 & 2 & 3 & 4 & \bigm| & 10 \\ 0 & -4 & -16 & -26 & \bigm| & -30 \\ 0 & 0 & 43 & 75 & \bigm| & 54 \\ 0 & 0 & 7 & 7 & \bigm| & 16 \end{bmatrix} \quad R_{32}(-5),\ R_{42}(-2)$$

$$\sim \begin{bmatrix} 1 & 2 & 3 & 4 & 10 \\ 0 & -4 & -16 & -26 & -30 \\ 0 & 0 & 43 & 75 & 54 \\ 0 & 0 & 7 & -\dfrac{224}{43} & \dfrac{224}{43} \end{bmatrix}$$

The equivalent system is

$$x_1 + 2x_2 + 3x_3 + x_4 = 10$$
$$-4x_2 - 16x_3 - 26x_4 = -30$$
$$43x_3 + 75x_4 = 54$$
$$-\frac{224}{43}x_4 = \frac{224}{43}$$

Solving the above relations using back substitution, we get $x_4 = -1$, $x_3 = 3$, $x_2 = 2$, $x_1 = 1$

Example 4: Solve $10x - 7y + 3z + 5u = 6$, $-6x + 8y - z - 4u = 5$, $3x + y + 4z + 11u = 2$ and $5x - 9y - 2z + 4u = 7$ by Gauss-elimination method.

Solution: We can solve problem by the following way. The given system can be written as

$$x - 0.7y + 0.3z + 0.5u = 0.6 \tag{i}$$
$$-6x + 8y - z - 4u = 5 \tag{iii}$$
$$3x + y + 4z + 11u = 2 \tag{iii}$$
$$5x - 9y - 2z + 4u = 7 \tag{iv}$$

To eliminate the variable x, subtract (-6) times Eq. (i) from Eq. (ii), 3 times Eq. (i) from Eq. (iii), 5 times Eq. (i) from Eq. (iv).

Then the new system will be

$$x - 0.7y + 0.3z + 0.5u = 0.6 \tag{v}$$
$$3.8y + 0.8z - u = 8.6 \tag{vi}$$
$$3.1y + 3.1z + 9.5u = 0.2 \tag{vii}$$
$$-5.5y - 3.5z + 1.5u = 4 \tag{viii}$$

In this system, coefficient of y is maximum (numerically) in Eq. (viii), so, we shall interchange the Eqs (ii) and (iv).

$$x - 0.7y + 0.3z + 0.5u = 0.6 \tag{ix}$$
$$5.5y + 3.5z - 1.5u = -4 \tag{x}$$
$$3.1y + 3.1z + 9.5u = 0.2 \tag{xi}$$
$$3.8y + 0.8z - u = 8.6 \tag{xii}$$
or
$$x - 0.7y + 0.3z + 0.5u = 0.6 \tag{xiii}$$
$$y + 0.63636z - 0.27275u = -0.72727 \tag{xiv}$$
$$3.1y + 3.1z + 9.5u = 0.2 \tag{xv}$$
$$3.8y + 0.8z - u = 8.6 \tag{xvi}$$

To eliminate the variable y in Eqs (ii), (iii) and (iv), subtract (3.1) times the Eq. (xiv) from Eq. (xv), (3.8) times the Eq. (xiv) from Eq. (xvi).

We have

$$x - 0.7y + 0.3z + 0.5u = 0.6 \tag{xvii}$$

$$y + 0.63636z - 0.27275u = -0.72727 \tag{xviii}$$

$$1.12728z + 10.34552u = 2.454537 \tag{xix}$$

$$-1.61816z + 0.03645u = 11.363626 \tag{xx}$$

In this system, the coefficient of z is maximum (numerically) in Eq. (xx). Interchanging Eq. (xix) and (xx), we have

$$x - 0.7y + 0.3z + 0.5u = 0.6 \tag{xxi}$$

$$y + 0.63636z - 0.27275u = -0.72727 \tag{xxii}$$

$$-1.61816z + 0.03645u = 11.363626 \tag{xxiii}$$

$$1.12728z + 10.34552u = 2.454537 \tag{xxiv}$$

or

$$x - 0.7y + 0.3z + 0.5u = 0.6 \tag{xxv}$$

$$y + 0.63636z - 0.27275u = -0.72727 \tag{xxvi}$$

$$z - 0.02247u = -7.02256 \tag{xxvii}$$

$$1.12728z + 10.34552u = 2.454537 \tag{xxviii}$$

To eliminate the variable z in Eqs (xxvii) and (xxviii), subtract 1.12728 times Eq. (xxvii) from Eq. (xxviii). We have

$$10.37084998u = 10.37092843$$

which gives

$$u \approx 1$$

From Eq. (xxvii) $z = -7$.

From Eq (xxvi), $y = 4$ and from Eq. (xxv) $x \approx 5$.

So, the solution is $x = 5$, $y = 4$, $z = -7$, $u = 1$.

Example 5: Solve the following system of equations by Gauss-elimination method

$$x + 2y + z = 3, 2x + 3y + 3z = 10, 3x - y + 2z = 13$$

Solution: The given system is

$$x + 2y + z = 3 \tag{i}$$

$$2x + 3y + 3z = 10 \tag{ii}$$

$$3x - y + 2z = 13 \tag{iii}$$

Eliminating x from Eq. (ii) and (iii), we get

$$x + 2y + z = 3 \tag{iv}$$

$$y - z = -4 \tag{v}$$

$$7y + z = -4 \tag{vi}$$

Now, eliminating y from Eq. (vi), we get

$$x + 2y + z = 3 \tag{vii}$$

$$y - z = -4 \tag{viii}$$

$$-8z = -24 \tag{ix}$$

which is upper triangular form.

Solving the above system using back substitution, we get $z = 3$, $y = -1$, $x = 2$ i.e. $x = 2$, $y = -1$, $z = 3$.

EXERCISE 9.1

Solve the following system by Gauss-elimination method.

1. $3x - y + 2z = 12, x + 2y + 3z = 11, 2x - 2y - z = 2.$ [Ans. $x = 3, y = 1, z = 2$]

2. $2x + 3y + z = 9, x + 2y + 3z = 6, 3x + y + 2z = 8.$ $\left[\text{Ans. } x = \dfrac{35}{18}, y = \dfrac{29}{18}, z = \dfrac{5}{18}\right]$

3. $2x - 3y + z = -1, x + 4y + 5z = 25, 3x - 4y + z = 2.$ [Ans. $x = 8.7, y = 5.7, z = -1.3$]

4. $6x + 3y + 2z = 6, 6x + 4y + 3z = 0, 20x + 15y + 12z = 0.$ [Ans. $x = 9, y = -36, z = 30$]

5. $3x + 4y + 5z = 18, 2x - y + 8z = 13, 5x - 2y + 7z = 20.$ [Ans. $x = 3, y = 1, z = 1$]

6. $x + 2y + z - w = -2, 2x + 3y - z + 2w = 7, x + y + 3z - 2w = -6, x + y + z + w = 2.$
  [Ans. $x = 1, y = 0, z = -1, w = 2$]

7. $2x - y + 3z + w = 9, 3x + y - 4z + 3w = 3, 5x - 4y + 3z - 6w = 2, x - 2y - z + 2w = -2.$
  [Ans. $x = 1, y = 2, z = 2, w = 2$]

8. $x + 4y - z = -5, x + y - 6z = -12, 3x - y - z = 4.$ $\left[\text{Ans. } x = \dfrac{117}{71}, y = -\dfrac{81}{71}, z = \dfrac{148}{71}\right]$

9. $5x - y - 2z = 142, x - 3y - z = -30, 2x - y - 3z = -5.$
  [Ans. $x = 41.06, y = 15.76, z = 23.79$]

10. $2x - y = 0, -x + 2y - z = 0, -y + 2z - u = 0, -z + 2u = 1.$
  [Ans. $x = 0.2, y = 0.4, z = 0.6, u = 0.8$]

9.5 GAUSS-JORDAN ELIMINATION METHOD

This method is a modification of Gauss-elimination method. In this method, the coefficient matrix is reduced to a unit matrix.

Consider the system of equations

$$a_{11}x_1 + a_{12}x_2 + a_{13}x_3 = b_1$$
$$a_{21}x_1 + a_{22}x_2 + a_{23}x_3 = b_2$$
$$a_{31}x_1 + a_{32}x_2 + a_{33}x_3 = b_3$$

The matrix form of the system is

$$\begin{bmatrix} a_{11} & a_{12} & a_{13} \\ a_{21} & a_{22} & a_{23} \\ a_{31} & a_{32} & a_{33} \end{bmatrix} \begin{bmatrix} x \\ y \\ z \end{bmatrix} = \begin{bmatrix} b_1 \\ b_2 \\ b_3 \end{bmatrix}$$

The augmented matrix of the system is

$$[A : B] = \begin{bmatrix} a_{11} & a_{12} & a_{13} & b_1 \\ a_{21} & a_{22} & a_{23} & b_2 \\ a_{31} & a_{32} & a_{33} & b_3 \end{bmatrix}$$

Applying elementary row transformation to augmented matrix to reduce coefficient matrix into unit matrix.

$$[A : B] \xrightarrow[\text{row transformation}]{\text{elementary}} \begin{bmatrix} 1 & 0 & 0 & d_1 \\ 0 & 1 & 0 & d_2 \\ 0 & 0 & 1 & d_3 \end{bmatrix}$$

The corresponding system of equations is
$$x_1 = d_1, x_2 = d_2, x_3 = d_3$$
Hence, the solution is $x_1 = d_1, x_2 = d_2, x_3 = d_3$.

Example 6: Apply Gauss-Jordan method to find the solution of the following system:
$$10x + y + z = 12; 2x + 10y + z = 13; x + y + 5z = 7.$$

Solution: Since the coefficient of x in the last equation is unity, we rewrite the equations by interchanging the first and the last equation.
$$x + y + 5z = 7; 2x + 10y + z = 13; 10x + y + z = 12$$
The matrix form of the above system is
$$AX = B$$

$$\begin{bmatrix} 1 & 1 & 5 \\ 2 & 10 & 1 \\ 10 & 1 & 1 \end{bmatrix} \begin{bmatrix} x \\ y \\ z \end{bmatrix} = \begin{bmatrix} 9 \\ 13 \\ 12 \end{bmatrix}$$

The augmented matrix of the system is

$$[A:B] = \begin{bmatrix} 1 & 1 & 5 & | & 9 \\ 2 & 10 & 1 & | & 13 \\ 10 & 1 & 1 & | & 12 \end{bmatrix}$$

Applying elementary row transformations to the augmented matrix
$$R_2 + (-2)R_1 \text{ and } R_3 + (-10R_1)$$

$$[A:B] \sim \begin{bmatrix} 1 & 1 & 5 & | & 7 \\ 0 & 8 & -9 & | & -1 \\ 0 & -9 & -49 & | & -58 \end{bmatrix}$$

Applying the $R_2/8$

$$\sim \begin{bmatrix} 1 & 1 & 5 & | & 7 \\ 0 & 1 & -\dfrac{9}{8} & | & -\dfrac{1}{8} \\ 0 & -9 & -49 & | & -58 \end{bmatrix}$$

Operating $R_3 + 9R_2$

$$\sim \begin{bmatrix} 1 & 1 & 5 & | & 7 \\ 0 & 1 & -\dfrac{9}{8} & | & -\dfrac{1}{8} \\ 0 & 0 & -\dfrac{473}{8} & | & -\dfrac{473}{8} \end{bmatrix}$$

Operating $\left(-\dfrac{8}{473}\right)R_3$

$$\sim \begin{bmatrix} 1 & 1 & 5 & | & 7 \\ 0 & 1 & -\dfrac{9}{8} & | & -\dfrac{1}{8} \\ 0 & 0 & 1 & | & 1 \end{bmatrix}$$

Operating $R_1 + (-1) R_2$

$$\sim \begin{bmatrix} 1 & 0 & \dfrac{49}{8} & \dfrac{57}{8} \\ 0 & 1 & -\dfrac{9}{8} & -\dfrac{1}{8} \\ 0 & 0 & 1 & 1 \end{bmatrix}$$

Operating $R_2 + \left(\dfrac{9}{8}\right) R_3, R_1 + \left(\dfrac{-49}{8}\right) R_3$

$$\sim \begin{bmatrix} 1 & 0 & 0 \\ 0 & 1 & 0 \\ 0 & 0 & 1 \end{bmatrix} \begin{bmatrix} 1 \\ 1 \\ 1 \end{bmatrix}$$

$\therefore x = 1, y = 1, z = 1$.

Example 7: Solve the following system of equations:
$$x - 2y = -4; -5y + z = -9; 4x - 3z = -10.$$

Solution: The matrix form of the system is
$$AX = B$$

$$\begin{bmatrix} 1 & -2 & 0 \\ 0 & -5 & 1 \\ 4 & 0 & -3 \end{bmatrix} \begin{bmatrix} x \\ y \\ z \end{bmatrix} = \begin{bmatrix} -4 \\ -9 \\ -10 \end{bmatrix}$$

The augmented matrix of the system is

$$[A : B] = \begin{bmatrix} 1 & -2 & 0 & -4 \\ 0 & -5 & 1 & -9 \\ 4 & 0 & -3 & -10 \end{bmatrix}$$

Operating $R_3 - 4R_1$

$$\sim \begin{bmatrix} 1 & -2 & 0 & -4 \\ 0 & -5 & 1 & -9 \\ 0 & 8 & -3 & 6 \end{bmatrix}$$

Operating $\left(-\dfrac{1}{5}\right) R_2$

$$\sim \begin{bmatrix} 1 & -2 & 0 & -4 \\ 0 & 1 & -\dfrac{1}{5} & \dfrac{9}{5} \\ 0 & 8 & -3 & 6 \end{bmatrix}$$

Operating $R_1 + 2R_2, R_3 - 8R_2$

$$\sim \begin{bmatrix} 1 & 0 & -\dfrac{2}{5} & -\dfrac{2}{5} \\ 0 & 1 & -\dfrac{1}{5} & \dfrac{9}{5} \\ 0 & 0 & -\dfrac{7}{5} & -\dfrac{42}{5} \end{bmatrix}$$

Operating $\left(-\dfrac{5}{7}\right)R_3$

$$\sim \begin{bmatrix} 1 & 0 & -\dfrac{2}{5} & -\dfrac{2}{5} \\ 0 & 1 & -\dfrac{1}{5} & \dfrac{9}{5} \\ 0 & 0 & 1 & 6 \end{bmatrix}$$

Operating $R_1 + \left(\dfrac{2}{5}\right)R_3, R_2 + \left(\dfrac{1}{5}\right)R_3$

$$\sim \begin{bmatrix} 1 & 0 & 0 & 2 \\ 0 & 1 & 0 & 3 \\ 0 & 0 & 1 & 6 \end{bmatrix}$$

$\therefore x = 2, y = 3, z = 6.$

EXERCISE 9.2

Solve the following system(s) of linear equations by Gauss-Jordan elimination method:

1. $2x - 6y + 8z = 24; 5x + 4y - 3z = 2; 3x + y + 2z = 16.$ [Ans. $x = 1, y = 3, z = 5$]
2. $2x_1 + 8x_2 + 2x_3 = 14; 6x_1 + 6x_2 - x_3 = 13; 2x_1 - x_2 + 2x_3 = 5.$

         [Ans. $x_1 = 10/7, x_2 = 1, x_3 = 11/7$]

3. $x + 2y + z = 3; 2x + 3y + 3z = 10; 3x - y + 2z = 13.$ [Ans. $x = 2, y = -1, z = 3$]
4. $x + y + z - w = 2; x - y - z + 2w = 0; 4x + 4y + z + w = 11; 2x + y + 2z - 2w = 2.$

         [Ans. $x = 1, y = 2, z = -1, w = 0$]

5. $10x_1 + x_2 - x_3 = 11.19; x_1 + 10x_2 + x_3 = 20.08; -x_1 + x_2 + 10x_3 = 35.61.$

         [Ans. $x_1 = 1.32, x_2 = 1.5219, x_3 = 3.5408$]

9.6 TRIANGULARIZATION METHOD

This method is also known as the LU decomposition method or the method of factorization.

This method is based upon the fact that every square matrix A can be decomposed as the product of a lower triangular matrix (L) and an upper triangular matrix (U). We will explain this method in the case of three equations in three unknowns.

Consider the system of equations

$$a_{11}x_1 + a_{12}x_2 + a_{13}x_3 = b_1$$
$$a_{21}x_1 + a_{22}x_2 + a_{23}x_3 = b_2$$
$$a_{31}x_1 + a_{32}x_2 + a_{33}x_3 = b_3$$

The matrix form of the system is

$$AX = B \tag{9.9}$$

where
$$A = \begin{bmatrix} a_{11} & a_{12} & a_{13} \\ a_{21} & a_{22} & a_{23} \\ a_{31} & a_{32} & a_{33} \end{bmatrix}, x = \begin{bmatrix} x_1 \\ x_2 \\ x_3 \end{bmatrix} \text{ and } B = \begin{bmatrix} b_1 \\ b_2 \\ b_3 \end{bmatrix}$$

Let $A = LU$

where

$$L = \begin{bmatrix} l_{11} & 0 & 0 \\ l_{21} & l_{22} & 0 \\ l_{31} & l_{32} & l_{33} \end{bmatrix}, U = \begin{bmatrix} u_{11} & u_{12} & u_{13} \\ 0 & u_{22} & u_{23} \\ 0 & 0 & u_{33} \end{bmatrix}$$

Now

$$LU = \begin{bmatrix} l_{11}u_{11} & l_{11}u_{12} & l_{11}u_{13} \\ l_{21}u_{11} & l_{21}u_{12} + l_{22}u_{22} & l_{21}u_{13} + l_{22}u_{23} \\ l_{31}u_{11} & l_{31}u_{12} + l_{32}u_{22} & l_{31}u_{13} + l_{32}u_{23} + l_{33}u_{33} \end{bmatrix}$$

So, putting the values of A, L and U in equation $A = LU$

$$\begin{bmatrix} a_{11} & a_{12} & a_{13} \\ a_{21} & a_{22} & a_{23} \\ a_{31} & a_{32} & a_{33} \end{bmatrix} = \begin{bmatrix} l_{11}u_{11} & l_{11}u_{12} & l_{11}u_{13} \\ l_{21}u_{11} & l_{21}u_{12} + l_{22}u_{22} & l_{21}u_{13} + l_{22}u_{23} \\ l_{31}u_{11} & l_{31}u_{12} + l_{32}u_{22} & l_{31}u_{13} + l_{32}u_{23} + l_{33}u_{33} \end{bmatrix}$$

Equating corresponding coefficients, we get the value of all the unknowns, i.e. l_{11}, l_{21}, l_{31}, l_{22}, l_{32}, l_{33}, u_{11}, u_{12}, u_{13}, u_{22}, u_{23} and u_{33}.

To produce a unique solution, it is convenient to choose either $u_{ii} = 1$ or $l_{ii} = 1$; $1 \leq i \leq 3$.

1. When we select $l_{ii} = 1$, the method is called the Doolittle's method.
2. When we select $u_{ii} = 1$, the method is called the Crout's method.

The given system of equation is

$$AX = B$$

Putting $A = LU$ in above equations, we get

$$LUX = B \tag{9.10}$$

Let

$$UX = Y \tag{9.11}$$

Then Eq. (9.10) becomes

$$LY = B \tag{9.12}$$

Equation (9.12) is solved for Y. Then by putting y in Eq. (9.11), X is obtained. The limitation of this method is that the method fails if any of the diagonal elements l_{ii} or u_{ii} is zero.

Example 8: Solve the following equations by Crout's method:

$$x_1 + 2x_2 + 3x_3 = 14$$
$$2x_1 + 5x_2 + 2x_3 = 18$$
$$3x_1 + x_2 + 5x_3 = 20$$

Solution: The given system is written in the form $AX = B$, where

$$A = \begin{bmatrix} 1 & 2 & 3 \\ 2 & 5 & 2 \\ 3 & 1 & 5 \end{bmatrix}, X = \begin{bmatrix} x_1 \\ x_2 \\ x_3 \end{bmatrix} \text{ and } B = \begin{bmatrix} 14 \\ 18 \\ 20 \end{bmatrix}$$

Now $\qquad A = LU$

$$\therefore \quad \begin{bmatrix} 1 & 2 & 3 \\ 2 & 5 & 2 \\ 3 & 1 & 5 \end{bmatrix} = \begin{bmatrix} l_{11} & 0 & 0 \\ l_{21} & l_{22} & 0 \\ l_{31} & l_{32} & l_{33} \end{bmatrix} \begin{bmatrix} 1 & u_{12} & u_{13} \\ 0 & 1 & u_{23} \\ 0 & 0 & u_{33} \end{bmatrix}$$

$$= \begin{bmatrix} l_{11} & l_{11}u_{12} & l_{11}u_{13} \\ l_{21} & l_{21}u_{12} + l_{22} & l_{21}u_{13} + l_{22}u_{23} \\ l_{31} & l_{31}u_{12} + l_{32} & l_{31}u_{13} + l_{32}u_{23} + l_{33} \end{bmatrix}$$

Equating, we get

$$l_{11} = 1, l_{21} = 2, l_{31} = 3$$
$$l_{11}u_{12} = 2 \Rightarrow u_{12} = 2$$
$$l_{11}u_{13} = 3 \Rightarrow u_{13} = 3$$
$$l_{21}u_{12} + l_{22} = 5 \Rightarrow l_{22} = 1$$
$$l_{31}u_{12} + l_{32} = 1 \Rightarrow l_{32} = -5$$
$$l_{21}u_{13} + l_{22}u_{23} = 2 \Rightarrow u_{23} = -4$$
$$l_{31}u_{13} + l_{32}u_{23} + l_{33} = 5 \Rightarrow l_{33} = -24$$

Thus, we get $\quad A = LU = \begin{bmatrix} 1 & 0 & 0 \\ 2 & 1 & 0 \\ 3 & -5 & -24 \end{bmatrix} \begin{bmatrix} 1 & 2 & 3 \\ 0 & 1 & -4 \\ 0 & 0 & 1 \end{bmatrix}$

The given system is $AX = B$

$\Rightarrow \qquad\qquad LUX = B$ $\qquad\qquad$ (i)

Let $\qquad\qquad UX = Y$ $\qquad\qquad$ (ii)

$\therefore \qquad\qquad LY = B$ $\qquad\qquad$ (iii)

Then, we have

$$\begin{bmatrix} 1 & 0 & 0 \\ 2 & 1 & 0 \\ 3 & -5 & -24 \end{bmatrix} \begin{bmatrix} y_1 \\ y_2 \\ y_3 \end{bmatrix} = \begin{bmatrix} 14 \\ 18 \\ 20 \end{bmatrix}$$

which gives

$$y_1 = 14$$
$$2y_1 + y_2 = 18$$
$$3y_1 - 5y_2 - 24y_3 = 20$$

$\Rightarrow y_1 = 14, y_2 = -10, y_3 = 3$

Now $\qquad\qquad UX = Y$

$$\begin{bmatrix} 1 & 2 & 3 \\ 0 & 1 & -4 \\ 0 & 0 & 1 \end{bmatrix} \begin{bmatrix} x_1 \\ x_2 \\ x_3 \end{bmatrix} = \begin{bmatrix} 14 \\ -10 \\ 3 \end{bmatrix}$$

which gives $x_1 + 2x_2 + 3x_3 = 14$

$$x_2 - 4x_3 = -10$$
$$x_3 = 3$$

By back substitution $x_1 = 1, x_2 = 2, x_3 = 3$.

Example 9: Solve the following system of equations by the LU factorization method:

$$2x + 3y + z = 9$$
$$x + 2y + 3z = 6$$
$$3x + y + 2z = 8.$$

Solution: The given system of equation can be written in the form $AX = B$,

$$\begin{bmatrix} 2 & 3 & 1 \\ 1 & 2 & 3 \\ 3 & 1 & 2 \end{bmatrix} \begin{bmatrix} x \\ y \\ z \end{bmatrix} = \begin{bmatrix} 9 \\ 6 \\ 8 \end{bmatrix}$$

where

$$A = \begin{bmatrix} 2 & 3 & 1 \\ 1 & 2 & 3 \\ 3 & 1 & 2 \end{bmatrix}, \quad X = \begin{bmatrix} x \\ y \\ z \end{bmatrix}, \quad B = \begin{bmatrix} 9 \\ 6 \\ 8 \end{bmatrix}$$

Now
$$A = LU$$

$$\therefore \quad \begin{bmatrix} 2 & 3 & 1 \\ 1 & 2 & 3 \\ 3 & 1 & 2 \end{bmatrix} = \begin{bmatrix} 1 & 0 & 0 \\ l_{21} & l_{22} & 0 \\ l_{31} & l_{32} & l_{33} \end{bmatrix} \begin{bmatrix} u_{11} & u_{12} & u_{13} \\ 0 & u_{22} & u_{23} \\ 0 & 0 & u_{33} \end{bmatrix}$$

$$= \begin{bmatrix} u_{11} & u_{12} & u_{13} \\ l_{21}u_{11} & l_{21}u_{12} + u_{22} & l_{21}u_{13} + u_{23} \\ l_{31}u_{11} & l_{31}u_{12} + l_{32}u_{22} & l_{31}u_{13} + l_{32}u_{23} + u_{33} \end{bmatrix}$$

Equating, we get

$$u_{11} = 2, u_{12} = 3, u_{13} = 1$$
$$l_{21}u_{11} = 1 \Rightarrow l_{21} = 1/2$$
$$l_{31}u_{11} = 3 \Rightarrow l_{31} = 3/2$$
$$l_{21}u_{12} + u_{22} = 2 \Rightarrow u_{22} = 1/2$$
$$l_{31}u_{12} + l_{32}u_{22} = 1 \Rightarrow l_{32} = -7$$
$$l_{21}u_{13} + u_{23} = 3 \Rightarrow u_{23} = 5/2$$
$$l_{31}u_{13} + l_{32}u_{23} + u_{33} = 2 \Rightarrow u_{33} = 18$$

Thus, we get

$$A = LU = \begin{bmatrix} 1 & 0 & 0 \\ 1/2 & 1 & 0 \\ 3/2 & -7 & 1 \end{bmatrix} \begin{bmatrix} 2 & 3 & 1 \\ 0 & 1/2 & 5/2 \\ 0 & 0 & 18 \end{bmatrix}$$

The given system is

$$AX = B \quad \Rightarrow LUX = B \tag{i}$$

Let $UX = Y$, then Eq. (i) becomes

$$LY = B \qquad \text{(ii)}$$

$$\begin{bmatrix} 1 & 0 & 0 \\ 1/2 & 1 & 0 \\ 3/2 & -7 & 1 \end{bmatrix} \begin{bmatrix} y_1 \\ y_2 \\ y_3 \end{bmatrix} = \begin{bmatrix} 9 \\ 6 \\ 8 \end{bmatrix}$$

which gives

$$y_1 = 9$$

$$\frac{1}{2}y_1 + y_2 = 6$$

$$\frac{3}{2}y_1 - 7y_2 + y_3 = 8$$

$\Rightarrow y_1 = 9, y_2 = 3/2, y_3 = 5$

Now, $UX = Y$

$$\begin{bmatrix} 2 & 3 & 1 \\ 0 & 1/2 & 5/2 \\ 0 & 0 & 18 \end{bmatrix} \begin{bmatrix} x \\ y \\ z \end{bmatrix} = \begin{bmatrix} 9 \\ 3/2 \\ 5 \end{bmatrix}$$

which gives $\quad 2x + 3y + z = 9$

$$\frac{1}{2}y + \frac{5}{2}z = \frac{3}{2}$$

$$18z = 5$$

By back substitution, we get $x = 35/18$, $y = 29/18$, $z = 5/18$.

EXERCISE 9.3

Solve the system of following equations by Crout's method:

1. $3x - y + 2z = 12$; $x + 2y + 3z = 11$; $2x - 2y - z = 2$.   [Ans. $x = 3, y = 1, z = 2$]
2. $x_1 + x_2 + x_3 = 1$; $3x_1 + x_2 - 3x_3 = 5$; $x_1 - 2x_2 - 5x_3 = 10$.   [Ans. $x_1 = 6, x_2 = -7, x_3 = 2$]
3. $2x + 4y + z = 5$; $5x + 4y + 3z = 8$; $4x + 8y + z = 9$.   [Ans. $x = 1/2, y = 3/4, z = 1$]
4. $4x_1 + 3x_2 + x_3 - x_4 = 14$; $2x_1 + 5x_2 + 2x_3 + x_4 = 17$; $x_1 + 4x_2 + 4x_3 + 6x_4 = 20$, $3x_1 + x_2 - x_3 + 5x_4 = 12$.   [Ans. $x_1 = 2, x_2 = 2, x_3 = 1, x_4 = 1$]

Solve the system of following equations by Doolittle's method:

5. $2x_1 + x_2 + x_3 = 2$; $x_1 + 3x_2 + 2x_3 = 2$; $3x_1 + x_2 + 2x_3 = 2$ [Ans. $x_1 = 1, x_2 = 1, x_3 = -1$]
6. $3x + 2y + 7z = 4$; $2x + 3y + z = 5$; $3x + 4y + z = 7$.   [Ans. $x = 7/8, y = 9/8, z = -1/8$]
7. $4x + y - z = 13$; $3x + 5y + 2z = 21$; $2x + y + 6z = 14$.   [Ans. $x = 3.8421, y = 1.5789, z = 0.7895$]
8. $10x - 7y + 3z + 56u = 6$; $-6x + 8y - z - 4u = 5$; $3x + y + 4z + 11u = 2$.   [Ans. $x = 5, y = 4, z = -7, u = 1$]

9.7 ILL CONDITIONED SYSTEM OF EQUATIONS

A system of linear equations $AX = B$ is termed as ill conditioned or unstable if it is highly sensitive to small changes in their coefficients A and constants B. In other words, small changes in the coefficients and constants result a large change in the solution X.

On the other hand, if small changes in the coefficients of the equations result in small changes in the value of unknowns, then the system is said to be well-conditioned.

For example, consider the following systems

$$x - y = 1, \quad x - 1.00001y = 0$$

has the solution (100001, 100000).

Again if we take the new system which is almost identical to the above systems as

$$x - y = 1, \quad x - 0.99999y = 0$$

we get the solution $(-99999, -100000)$.

We see that the two solutions are very much different. Therefore the system is ill conditioned.

9.7.1 Ill Conditioned Matrix

A matrix is said to be ill conditioned if it is very sensitive to small changes. We see that such type of problems arise in some power systems. The power flow problem is said to be ill conditioned if the Jacobian matrix is ill conditioned.

For example
$$A = \begin{bmatrix} -73 & 78 & 24 \\ 92 & 66 & 25 \\ -80 & 37 & 10 \end{bmatrix}$$

is ill conditioned matrix because $|A| = 1$, while

$$\begin{bmatrix} -73 & 78.001 & 24 \\ 92 & 66 & 25 \\ -80 & 37 & 10 \end{bmatrix} = -1.92 \text{ is not an ill conditioned matrix.}$$

In solving simultaneous equations, we come across with inherent and induced instabilities. The inherent instability of a system is the property of the given problem and occurs due to problem being ill conditioned. By proper reformation of the problem, it can be avoided. Incorrect choice of the method is responsible for induced instability.

9.7.2 Improve Accuracy of ill Conditioned System

Consider the system of three unknowns

$$\left. \begin{array}{l} a_{11}x_1 + a_{12}x_2 + a_{13}x_3 = b_1 \\ a_{21}x_1 + a_{22}x_2 + a_{23}x_3 = b_2 \\ a_{31}x_1 + a_{32}x_2 + a_{33}x_3 = b_3 \end{array} \right] \tag{9.13}$$

Substituting x_1', x_2', x_3' for x_1, x_2, x_3 respectively in the above system, we get new values b_1', b_2', b_3' of b_1, b_2, b_3 respectively. So new system becomes

$$\left.\begin{array}{l} a_{11}x_1' + a_{12}x_2' + a_{13}x_3' = b_1' \\ a_{21}x_1' + a_{22}x_2' + a_{23}x_3' = b_2' \\ a_{31}x_1' + a_{32}x_2' + a_{33}x_3' = b_3' \end{array}\right] \qquad (9.14)$$

Subtracting each in Eq. (9.14) from the corresponding equations in Eq. (9.13), we get

$$\left.\begin{array}{l} a_{11}x_1^e + a_{12}x_2^e + a_{13}x_3^e = k_1 \\ a_{21}x_1^e + a_{22}x_2^e + a_{23}x_3^e = k_2 \\ a_{31}x_1^e + a_{32}x_2^e + a_{33}x_3^e = k_3 \end{array}\right] \qquad (9.15)$$

where $k_1 = b_1 - b_1', k_2 = b_2 - b_2', k_3 = b_3 - b_3', x_1^e = x_1 - x_1', x_2^e = x_2 - x_2', x_3^e = x_3 - x_3'$. We now solve the system in Eq. (9.15) for x_1^e, x_2^e, x_3^e so that $x_1 = x_1' + x_1^e, x_2 = x_2' + x_2^e, x_3 = x_3' + x_3^e$. It will be a better approximation for x_1, x_2, x_3. We can repeat the process for improving the accuracy.

Example 10: A system is described by the following simultaneous equations:
$$2x + y = 2; \quad 2x + 1.01y = 2.01$$
Discuss, whether the system is ill conditioned system.

Solution: On solving given equations, we get
$$x = 0.5, y = 1$$
Now, we make slight changes in the given system of equations. The new system becomes
$$2x + y = 2; \quad 2x + 1.02y = 2.01$$
Here, we get the solution as
$$x = 0.75, \quad y = 0.5$$
Hence the given system is ill conditioned.

Example 11: Consider the following system of equations
$$100x - 200y = 100; \quad -200x + 401y = -100$$
Determine, whether the given system is ill conditioned or not.

Solution: On solving the given equations, we get
$$x = 201, y = 100$$
Now, we make small changes in the given system of equations. The new system becomes
$$100x - 200y = 100; \quad -200.01x + 401y = -100$$
We get the solution
$$x = 205.102, \quad y = 102.051$$
Hence the given system is ill conditioned.

Example 12: Show that the following system of linear equation is ill conditioned
$$6x - 9y = 1; \quad 4x + 6y = 0.5$$

Solution: On solving the above system of equations, we get
$$x = 0.1458, y = -0.0139$$

Now, we make small changes in the given system of equations. The new system becomes

$$6x - 9y = 1; \quad 4x + 6y = 0.52$$

We get the solution

$$x = 0.1483, \quad y = -0.0122$$

Hence the given system is ill conditioned.

Example 13: An approximate solution of the equations $x + 4y + 7z = 5$; $2x + 5y + 8z = 7$; $3x + 6y + 9.1z = 9.1$ is given by $x = 1.8, y = -1.2, z = 1$. Improve this solution by using the iterative method.

Solution: The approximate solution of the system is

$$x' = 1.8, y' = -1.2, z' = 1$$

Substituting x', y', z' for x, y, z respectively in the given system, we get new values k_1, k_2, k_3. New system becomes

$$x' + 4y' + 7z' = k_1$$
$$2x' + 5y' + 8z' = k_2$$
$$3x' + 6y' + 9.1z' = k_3$$

$$\Rightarrow \quad k_1 = 1.8 + 4 \times (-1.2) + 7 \times 1 = 4$$
$$k_2 = 2 \times 1.8 + 5 \times (-1.2) + 8 \times 1 = 5.6$$
$$k_3 = 3 \times 1.8 + 6 \times (-1.2) + 9.1 \times 1 = 7.3$$

Subtracting each equation of this system from corresponding equation of the system, we get

$$(x - x') + 4(y - y') + 7(z - z') = 5 - 4$$
$$x^e + 4y^e + 7z^e = 1$$

Similarly
$$2x^e + 5y^e + 8z^e = 1.4$$
$$3x^e + 6y^e + 9.1z^e = 1.8$$

where
$$x^e = x - x'; y^e = y - y'; z^e = z - z'$$
$$\Rightarrow \quad x^e = x - 1.8; y^e = y + 1.2; z^e = z - 1$$

Solving new system we get

$$x^e = 0.2, y^e = 0.2, z^e = 0$$

This gives $x = 2, y = -1.0, z = 1$ which is better solution and incidently it is the exact solution.

Example 14: An approximate solution of the system $2x + 2y - z = 6$; $x + y + 2z = 8$; $-x + 3y + 2z = 4$ is given by $x = 2.8, y = 1, z = 1.8$. Improve this solution by using the iterative method.

Solution: The approximate solution of the system is

$$x' = 2.8, y' = 1, z' = 1.8$$

Substituting x', y', z' for x, y, z respectively in the given system, we get new values k_1, k_2, k_3. New system becomes

$$2x' + 2y' - z' = k_1$$
$$x' + y' + 2z' = k_2$$
$$-x' + 3y' + 2z' = k_3$$

$$\Rightarrow \quad k_1 = 2 \times 2.8 + 2 \times 1 - 1.8 = 5.8$$

$$k_2 = 2.8 + 1 + 2 \times 1.8 = 7.4$$
$$k_3 = -2.8 + 3 \times 1 + 2 \times 1.8 = 3.8$$

Subtracting each equation of this system from corresponding equation of the system, we get

$$2(x - x') + 2(y - y') - (z - z') = 6 - 5.8$$
$$2x^e + 2y^e - z^e = 0.2$$

Similarly

$$x^e + y^e + 2z^e = 0.6 \tag{i}$$
$$-x^e + 3y^e + 2z^e = 0.2$$

where

$$x^e = x - 2.8; \ y^e = y - 1; \ z^e = z - 1.8$$

Solving the new system in Eq. (i), we get

$$x^e = 0.2, \ y^e = 0, \ z^e = 0.2$$

This gives $x = 3$, $y = 1$, $z = 2$ which incidently is the exact solution.

EXERCISE 9.4

1. Show that the following system of linear equation is ill conditioned
$$7x - 10y = 1; \ \ 5x + 7y = 0.7 \qquad \text{[Ans. Yes]}$$
2. Establish whether the system of equations
$$10x + 8y + 9z + 6w = 33; \ \ 6x + 7y + 5z + 5w = 23;$$
$$8x + 10y + 7z + 7w = 3; \ \ 9x + 7y + 10z + 5w = 31$$
   is well conditioned or not?
3. A system is described by the following simultaneous equations
$$x - 2y = -2; \ \ 0.45x - 0.91y = -1$$
   Discuss, whether the system is ill conditioned system. \qquad [Ans. Yes]
4. Show that the following system of linear equations is ill conditioned
$$7x - 10y = 1; \ \ 5x + 7y = 0.7 \qquad \text{[Ans. Yes]}$$
5. Establish whether the system $1.01x + 2y = 2.01; \ x + 2y = 2$ is well conditioned or not. \qquad [Ans. No]

9.8 VECTOR NORM

A function $\|\cdot\| \cdot \mathbb{R}^n \to \mathbb{R}$ is called a vector norm if it has the following properties:

i. $\|x\| \geq 0$ for any vector $x \in \mathbb{R}^n$, and $\|x\| = 0$ if and only if $x = 0$
ii. $\|\alpha x\| = |\alpha| \, \|x\|$ for any vector $x \in \mathbb{R}^n$ and any scaler $\alpha \in \mathbb{R}$
iii. $\|x + y\| \leq \|x\| + \|y\|$ for any vector $x, y \in \mathbb{R}^n$.

The property (iii) is called the triangle inequality. It should be noted that when $n = 1$, the absolute value function is a vector norm.

The most commonly used vector norms belong to the family of p norms, that is given by

$$\|x\|_p = \left(\sum_{i=1}^{n} |x_i|^p \right)^{\frac{1}{p}}.$$

It can be shown that for any $p > 0, \|\cdot\|_p$ defines a vector norm. The below p-norms are of particular interest.

a. $p = 1$: The absolute norm (l_1-norm)

$$\|x\|_1 = \sum_{i=1}^{n} |x_i|$$

b. $p = 2$: The Euclidean norm (l_2-norm)

$$\|x\|_2 = \left(\sum_{i=1}^{n} |x_i|^2 \right)^{\frac{1}{2}} = (x^T x)^{\frac{1}{2}}$$

c. $p = \infty$: Maximum norm (l_∞-norm)

$$\|x\|_\infty = \max_{1 \le i \le n} |x_i|$$

9.9 MATRIX NORM

A matrix norm of a matrix $\|A\|$ is a mapping from $\mathbb{R}^{n \times n} \to \mathbb{R}$ with the following properties:

   i. $\|A\| > 0$ if $A \ne 0$

   ii. $\|\alpha A\| = |\alpha| \, \|A\|$, for any $\alpha \in \mathbb{R}$

   iii. $\|A + B\| \le \|A\| + \|B\|$ (triangular inequality) for any matrix $A, B \in \mathbb{R}^{n \times n}$.

Matrix norm also satisfies

   iv. $\|AB\| \le \|A\| \cdot \|B\|$

   v. $\|Ax\| \le \|A\| \cdot \|x\|$ for any vector x.

The following matrix norm are commonly used

a. Schur or Frobenius norm

$$\|A\|_F = \left[\sum \sum |a_{ij}|^2 \right]^{\frac{1}{2}}$$

b. Maximum absolute row sum norm

$$\|A\|_\infty = \max_i \sum_j |a_{ij}|$$

c. Maximum absolute column sum norm

$$\|A\|_1 = \max_j \sum_i |a_{ij}|$$

d. Spectral norm or Hilbert norm

$$\|A\|_2 = (\lambda)^{\frac{1}{2}} \text{ where } \lambda = \rho(A^*T)$$

If A is Hermitain or real and symmetric then

$$\lambda = \rho(A^2) = \rho^2(A)$$

so that $\|A\|_2 = \rho(A)$

9.10 CONDITION NUMBER

Condition number of a matrix indicates if the solution of linear system is sensitive to small changes. It bring out that this sensitivity can be measured by condition number that is defined as

$$\text{Cond (A)} = k(A) = \|A^{-1}\| \cdot \|A\|$$

If Cond (A) is large, then small change in A or b produces large relative changes in x, and the system of equations $Ax = b$ is ill conditioned. If cond (A) = 1, the system $Ax = b$ is well conditioned

If $\|\cdot\|$ is the spectral norm, then

$$\text{Cond } (A) = k(A) = \|A\|_2 \cdot \|A^{-1}\|_2 = \sqrt{\frac{\lambda}{\mu}}$$

where λ and μ are the largest and smallest eigen values in modulus of A^*A. If A is Hermitain or real and symmetric then we have

$$\text{Cond } (A) = k(A) = \frac{\lambda^*}{\mu^*}$$

where λ^* and μ^* are the smallest and largest values in moduius of A.

9.10.1 Properties of Condition Number

We have the following properties of condition number:
1. For any matrix A, Cond $(A) \geq 1$
2. For identity matrix Cond $(I) = 1$
3. For any matrix A and scaler α, Cond $(\alpha A) = $ Cond (A)
4. For any diagonal matrix $D = $ diagonal (d_i), Cond $(D) = \dfrac{\max |d_i|}{\min |d_i|}$

Example 15: Determine the condition number of the matrix

$$A = \begin{bmatrix} 1 & 7 & -4 \\ 4 & -4 & 9 \\ 12 & -1 & 3 \end{bmatrix}$$

using the maximum absolute row sum norm.

Solution: We have $\quad |A| = 493$

$$\text{Cofactor of matrix } A = \begin{bmatrix} -3 & 96 & 44 \\ -17 & 51 & 85 \\ 47 & -25 & -32 \end{bmatrix}$$

adj A = Transpose of cofactor of matrix A

$$= \begin{bmatrix} -3 & -17 & 47 \\ 96 & 51 & -25 \\ 44 & 85 & -32 \end{bmatrix}$$

$$A^- = \frac{\text{adj } A}{|A|} = \frac{1}{493} \begin{bmatrix} -3 & -17 & 47 \\ 96 & 51 & -25 \\ 44 & 85 & -32 \end{bmatrix}$$

$\|A\|_\infty = $ max. absolute row sum norm for A

$\qquad = $ max. $(12, 17, 16) = 17$

$$\|A^{-1}\|_{\infty} = \text{max. absolute row sum norm for } A^{-1}$$

$$= \max.\left(\frac{3}{493}+\frac{17}{493}+\frac{47}{493}\right),\left(\frac{96}{493}+\frac{51}{493}+\frac{25}{493}\right),\left(\frac{44}{493}+\frac{85}{493}+\frac{32}{493}\right)$$

$$= \max.\left(\frac{67}{493},\frac{172}{493},\frac{161}{493}\right)=\frac{172}{493}$$

$\therefore$ $\text{Cond }(A) = \|A\|_{\infty}\cdot\|A^{-1}\|_{\infty} = 17\times\dfrac{172}{493}=5.93$

Example 16: Let $A(\alpha)=\begin{bmatrix}0.1\alpha & 0.1\alpha \\ 1.0 & 1.5\end{bmatrix}$

Determine α such that Cond $[A(\alpha)]$ is minimized. Use the maximum norm.

Solution: We have $A(\alpha) = \begin{bmatrix}0.1\alpha & 0.1\alpha \\ 1.0 & 1.5\end{bmatrix}$

and its $A^{-1}(\alpha) = \dfrac{1}{0.05\alpha}\begin{bmatrix}1.5 & -0.1\alpha \\ -1.0 & 0.1\alpha\end{bmatrix}$

Using maximum norm, we get

$$\|A(\alpha)\| = \max.\,[0.2|\alpha|,\,2.5]$$

$$\|A^{-1}(\alpha)\| = \max.\left[\frac{\dfrac{1.5}{0.05}+\dfrac{0.1|\alpha|}{0.05}}{|\alpha|},\ \frac{\dfrac{1.0}{0.05}+\dfrac{0.1|\alpha|}{0.05}}{|\alpha|}\right]$$

$$= \max.\left[\frac{2|\alpha|+30}{|\alpha|},\frac{2|\alpha|+20}{|\alpha|}\right]=\frac{2|\alpha|+30}{|\alpha|}$$

We have   Cond $[A(\alpha)] = \dfrac{1}{|\alpha|}\,[2|\alpha|+30]\,[\max.\,\{0.2|\alpha|,\,2.5\}]$

Now we want to determine α such that Cond $(A(\alpha)$ is minimum. We have

$$\text{Cond }[A(\alpha)] = \max.\left[0.4|\alpha|+6,\,5+\frac{75}{|\alpha|}\right]=\text{minimum} \qquad\qquad \text{(i)}$$

Select α such that

$$0.4|\alpha|+6 = 5+\frac{75}{|\alpha|}$$

which gives $|\alpha| = 12.5$.

From Eq. (i), the minimum value of Cond $[A(\alpha)] = 11$.

Example 17: Determine the condition number of matrix

$$A = \begin{bmatrix}1 & 4 & 9 \\ 4 & 9 & 16 \\ 9 & 16 & 25\end{bmatrix}$$

using the (i) maximum absolute row sum norm (ii) spectral norm.

Solution: i. We have $|A| = -8$

$$\text{Cofactor of matrix } A = \begin{bmatrix} -31 & 44 & -17 \\ 44 & -56 & 20 \\ -17 & 20 & -7 \end{bmatrix}$$

adj A = Transpose of cofactor matrix A

$$= \begin{bmatrix} -31 & 44 & -17 \\ 44 & -56 & 20 \\ -17 & 20 & -7 \end{bmatrix}$$

$$A^{-1} = \frac{\text{adj } A}{|A|} = -\frac{1}{8} \begin{bmatrix} -31 & 44 & -17 \\ 44 & -56 & 20 \\ -17 & 20 & -7 \end{bmatrix}$$

$\|A\|_\infty$ = max. absolute row sum norm of A
$\qquad$ = max. $(14, 29, 50) = 50$
$\|A^{-1}\|_\infty$ = max. absolute row sum norm of A^{-1}

$$= \text{max.} \left[\left(\frac{31}{8} + \frac{44}{8} + \frac{17}{8} \right), \left(\frac{44}{8} + \frac{56}{8} + \frac{20}{8} \right), \left(\frac{17}{8} + \frac{20}{8} + \frac{7}{8} \right) \right]$$

$$= \text{max.} \left[\frac{92}{8}, 15, \frac{44}{8} \right] = 15$$

$\therefore$ Cond $(A) = \|A\|_\infty \cdot \|A\|_\infty^{-1} = 50 \times 15 = 750$.

ii. The given matrix is real and symmetric. So, Cond $(A) = \lambda^*/\mu^*$, where λ^* and μ^* are the largest and smallest eigen values in modulus of A.
Characteristic equation is

$$|A - \lambda I| = \begin{vmatrix} 1-\lambda & 4 & 9 \\ 4 & 9-\lambda & 16 \\ 9 & 16 & 25-\lambda \end{vmatrix} = 0$$

$$\Rightarrow (1-\lambda)[(9-\lambda)(25-\lambda) - 256] - 4[(100 - 4\lambda) - 144]$$
$$+ 9[64 - 81 + 9\lambda] = 0$$
$$\Rightarrow -\lambda^3 + 35\lambda^2 + 94\lambda - 8 = 0$$

Now we find root with the help of Newton-Raphson method.
Consider $\qquad f(\lambda) = -\lambda^3 + 35\lambda^2 + 94\lambda - 8$
then $\qquad\qquad \lambda(0) = -8$
and $\qquad\qquad \lambda(0.1) = -0.001 + 0.35 + 9.4 - 8 = 1.749$
So, the root lie between $(0, 0.1)$.

Now $\qquad\qquad \lambda_{n+1} = \lambda_n - \dfrac{f(\lambda_n)}{f'(\lambda_n)} = \lambda_n - \dfrac{(-\lambda_n^3 + 35\lambda_n^2 + 94\lambda_n - 8)}{-3\lambda_n^2 + 70\lambda_n + 94}$

$$= \lambda_n - \frac{\lambda_n^3 - 35\lambda_n^2 - 94\lambda_n + 8}{3\lambda_n^2 - 70\lambda_n - 94}$$

$$= \frac{(3\lambda_n^3 - 70\lambda_n^2 - 94\lambda_n) - (\lambda_n^3 - 35\lambda_n^2 - 94\lambda_n + 8)}{3\lambda_n^2 - 70\lambda_n - 94}$$

$$= \frac{2\lambda_n^3 - 35\lambda_n^2 - 8}{3\lambda_n^2 - 70\lambda_n - 94}$$

Put $n = 0, 1, 2, \ldots$ and we take $\lambda_0 = 0.1$.

$$\lambda_1 = \frac{2\lambda_0^3 - 35\lambda_0^2 - 8}{3\lambda_0^2 - 70\lambda_0 - 94} = \frac{2(0.1)^3 - 35(0.1)^2 - 8}{3(0.1)^2 - 70 \times 0.1 - 94}$$

$$= 0.082648$$

$$\lambda_2 = 0.082574$$

$$\lambda_3 = 0.082574$$

Therefore the root correct to five decimal places is 0.08257. Dividing the characteristic equation by $(\lambda - 0.08257)$, we get the deflated polynomials as

$$\lambda^2 - 34.91743\lambda - 96.88313 = 0$$

whose roots are 37.50092, −2.58349.

Hence $$\text{Cond }(A) = \frac{37.50092}{0.08257} = 454.17$$

Example 18: Determine the condition number of matrix $A = \begin{bmatrix} 2 & -1 & 1 \\ 1 & 0 & 1 \\ 3 & -1 & 4 \end{bmatrix}$ using the maximum absolute row sum norm and maximum absolute column sum norm.

Solution: i. We have $|A| = 2(0+1) + 1(4-3) + 1(-1-0)$

$$= 2 + 1 - 1 = 2$$

and $$|A^{-1}| = \begin{bmatrix} \dfrac{1}{2} & \dfrac{3}{2} & -\dfrac{1}{2} \\ -\dfrac{1}{2} & \dfrac{5}{2} & -\dfrac{1}{2} \\ -\dfrac{1}{2} & -\dfrac{1}{2} & \dfrac{1}{2} \end{bmatrix}.$$

$$\|A\|_\infty = \text{max. absolute row sum norm for } A$$

$$= \text{max. } (4, 2, 8) = 8$$

$$\|A^{-1}\|_\infty = \text{max. absolute row sum norm for } A^{-1}$$

$$= \text{max. } \left[\left(\frac{1}{2} + \frac{3}{2} + \frac{1}{2} \right), \left(\frac{1}{2} + \frac{5}{2} + \frac{1}{2} \right), \left(\frac{1}{2} + \frac{1}{2} + \frac{1}{2} \right) \right]$$

$$= \text{max. } \left(\frac{5}{2}, \frac{7}{2}, \frac{3}{2} \right) = \frac{7}{2}$$

$$\text{Cond}_\infty(A) = \|A\|_\infty \cdot \|A^{-1}\|_\infty = 8 \times \frac{7}{2} = 28$$

ii. $$\|A\|_1 = \text{max. absolute column sum norm for } A$$

$$= \text{max. } (6, 2, 6) = 6$$

$$\|A^{-1}\|_1 = \max. \left[\left(\frac{1}{2} + \frac{1}{2} + \frac{1}{2} \right), \left(\frac{3}{2} + \frac{5}{2} + \frac{1}{2} \right), \left(\frac{1}{2} + \frac{1}{2} + \frac{1}{2} \right) \right]$$

$$= \max. \left(\frac{3}{2}, \frac{9}{2}, \frac{3}{2} \right) = \frac{9}{2}$$

$$\text{Cond}_1(A) = \|A\|_1 \cdot \|A^{-1}\|_1 = 6 \times \frac{9}{2} = 27$$

EXERCISE 9.5

1. Determine the condition number of the matrix $A = \begin{bmatrix} 1.2969 & 0.8648 \\ 0.2161 & 0.1441 \end{bmatrix}$ using the maximum absolute row sum norm and maximum absolute column sum norm.

[Ans. 3.2707×10^8, 3.2706×10^8]

2. Determine the condition number of the matrix $A = \begin{bmatrix} 5 & 7 & 6 & 5 \\ 7 & 10 & 8 & 7 \\ 6 & 8 & 10 & 9 \\ 5 & 7 & 9 & 10 \end{bmatrix}$ using the maximum absolute row sum norm.

[Ans. 4488]

3. Determine the condition number of the matrix $A = \begin{bmatrix} 1 & 3 & 4 \\ 4 & 5 & 6 \\ -15 & 6 & 9 \end{bmatrix}$ using the maximum absolute row sum norm.

[Ans. 177.340]

4. Find the condition number of the system $\begin{bmatrix} 2.1 & 1.8 \\ 6.2 & 5.3 \end{bmatrix} \begin{bmatrix} x_1 \\ x_2 \end{bmatrix} = \begin{bmatrix} 2.1 \\ 6.2 \end{bmatrix}$.

[Ans. Cond (A) = 2472.73]

9.11 SUCCESSIVE OVER RELAXATION (SOR) METHOD

Consider a system of three equations given below for the sake of simplicity. The method is also applicable even for more number of equations.

Consider the system of equations

$$\left. \begin{array}{l} a_{11}x + a_{12}y + a_{13}z = d_1 \\ a_{21}x + a_{22}y + a_{23}z = d_2 \\ a_{31}x + a_{32}y + a_{33}z = d_3 \end{array} \right] \tag{9.16}$$

Now we define the residuals r_1, r_2, r_3 by the relation

$$\left. \begin{array}{l} r_1 = a_{11}x + a_{12}y + a_{13}z - d_1 \\ r_2 = a_{21}x + a_{22}y + a_{23}z - d_2 \\ r_3 = a_{31}x + a_{32}y + a_{33}z - d_3 \end{array} \right] \tag{9.17}$$

If we can find the values of x, y, z, so that $r_1 = r_2 = r_3 = 0$ then those values of x, y, z are the exact values of the system of Eq. (9.16). If it is not possible to make $r_1 = r_2 = r_3 = 0$, then we make simultaneously the values of r_1, r_2, r_3 to as nearer to zero as possible.

To understand this, we are giving below an operation table from which we can easily know the corresponding changes in r_1, r_2, r_3 when any one of the unknows x, y, z is given an increment 1, other unknowns remaining constant.

Let R_1, R_2, R_3 respectively denote the operations of increasing x, y, z by 1. Thus by R_1, x is increased by 1 in each equation while y, z remain constant. Due to this, r_1 is increased by a_{11}, r_2 by a_{21} and r_3 by a_{31}.

Similarly, by R_2, y is increased by 1 while x and z remain constant and hence r_1 is increased by a_{12}, r_2 by a_{22} and r_3 by a_{32}. Again R_3 increases the value of z by 1 while x and y remain constant and hence r_1 is increased by a_{13}, r_2 by a_{23} and r_3 by a_{33}.

						Operation table					
	x	y	z	r_1	r_2	r_3
R_1	1	0	0	a_{11}	a_{21}	a_{31}
R_2	0	1	0	a_{12}	a_{22}	a_{32}
R_3	0	0	1	a_{13}	a_{23}	a_{33}

Clearly the above table involves the unit matrix I and the transpose of the matrix A namely A', where A is the coefficient matrix of the system of equation

$$A = \begin{bmatrix} a_{11} & a_{12} & a_{13} \\ a_{21} & a_{22} & a_{23} \\ a_{31} & a_{32} & a_{33} \end{bmatrix}$$

9.11.1 Convergence of the Relaxation Method

If the method should converge, the diagonal elements of the coefficient matrix A should be dominant, that is A is diagonally dominant. Referring to the system of Eq. (9.16), the system can be solved by this method successfully only if

$$|a_{11}| \geq |a_{12}| + |a_{13}|$$
$$|a_{22}| \geq |a_{21}| + |a_{23}|$$
$$|a_{33}| \geq |a_{31}| + |a_{32}|$$

where at least once the strict inequality holds.

The relaxation method is explained through the following examples.

Example 19: Solve by relaxation method, the following equations:

$$x + 9y - z = 10$$
$$2x - y + 11z = 20$$
$$10x - 2y + z = 12$$

Solution: We observe that, the above system is not diagonally dominant. So, we rearrange the equations as follows

$$\left. \begin{array}{l} 10x - 2y + z = 12 \\ x + 9y - z = 10 \\ 2x - y + 11z = 20 \end{array} \right] \qquad (i)$$

Now, we can use relaxation method. The residuals are

$$\left.\begin{array}{l} r_1 = 10x - 2y + z - 12 \\ r_2 = x + 9y - z - 10 \\ r_3 = 2x - y + 11z - 20 \end{array}\right]$$ (ii)

Operation table						
	x	y	z	r_1	r_2	r_3
R_1	1	0	0	10	1	2
R_2	0	1	0	-2	9	-1
R_3	0	0	1	1	-1	11

Take initial approximation as $x = 0$, $y = 0$, $z = 0$. Relaxation procedure is given below:

	x	y	z	r_1	r_2	r_3	
Initial	0	0	0	-12	-10	-20	1
$2R_3$	0	0	2	-10	-12	2	2
$1R_2$	0	1	0	-12	-3	1	3
$1R_1$	1	0	0	-2	-2	3	4
	1	1	2	-2	-2	3	5

Multiply the line 5 by 10

	10	10	20	-20	-20	30	6
$-3R_3$	0	0	-3	-23	-17	-3	7
$2R_1$	2	0	0	-3	-15	1	8
$2R_2$	0	2	0	-7	3	-1	9
$1R_1$	1	0	0	3	4	1	
	13	12	17	3	4	1	10

Multiply the line 10 by 10.

	130	120	170	30	40	10	11
$-4R_2$	0	-4	0	38	4	14	12
$-4R_1$	-4	0	0	-2	0	6	13
$-1R_3$	0	0	-1	-3	1	-5	14
	126	116	169	-3	1	-5	15

Multiply the line 15 by 10.

	1260	1160	1690	-30	10	-50	16
$4R_3$	0	0	4	-26	6	-6	17
$2R_1$	2	0	0	-6	8	-2	18
$-1R_2$	0	-1	0	-4	-1	0	19
	1262	1159	1694	-4	-1	0	20

Explanation: Due to initial values, $r_1 = -12$, $r_2 = -10$, $r_3 = -20$, r_3 being numerically the largest. In step one, r_3 is reduced by a suitable multiple of R_3 i.e. by $2R_3$, giving new values of residuals r_1, r_2, r_3 as -10, -12, 2 respectively, $r_2 = -12$ being numerically largest new residual. In step 2, it is reduced by $1R_2$, giving new values of residuals r_1, r_2, r_3 as -12, -3, $+1$ respectively, $r_1 = -12$ being numerically largest new residual. In step 3, it is reduced by $1R_1$, giving new values of residuals r_1, r_2, r_3 as -2, -2, 3 respectively.

The line 5 is obtained by adding the increments in steps 1, 2, 3 to the initial approximations showing the residuals $r_1 = -2$, $r_2 = -2$, $r_3 = 3$ corresponding to these values. Thus for $x = 1$, $y = 1$, $z = 2$ residuals are $r_1 = -2$, $r_2 = -2$, $r_3 = 3$. This can be checked from given equations.

In line 5, all the residuals are less than their initial values. This indicate that we go for the next decimal point. At line 5, we multiply the line 5 by 10 and get line 6. In line 5, $x = 1$, $y = 1$, $z = 2$ make $r_1 = -2$, $r_2 = -2$, $r_3 = 3$. In line 6, it means that $10x = -20$, $10y = -20$, $10z = 30$ and $10r_1 = -20$, $10r_2 = -20$, $10r_3 = 30$. Again from line 6, we proceed as usual and at line 10, the residuals are again less than the initial residuals and hence we multiply by 10 against line 10 and get line 11. In line 11, it means $10^2x = 130$, $10^2y = 120$, $10^2z = 170$, $10^2r_1 = 30$, $10^2r_2 = 40$, $10^2r_3 = 10$. Again proceeding from 11 at line 15, again the residuals are less than initial residuals. Hence, we again multiply by 10 at line 15. In line 16, it means, $10^3x = 1260$, $10^3y = 1160$, $10^3z = 1690$, $10^3r_1 = -30$, $10^3r_2 = 10$, $10^3r_3 = -50$. Now again we proceed till line 20 where the residuals are again less than the initial residuals. At line 20, we have, $10^3x = 1262$, $10^3y = 1159$, $10^3z = 1694$, $10^3r_1 = -4$, $10^3r_2 = -1$, $10^3r_3 = 0$.

From these, we get $x = 1.262$, $y = 1.159$, $z = 1.694$, $r_1 = -0.004$, $r_2 = -0.001$, $r_3 = 0$, which can be checked by putting values of x, y, z in system (ii).

Thus the values $x = 1.262$, $y = 1.159$, $z = 1.694$ are correct to three decimal places.

Example 20: What is ill conditioned system of equation? Solve the following equations using relaxation method:

$$9x - 2y + z = 50$$
$$x + 5y - z = 18$$
$$-2x + 2y + 7z = 19$$

Solution: The residuals are given by

$$r_1 = 9x - 2y + z - 50$$
$$r_2 = x + 5y - 3z - 18$$
$$r_3 = -2x + 2y + 7z - 19$$

Evidently the coefficient matrix is diagonally dominant.

Operation table						
	x	y	z	r_1	r_2	r_3
R_1	1	0	0	9	1	-2
R_2	0	1	0	-2	5	2
R_3	0	0	1	1	-3	7

We will take the initial values of x, y, z as 0, 0, 0. Setting $x = y = z = 0$, we get $r_1 = -50$, $r_2 = -18$, $r_3 = -19$. We write these residuals below relax these values making changes in x, y, z as shown below:

	x	y	z	r_1	r_2	r_3	
Initial	0	0	0	-50	-18	-19	1
$5R_1$	5	0	0	-5	-13	-29	2
$4R_3$	0	0	4	-1	-25	-1	3
$5R_2$	0	5	0	-11	0	9	4
	5	5	4	-11	0	9	5

Multiply the line 5 by 10

Initial	50	50	40	-110	0	90	6
$12R_1$	12	0	0	-2	12	66	7
$-9R_3$	0	0	-9	-11	39	3	8
$-8R_2$	0	-8	0	5	-1	-13	9
	62	42	31	5	-1	-13	10

Multiply the line 10 by 10.

	620	420	310	50	-10	-130	11
$19R_3$	0	0	19	69	-67	3	12
$-8R_1$	-8	0	0	-3	-75	19	13
$15R_2$	0	15	0	-33	0	49	14
$-7R_3$	0	0	-7	-40	21	0	15
$4R_1$	4	0	0	-4	25	-8	16
$-5R_2$	0	-5	0	6	0	-18	17
$3R_3$	0	0	3	9	-9	3	18
	616	430	322	9	-9	3	19

Since we multiplied by 10 three times. $10^2x = 616$, $10^2y = 430$, $10^2z = 322$, $10^2r_1 = 9$, $10^2r_2 = -9$, $10^2r_3 = 3$ and hence $x = 6.16$, $y = 4.30$, $z = 3.22$ while the residuals are $r_1 = 0.09$, $r_2 = -0.09$, $r_3 = 0.03$.

Thus the values $x = 6.16$, $y = 4.30$, $z = 3.22$ are correct to two decimal place.

Explanation: Reader can explain the procedure according to Example 1.

Example 21: Use relaxation method to solve the system

$$8x + y + z + w = 14$$
$$2x + 10y + 3z + w = -8$$
$$x - 2y - 20z + 3w = 111$$
$$3x + 2y + 2z + 19w = 53$$

Solution: The coefficient matrix is diagonally dominant. Hence we will use relaxation method. The residuals are

$$r_1 = 8x + y + z + w - 14$$
$$r_2 = 2x + 10y + 3z + w + 8$$
$$r_3 = x - 2y - 20z + 3w - 111$$
$$r_4 = 3x + 2y + 2z + 19w - 53$$

| Operation table | | | | | | | |
	x	y	z	w	r_1	r_2	r_3	r_4
R_1	1	0	0	0	8	2	1	3
R_2	0	1	0	0	1	10	-2	2
R_3	0	0	1	0	1	3	-20	2
R_4	0	0	0	1	1	1	3	19

We will start with $x = 0, y = 0, z = 0, w = 0$ as initial values. Relaxation procedure is given below:

	x	y	z	w	r_1	r_2	r_3	r_4	
Initial	0	0	0	0	-14	8	**-111**	-53	1
$-5R_3$	0	0	-5	0	-19	-7	-11	-63	2
$3R_4$	0	0	0	3	-16	-4	-2	-6	3
$2R_1$	2	0	0	0	0	0	0	0	4
	2	0	-5	3	0	0	0	0	5

Since all the residuals are zero, the exact solution is $x = 2, y = 0, z = -5, w = 3$.

Explanation: In line 1, for $x = 0, y = 0, z = 0, w = 0$, the residuals are $-14, 8, -111, -53$. The numerically largest residual is -111 which is shown in **bold** letters.

First, we liquidate the numerically largest residual $r_3 = -111$ by a suitable multiple of R_3. In R_3, a change of 1 in z will cause an addition -20 in r_3. To nullify -111, we add -5 in z so that a change of $(+100)$ is effected in r_3. Therefore, using $-5R_3$, changes in r_1, r_2, r_3, r_4 are $-5, -15, 100, -10$. Hence, new $r_1 = -14 + (-5) = -19$, $r_2 = -15 + 8 = -7$, $r_3 = 100 + (-111) = -11$, $r_4 = -10 + (-53) = -63$, which is line number 2. In line 2, the numerically largest residual is -63. We liquidate this, by a proper multiple of R_4. By R_4, an increase of 1 in w makes changes 1 in r_1, 1 in r_2, 3 in r_3 and 19 in r_4. To liquidate (-63), we do operation $3R_4$. So that $3R_4$ causes changes 3, 3, 9, 57 in r_1, r_2, r_3, r_4 respectively. Hence, by $3R_4$, the new $r_1 = -19 + 3 = -16$, $r_2 = -7 + 3 = -4$, $r_3 = -11 + 9 = -2$, $r_4 = -63 + 57 = -6$. Therefore, in line 3, $r_1 = -16$, $r_2 = -4$, $r_3 = -2$, $r_4 = -6$. Now $r_1 = -16$ is the numerically largest residual. Hence, we now liquidate the value of $r_1 = -16$. An increase of 1 in x (i.e. R_1 operation) increases 8 in r_1. Hence, we do operation $2R_1$ to liquidate $r_1 = -16$. This causes for new $r_1 = 0, r_2 = 0, r_3 = 0, r_4 = 0$. Adding all these values of x, y, z, w we get $x = 2, y = 0, z = -5, w = 3$ as the exact solution.

Example 22: Solve the following system of equations by relaxation method.

$$8x + y - z = 8$$
$$x - 7y + 2z = -4$$
$$2x + y + 9z = 12$$

Solution: We observe that the coefficient matrix is diagonally dominant. So, we will use relaxation method. The residuals are

$$r_1 = 8x + y - z - 8$$
$$r_2 = x - 7y + 2z + 4$$
$$r_3 = 2x + y + 9z - 12$$

	x	y	z	r_1	r_2	r_3
			Operation table			
R_1	1	0	0	8	1	2
R_2	0	1	0	1	-7	1
R_3	0	0	1	-1	2	9

We will start with the initial values $x = 0, y = 0, z = 0$. Relaxation procedure is given below:

	x	y	z	r_1	r_2	r_3
Initial	0	0	0	-8	4	-12
$1 \cdot R_3$	0	0	1	-9	6	-3
$1 \cdot R_1$	1	0	0	-1	7	-1
$1 \cdot R_2$	0	1	0	0	0	0
	1	1	1	0	0	0

Since all the residuals are made to zero, the solution is $x = 1, y = 1, z = 1$.

Explanation: We explain the procedure according to example 3.

EXERCISE 9.6

Solve the following equations using relaxation method:

1. $10x - 2y - 2z = 6; -x + 10y - 2z = 7; -x - y + 10z = 8$ [Ans. $x = 1, y = 1, z = 1$]
2. $12x + y + z = 31; 2x + 8y - z = 24; 3x + 4y + 10z = 58$ [Ans. $x = 2, y = 3, z = 4$]
3. $9x - y + 2z = 9; x + 10y - 2z = 15; 2x - 2y - 13z = -17$

                                   [Ans. $x = 0.917, y = 1.647, z = 1.195$]
4. $10x - 2y - z - u = 3; -2x + 10y - z - u = 15; -x - y + 10z - 2u = 27; -x - y - 2z + 10u = -9$ [Ans. $x = 1, y = 2, z = 3, u = 0$]
5. $6x - 3y + z = 11; 2x + y - 8z = -15; x - 7y + z = 10$ [Ans. $x = 1, y = -1, z = 2$]
6. $27x + 6y - z = 85; 6x + 15y + 2z = 72; x + y + 54z = 110$

                                     [Ans. $x = 2.426, y = 3.572, z = 1.926$]
7. $3x + 9y - 2z = 11; 4x + 2y + 13z = 24; 4x - 4y + 3z = -8$

                                     [Ans. $x = 1.35, y = 2.103, z = 2.845$]
8. $2x - 3y + 10z = 3; -x + 4y + 2z = 20; 5x + 2y + z = -12$ [Ans. $x = -4, y = 3, z = 2$]
9. $3x + y - z - w = 0; x + 3y - z + 2w = -3; -2x + 2y + 3z - 2w = 4; x + 2y + z - 5w = -1$

                                     [Ans. $x = 2, y = -1, z = 4, w = 1$]
10. $10x + y + z = 12; x + 10y + z = 12; x + y + 10z = 12$ [Ans. $x = 1, y = 1, z = 1$]

In the beginning of this chapter, we have discussed direct method (Gauss elimination method) for the solution of the system of linear equations. Now, we will discuss indirect method or iterative method (Gauss-Siedel iterative method, Jacobi's iterative method).

Gauss-Siedel method is a modification of Jacobi's method. But this iterative method is not always successful to all the systems of equations.

Remark: *The solution will exist if the absolute value of the largest coefficient is greater than the sum of the absolute values of all remaining coefficients in each equation. If these are not so, then on interchanging the equation, we can make the leading diagonal dominant diagonal.*

9.12 JACOBI ITERATIVE METHOD OR METHOD OF SIMULTANEOUS DISPLACEMENTS

This method is applicable to the system of equations in which leading diagonal elements of the coefficient matrix are dominant (larger in magnitude) in their respective rows.

Consider the system

$$\left.\begin{array}{l} a_1 x + b_1 y + c_1 z = d_1 \\ a_2 x + b_2 y + c_2 z = d_2 \\ a_3 x + b_3 y + c_3 z = d_3 \end{array}\right] \tag{9.18}$$

with assumption that $|a_1|$, $|b_2|$, $|c_3|$ are large as compared to other coefficients in the corresponding row and satisfy the condition of convergence as follows.

$$|a_1| > |b_1| + |c_1|$$
$$|b_2| > |a_2| + |c_2|$$
$$|c_3| > |a_3| + |b_3|$$

System in Eq. (9.18) can be written as

$$\left.\begin{array}{l} x = \dfrac{1}{a_1}(d_1 - b_1 y - c_1 z) \\[2mm] y = \dfrac{1}{b2}(d_2 - a_2 x - c_2 z) \\[2mm] z = \dfrac{1}{c_3}(d_3 - a_3 x - b_3 y) \end{array}\right] \tag{9.19}$$

We start the process by considering the initial approximation as $x = x_0$, $y = y_0$, $z = z_0$ and substituting in Eq. (9.19)

$$x^{(1)} = \frac{1}{a_1}[d_1 - b_1 y^{(0)} - c_1 z^{(0)}]$$

$$y^{(1)} = \frac{1}{b_2}[d_2 - a_2 x^{(0)} - c_2 z^{(0)}]$$

$$z^{(1)} = \frac{1}{c_3}[d_3 - a_3 x^{(0)} - b_3 y^{(0)}]$$

Now, we substitute the value $x^{(1)}, y^{(1)}, z^{(1)}$ in Eq. (9.19), we get the next approximation. The above iteration process is continued until two successive approximations are nearly equal.

Example 23: Solve the following system of equations:
$$4x + y + 3z = 17, \ x + 5y + z = 14, \ 2x - y + 8z = 12.$$

Solution: The given system of linear equation is diagonally dominant. So, the equation can be written as

$$\left.\begin{array}{l} x = \dfrac{1}{4}(17 - y - 3z) \\[2mm] y = \dfrac{1}{5}(14 - x - z) \\[2mm] z = \dfrac{1}{8}(12 - 2x + y) \end{array}\right] \tag{i}$$

We start with an approximation $x_0 = y_0 = z_0 = 0$. Substituting these in Eq. (i), we get the first approximation as

$$x^{(1)} = \frac{17}{4} = 4.25$$

$$y^{(1)} = \frac{14}{5} = 2.8$$

$$z^{(1)} = \frac{12}{8} = 1.5$$

Again putting the values of first approximation in Eq. (i), we get second approximation as

$$x^{(2)} = \frac{1}{4}[17 - y^{(1)} - 3z^{(1)}] = \frac{1}{4}[17 - 2.8 - 3 \times 1.5] = 2.425$$

$$y^{(2)} = \frac{1}{5}[14 - x^{(1)} - z^{(1)}] = \frac{1}{5}[14 - 4.25 - 1.5] = 1.65$$

$$z^{(2)} = \frac{1}{8}[12 - 2x^{(1)} + y^{(1)}] = \frac{1}{8}[12 - 2 \times 4.25 + 2.8] = 0.7875$$

Again, putting the values of second approximation in Eq. (i), we get third approximation as

$$x^{(3)} = \frac{1}{4}[17 - y^{(2)} - 3z^{(2)}] = \frac{1}{4}[17 - 1.65 - 3 \times 0.7875] = 3.246875$$

$$y^{(3)} = \frac{1}{5}[14 - x^{(2)} - z^{(2)}] = \frac{1}{5}[14 - 2.425 - 0.7875] = 2.1575$$

$$z^{(3)} = \frac{1}{8}[12 - 2x^{(2)} + y^{(2)}] = \frac{1}{8}[12 - 2 \times 2.425 + 1.65] = 1.1$$

Putting the values of third approximation in Eq. (i), we get fourth approximation as

$$x^{(4)} = \frac{1}{4}[17 - y^{(3)} - 3z^{(3)}] = \frac{1}{4}[17 - 2.1575 - 3 \times 1.1] = 2.885625$$

$$y^{(4)} = \frac{1}{5}[14 - x^{(3)} - z^{(3)}] = \frac{1}{5}[14 - 3.246875 - 1.1] = 1.930625$$

$$z^{(4)} = \frac{1}{8}[12 - 2x^{(3)} + y^{(3)}] = \frac{1}{8}[12 - 2 \times 3.246875 + 2.1575] = 0.95796875$$

Putting the values of fourth approximation in Eq. (i), we get fifth approximation as

$$x^{(5)} = \frac{1}{4}[17 - y^{(4)} - 3z^{(4)}] = \frac{1}{4}[17 - 1.930625 - 3 \times 0.95796875] = 3.048867188$$

$$y^{(5)} = \frac{1}{5}[14 - x^{(4)} - z^{(4)}] = \frac{1}{5}[14 - 2.885625 - 0.95796875] = 2.03128125$$

$$z^{(5)} = \frac{1}{8}[12 - 2x^{(4)} + y^{(4)}] = \frac{1}{8}[12 - 2 \times 2.885625 + 1.930625] = 1.019921875$$

Putting the values of fifth approximation in Eq. (i), we get sixth approximation as

$$x^{(6)} = \frac{1}{4}[17 - y^{(5)} - 3z^{(5)}] = \frac{1}{4}[17 - 2.03128125 - 3 \times 1.019921875] = 2.977238281$$

$$y^{(6)} = \frac{1}{5}[14 - x^{(5)} - z^{(5)}] = \frac{1}{5}[14 - 3.048867188 - 1.019921875] = 1.986242187$$

$$z^{(6)} = \frac{1}{8}[12 - 2x^{(5)} + y^{(5)}] = \frac{1}{8}[12 - 2 \times 3.048867188 + 2.03128125] = 0.991693359$$

Similarly

$$x^{(7)} = \frac{1}{4}[17 - 1.986242187 - 3 \times 0.991693359] = 3.009669434$$

$$y^{(7)} = \frac{1}{5}[14 - 2.977238281 - 0.991693359] = 2.006213672$$

$$z^{(7)} = \frac{1}{8}[12 - 2 \times 2.977238281 + 1.986242187] = 1.003970703$$

and

$$x^{(8)} = \frac{1}{4}[17 - 2.006213672 - 3 \times 1.003970703] = 3.009669434$$

$$y^{(8)} = \frac{1}{5}[14 - 3.009669434 - 1.003970703] = 2.006213672$$

$$z^{(8)} = \frac{1}{8}[12 - 2 \times 3.009669434 + 2.006213672] = 1.003970703$$

Since the seventh and eighth iteration values are nearly equal, the approximate solution is $x = 3, y = 2, z = 1$.

EXERCISE 9.7

Use Jacobi iterative method to solve the following system of linear equations:
1. $83x + 11y - 4z = 95; 7x + 52y + 13z = 104; 3x + 8y + 29z = 71$.

[Ans. $x = 1.05749, y = 1.36609, z = 1.96032$]
2. $27x + 6y - z = 85; 6x + 15y + 2z = 72; x + y + 54z = 110$

[Ans. $x = 2.42, y = 3.57, z = 1.92$]
3. $20x + y - 2z = 17; 3x + 20y - z = -18; 2x - 3y + 20z = 25$.   [Ans. $x = 1, y = -1, z = 1$]
4. $8x - y + 2z = 13; x - 10y + 3z = 17; 3x + 2y - 12z = 25$.   [Ans. $x = 1, y = -1, z = 2$]
5. $6x + 2y - z = 4; x + 5y + z = 3; 2x + y + 4z = 27$.   [Ans. $x = 2, y = -1, z = 6$]
6. $5x - y + z = 10; 2x + 4y = 12; x + y + 5z = -1$.   [Ans. $x = 2.56, y = 1.72, z = -1.06$]
7. $5x + 2y + z = 12; x + 4y + 2z = 15; x + 2y + 5z = 20$. [Ans. $x = 1.08, y = 1.95, z = 3.16$]
8. $10x + 2y + z = 9; 2x + 20y - 2z = -14; -2x + 3y + 10z = 22$.

[Ans. $x = 1.103, y = -1.9969, z = 3.001$]

9.13 GAUSS-SEIDEL METHOD

For this method, we consider a system of simultaneous equations

$$a_1x + b_1y + c_1z = d_1$$
$$a_2x + b_2y + c_2z = d_2$$
$$a_3x + b_3y + c_3z = d_3$$

With assumption that a_1, b_2, c_3 are large as compared to other coefficients.

The above system can be written as

$$x = \frac{1}{a_1}(d_1 - b_1 y - c_1 z) \tag{9.20}$$

$$y = \frac{1}{b_2}(d_2 - a_2 x - c_2 z) \tag{9.21}$$

$$z = \frac{1}{c_3}(d_3 - a_3 x - b_3 y) \tag{9.22}$$

We start the process by considering the initial approximation as $y = y_0$ and $z = z_0$ then

$$x_1 = \frac{1}{a_1}(d_1 - b_1 y_0 - c_1 z_0)$$

$$y_1 = \frac{1}{b_2}(d_2 - a_2 x_1 - c_2 z_0)$$

$$z_1 = \frac{1}{c_3}(d_3 - a_3 x_1 - b_3 y_1)$$

This completes the first iteration

For second iteration: Now, we substitute $y = y_1$ and $z = z_1$ in Eq. (9.20), then

$$x_2 = \frac{1}{a_1}(d_1 - b_1 y_1 - c_1 z_1)$$

$$y_2 = \frac{1}{b_2}(d_2 - a_2 x_2 - c_2 z_1)$$

$$z_3 = \frac{1}{c_3}(d_3 - a_3 x_2 - b_3 y_2)$$

Continuing this process till the difference between two consecutive approximations is as small as we desire.

Note: 1. Iteration method is self-correcting method. That is, any error made in computation, is corrected in the subsequent iterations.

2. The rate of convergence in Gauss-Seidel method is roughly two times that of Gauss-Jacobi method.

Example 24: Solve the system of equations
$$27x + 6y - z = 85$$
$$6x + 15y + 2z = 72$$
and $$x + y + 54z = 110$$
by Gauss-Seidel iteration method.

Solution: The given equations are:

$$27\ \ x\ + 6y - z = 85$$
$$6x\ +\ 15\ \ y + 2z = 72$$
$$x + y +\ \ 54\ \ z = 110$$

Here, we see that the diagonal elements are dominant. Hence, the iteration process can be applied.

Given system of equations can be written as

$$x = \frac{1}{27}(85 - 6y + z) \tag{i}$$

$$y = \frac{1}{15}(72 - 6x - 2z) \tag{ii}$$

$$z = \frac{1}{54}(110 - x - y) \tag{iii}$$

First iteration: Putting $y = y_0 = 0$ and $z = z_0 = 0$ in Eq. (i), we get

$$x_1 = \frac{1}{27}[85 - 6y_0 + z_0] = \frac{1}{27}[85 - 0 + 0] = \frac{85}{27} = 3.148$$

Putting $x = x_1$ and $z = z_0$ in Eq. (ii), we get

$$y_1 = \frac{1}{15}[72 - 6x_1 - 2z_0] = \frac{1}{15}[72 - 6 \times (3.148) - 0] = 3.541$$

Putting $x = x_1$ and $y = y_1$ in Eq. (iii), we get

$$z_1 = \frac{1}{54}[110 - x_1 - y_1] = \frac{1}{54}[110 - 3.148 - 3.541] = 1.913$$

Second iteration: Putting $y = y_1$ and $z = z_1$ in Eq. (i), we get

$$x_2 = \frac{1}{27}[85 - 6y_1 + z_1] = \frac{1}{27}[85 - 6 \times (3.541) + 1.913] = 2.43$$

Putting $x = x_2 = 2.43$ and $z = z_1 = 1.913$ in Eq. (ii), we get

$$y_2 = \frac{1}{15}[72 - 6x_2 - 2z_1] = \frac{1}{15}[72 - 6 \times (2.43) - 2 \times (1.913)]$$

$$= \frac{1}{15}[72 - 14.58 - 3.826] = 3.573$$

Putting $x = x_2$ and $y = y_2$ in Eq. (iii), we get

$$z_2 = \frac{1}{54}[110 - x_2 - y_2] = \frac{1}{54}[110 - 2.43 - 3.573] = 1.926$$

Third iteration: Putting $y = y_2$ and $z = z_2$ in Eq. (i), we get

$$x_3 = \frac{1}{27}[85 - 6y_2 + z_2] = \frac{1}{27}[85 - 6 \times (3.573) + 1.926]$$

$$= \frac{1}{27}[85 - 21.438 + 1.926] = 2.43$$

Putting $x = x_3$ and $z = z_2$ in Eq. (ii), we get

$$y_3 = \frac{1}{15}[72 - 6x_3 - 2z_2] = \frac{1}{15}[72 - 6 \times (2.43) - 2 \times (1.926)]$$

$$= \frac{1}{15}[72 - 14.58 - 3.852] = 3.571$$

Putting $x = x_3$ and $y = y_3$ in Eq. (iii), we get

$$z_3 = \frac{1}{54}[110 - x_3 - y_3] = \frac{1}{54}[110 - 2.43 - 3.571] = 1.926$$

Fourth iteration: Putting $y = y_3$ and $z = z_3$ in Eq. (i), we get

$$x_4 = \frac{1}{27}[85 - 6y_3 + z_3] = \frac{1}{27}[85 - 6 \times (3.571) + 1.926]$$

$$= \frac{1}{27}[85 - 21.426 + 1.926] = 2.43$$

Putting $x = x_4$ and $z = z_3$ in Eq. (ii), we get

$$y_4 = \frac{1}{15}[72 - 6x_4 - 2z_3] = \frac{1}{15}[72 - 6 \times (2.43) - 2 \times (1.926)]$$

$$= \frac{1}{15}[72 - 14.58 - 3.852] = 3.571$$

Putting $x = x_4$ and $y = y_4$ in Eq. (iii), we get

$$z_4 = \frac{1}{54}[110 - x_4 - y_4] = \frac{1}{54}[110 - 2.43 - 3.571] = 1.926$$

Thus, values have an accuracy upto three decimal places and final solution is
$$x = 2.43, y = 3.571, z = 1.926.$$

Example 24: Obtain the solution of the following system using Gauss-Siedel iteration method:

$$\begin{array}{l} 2\ x + y + z = 5 \\ 3x + 5\ y + 2z = 15 \\ 2x + y + 4\ z = 8 \end{array}$$

Solution: The given equations are:

$$\begin{array}{l} 2\ x + y + z = 5 \\ 3x + 5\ y + 2z = 15 \\ 2x + y + 4\ z = 8 \end{array}$$

Here, we see that the diagonal elements are dominant. Hence, the iteration process can be applied.

Given system of equations can be written as

$$x = \frac{1}{2}(5 - y - z) \tag{i}$$

$$y = \frac{1}{5}(15 - 3x - 2z) \tag{ii}$$

$$z = \frac{1}{4}(8 - 2x - y) \tag{iii}$$

First iteration: Putting $y = y_0 = 0$ and $z = z_0 = 0$ in Eq. (i), we get

$$x_1 = \frac{1}{2}[5 - y_0 - z_0] = \frac{1}{2}[5 - 0 - 0] = \frac{5}{2} = 2.5$$

Putting $x = x_1$ and $z = z_0$ in Eq. (ii), we get

$$y_1 = \frac{1}{5}[15 - 3x_1 - 2z_0] = \frac{1}{5}[15 - 3 \times (2.5) - 0]$$

$$= \frac{1}{5}[15 - 7.5] = \frac{7.5}{5} = 1.5$$

Putting $x = x_1$ and $y = y_1$ in Eq. (iii), we get

$$z_1 = \frac{1}{4}[8 - 2x_1 - y_1] = \frac{1}{4}[8 - 2 \times (2.5) - 1.5]$$

$$= \frac{1}{4}[8 - 5 - 1.5] = \frac{1.5}{4} = 0.375$$

Second iteration: Putting $y = y_1$ and $z = z_1$ in Eq. (i), we get

$$x_2 = \frac{1}{2}[5 - y_1 - z_1] = \frac{1}{2}[5 - 1.5 - 0.375] = 1.5625$$

Putting $x = x_2$ and $z = z_1$ in Eq. (ii), we get

$$y_2 = \frac{1}{5}[15 - 3x_2 - 2y_1]$$

$$= \frac{1}{5}[15 - 3 \times (1.5625) - 2 \times (0.375)] = 1.9125$$

Putting $x = x_2$ and $y = y_2$ in Eq. (iii), we get

$$z_2 = \frac{1}{4}[8 - 2x_2 - y_2] = \frac{1}{4}[8 - 2 \times (1.5625) - 1.9125] = 0.7406$$

Third iteration: Putting $y = y_2$ and $z = z_2$ in Eq. (i), we get

$$x_3 = \frac{1}{2}[5 - y_2 - z_2] = \frac{1}{2}[5 - 1.9125 - 0.740625] = 1.17344$$

Putting $x = x_3$ and $z = z_2$ in Eq. (ii), we get

$$y_3 = \frac{1}{5}[15 - 3x_3 - 2z_2] = \frac{1}{5}[15 - 3 \times (1.17344) - 2 \times (0.7406)]$$

$$= \frac{1}{5}[15 - 5.00152] = 1.9997$$

Putting $x = x_3$ and $y = y_3$ in Eq. (iii), we get

$$z_3 = \frac{1}{4}[8 - 2x_3 - y_3] = \frac{1}{4}[8 - 2 \times (1.17344) - 1.9997]$$

$$= \frac{1}{4}[8 - 2.34688 - 1.9997] = 0.913355$$

Fourth iteration: Putting $y = y_3$ and $z = z_3$ in Eq. (i), we get

$$x_4 = \frac{1}{2}[5 - y_3 - z_3] = \frac{1}{2}[5 - 1.9997 - 0.913355] = 1.0435$$

Putting $x = x_4$ and $z = z_3$ in Eq. (ii), we get

$$y_4 = \frac{1}{5}[15 - 3x_4 - 2z_3]$$

$$= \frac{1}{5}[15 - 3 \times (1.0435) - 2 \times (0.913355)] = 2.00856$$

Putting $x = x_4$ and $y = y_4$ in Eq. (iii), we get

$$z_4 = \frac{1}{4}[8 - 2x_4 - y_4]$$

$$= \frac{1}{4}[8 - 2 \times (1.0435) - 2.00856] = 0.97611$$

Fifth iteration: Putting $y = y_4$ and $z = z_4$ in Eq. (i), we get

$$x_5 = \frac{1}{2}[5 - y_4 - z_4] = \frac{1}{2}[5 - 2.00856 - 0.97611] = 1.0077$$

Putting $x = x_5$ and $z = z_4$ in Eq. (ii), we get

$$y_5 = \frac{1}{5}[15 - 3x_5 - 2z_4]$$

$$= \frac{1}{5}[15 - 3 \times (1.0077) - 2 \times (0.97611)] = 2.005$$

Putting $x = x_5$ and $y = y_5$ in Eq. (iii), we get

$$z_5 = \frac{1}{4}[8 - 2x_5 - y_5] = \frac{1}{4}[8 - 2 \times (1.0077) - 2.005] = 0.9949$$

Hence, the solution is $x = 1$, $y = 2$, $z = 1$.

Example 26: Solve the system by using Gauss-Seidel iteration method:

$$x_1 + 6x_2 - 3x_3 = 4$$
$$5x_1 + 2x_2 - x_3 = 6$$
$$2x_1 + x_2 + 4x_3 = 7$$

Solution: Here, we see that the coefficient matrix of the given system is not diagonally dominant. Hence, we rearrange the equations as follows, such that the element in the coefficient matrix are diagonally dominant.

$$5x_1 + 2x_2 - x_3 = 6$$
$$x_1 + 6x_2 - 3x_3 = 4$$
$$2x_1 + x_2 + 4x_3 = 7$$

Now, the coefficient matrix of the given system is diagonally dominant.
The above system of equations can be written as

$$x_1 = \frac{1}{5}[6 - 2x_2 + x_3] \tag{i}$$

$$x_2 = \frac{1}{6}[4 - x_1 - 3x_3] \tag{ii}$$

$$x_3 = \frac{1}{4}[7 - 2x_1 - x_2] \tag{iii}$$

In this problem, we shall use the notation $x_1^{(1)}, x_2^{(1)}, x_3^{(1)}$ for first iteration and $x_1^{(2)}, x_2^{(2)}, x_3^{(2)}$ for second iteration etc.

First iteration: Putting $x_2 = 0$ and $x_3 = 0$ in Eq. (i), we get

$$x_1^{(1)} = \frac{1}{6}[6 - 2x_2 + x_3] = \frac{1}{5}[6 - 0 + 0] = \frac{6}{5} = 1.2$$

Putting $x_1 = x_1^{(1)}$ and $x_3 = 0$ in Eq. (ii), we get

$$x_2^{(1)} = \frac{1}{6}[4 - x_1^{(1)} + 3x_3] = \frac{1}{6}[4 - 1.2 + 0] = \frac{2.8}{6} = 0.47$$

Putting $x_1 = x_1^{(1)}$ and $x_2 = x_2^{(1)}$ in Eq. (iii), we get

$$x_3^{(1)} = \frac{1}{4}[7 - 2x_1^{(1)} - x_2^{(1)}] = \frac{1}{4}[7 - 2 \times (1.2) - 0.47]$$

$$= \frac{1}{4}[7 - 2.4 - 0.47] = 1.03$$

Second iteration: Putting $x_2 = x_2^{(1)}$ and $x_3 = x_3^{(1)}$ in Eq. (i), we get

$$x_1^{(2)} = \frac{1}{5}[6 - 2x_2^{(1)} + x_3^{(1)}] = \frac{1}{5}[6 - 2 \times (0.47) + 1.03] = 1.218$$

Putting $x_1 = x_1^{(2)}$ and $x_3 = x_3^{(1)}$ in Eq. (ii), we get

$$x_2^{(2)} = \frac{1}{6}[4 - x_1^{(2)} + 3x_3^{(1)}] = \frac{1}{6}[4 - 1.218 + 3 \times (1.03)] = 0.9787$$

Putting $x_1 = x_1^{(2)}$ and $x_2 = x_2^{(2)}$ in Eq. (iii), we get

$$x_3^{(2)} = \frac{1}{4}[7 - 2x_1^{(2)} - x_2^{(2)}] = \frac{1}{4}[7 - 2 \times (1.218) - 0.9787] = 0.8963$$

Third iteration: Putting $x_1 = x_2^{(2)}$ and $x_3 = x_3^{(2)}$ in Eq. (i), we get

$$x_1^{(3)} = \frac{1}{5}[6 - 2x_2^{(2)} + x_3^{(2)}] = \frac{1}{5}[6 - 2 \times (0.9787) + 0.8963] = 0.98778$$

Putting $x_1 = x_1^{(3)}$ and $x_3 = x_3^{(2)}$ in Eq. (ii), we get

$$x_2^{(3)} = \frac{1}{6}[4 - x_1^{(3)} + 3x_3^{(2)}] = \frac{1}{6}[4 - 0.98778 + 3 \times (0.8963)] = 0.9502$$

Putting $x_1 = x_1^{(3)}$ and $x_2 = x_2^{(3)}$ in Eq. (iii), we get

$$x_3^{(3)} = \frac{1}{4}[7 - 2x_1^{(3)} - x_2^{(3)}] = \frac{1}{4}[7 - 2 \times (0.98778) - 0.9502] = 1.02$$

Fourth iteration: Putting $x_2 = x_2^{(3)}$ and $x_3 = x_3^{(3)}$ in Eq. (i), we get

$$x_1^{(4)} = \frac{1}{5}[6 - 2x_2^{(3)} + x_3^{(3)}] = \frac{1}{5}[6 - 2 \times (0.9502) + 1.02] = 1.02$$

Putting $x_1 = x_1^{(4)}$ and $x_3 = x_3^{(3)}$ in Eq. (ii), we get

$$x_2^{(4)} = \frac{1}{6}[4 - x_1^{(4)} + 3x_3^{(3)}] = \frac{1}{6}[4 - 1.02 + 3 \times (1.02)] = 1.00667$$

Putting $x_1 = x_1^{(4)}$ and $x_2 = x_2^{(4)}$ in Eq. (iii), we get

$$x_3^{(4)} = \frac{1}{4}[7 - 2x_1^{(4)} - x_2^{(4)}] = \frac{1}{4}[7 - 2 \times (1.02) - 1.0067] = 0.9883$$

Fifth iteration: Putting $x_2 = x_2^{(4)}$ and $x_3 = x_3^{(4)}$ in Eq. (i), we get

$$x_1^{(5)} = \frac{1}{5}[6 - 2x_2^{(4)} + x_3^{(4)}] = \frac{1}{5}[6 - 2 \times (1.0067) + 0.9883] = 0.99498$$

Putting $x_1 = x_1^{(5)}$ and $x_3 = x_3^{(4)}$ in Eq. (ii), we get

$$x_2^{(5)} = \frac{1}{6}[4 - x_1^{(5)} + 3x_3^{(4)}] = \frac{1}{6}[4 - 0.99498 + 3 \times (0.9883)] = 0.99498$$

Putting $x_1 = x_1^{(5)}$ and $x_2 = x_2^{(5)}$ in Eq. (iii), we get

$$x_3^{(5)} = \frac{1}{4}[7 - 2x_1^{(5)} - x_2^{(5)}] = \frac{1}{4}[7 - 2 \times (0.99498) - 0.99498] = 1.003$$

Hence, all the roots approximate to 1.

Thus, $x_1 = 1$, $x_2 = 1$, $x_3 = 1$.

EXERCISE 9.8

Solve the following system of equations by Gauss-Seidel method:

1. $2x + 10y + z = 51$; $10x + y + 2z = 44$; $x + 2y + 10z = 61$
   [Ans. $x = 2.9995$, $y = 4.000016$, $z = 5.0000018$]
2. With the system of equations $3x + 2y = 4.5$; $2x + 3y - z = 5$; $-y + 2z = -0.5$, set up the Gauss-Seidel iteration method. Iterate two times, using the initial approximation as $x_0 = 0.4$, $y_0 = 1.6$, $z_0 = 0.4$.   [Ans. $x = 0.5$, $y = 1.5$, $z = 0.5$]
3. $5x + 2y + z = 12$; $x + 4y + 2z = 15$; $x + 2y + 5z = 20$.
   [Ans. $x = 1.000121799$, $y = 1.999852036$, $z = 3.000034826$]
4. $1.2x + 2.1y + 4.2z = 9.9$; $5.3x + 6.1y + 4.7z = 21.6$; $9.2x + 8.3y + z = 15.2$.
   [Ans. $x = -13.22$, $y = 16.76$, $z = -2.30$]
5. $10x_1 - 2x_2 - x_3 - x_4 = 3$; $-2x_1 + 10x_2 - x_3 - x_4 = 15$; $-x_1 - x_2 + 10x_3 - 2x_4 = 27$; $-x_1 - x_2 - 2x_3 + 10x_4 = -9$.   [Ans. $x_1 = 0.9999$, $x_2 = 1.9999$, $x_3 = 0.00004$]

10 Solution of Ordinary Differential Equations

10.1 INTRODUCTION

Ordinary differential equations frequently occur as mathematical models in many branches of science, engineering and economy. For example, simple harmonic motion, deflection of beam etc., are represented by differential equations. So, the solution of differential equation is necessary in such studies. Unfortunately it is seldom that these equations have solutions that can be expressed in closed form, so it is common to seek approximate solutions by means of numerical methods. Hence, we will study here some of the methods of numerical solutions of the ordinary differential equations.

10.2 PICARD'S METHOD

Let the initial value problem is

$$\frac{dy}{dx} = f(x, y), \, y(x_0) = y_0 \tag{10.1}$$

This can be written as

$$dy = f(x, y) \, dx$$

On integrating it between x_0 to x, we find

$$\int_{y_0}^{y} dy = \int_{x_0}^{x} f(x, y) \, dx$$

$$\Rightarrow \qquad y = y_0 + \int_{x_0}^{x} f(x, y) \, dx \tag{10.2}$$

Equation (10.2) is known as integral equation corresponding to the initial value problem in Eq. (10.1). Integrand in Eq. (10.2) is a function of x and y both which complicates the integration with respect to x. So the method of successive approximations is used to solve this integral Eq. (10.2).

In order to get the first approximated value y_1 of y, we use y_0 for y in the integrand of Eq. (10.2). So, we get

$$y_1 = y_0 + \int_{x_0}^{x} f(x, y_0) \, dx \tag{10.3}$$

If we substitute y_1 for y in Eq. (10.2), we shall get the second approximate value y_2 of y as

$$y_2 = y_0 + \int_{x_0}^{x} f(x, y_1) \, dx \tag{10.4}$$

We can continue this process of substitution as many times as may be necessary. The nth approximation is given by

$$y_n = y_0 + \int_{x_0}^{x} f(x, y_{n-1})\, dx \tag{10.5}$$

In this way, we get a sequence of approximations

$$<y_n> = \{y_1, y_2, y_3, ..., y_n, ...\}$$

which converges to the solution of IVP in Eq. (10.1) if the function $f(x, y)$ is bounded in some region about the point (x_0, y_0) and also the function $f(x, y)$ satisfies the Lipschitz condition

$$|f(x, y_n) - f(x, y_{n-1})| < L\,|y_n - y_{n-1}|$$

$\rightarrow n \geq n_0$ where n_0 is a positive integer and L is a constant. This constant L is called a Lipschitz constant.

10.3 TAYLOR'S SERIES METHOD

Consider the differential equation is

$$\frac{dy}{dx} = f(x, y) \tag{10.6}$$

with initial condition $y(x_0) = y_0$ $\tag{10.7}$

On differentiating Eq. (10.6) with respect to x, we have

$$\frac{d^2 y}{dx^2} = y'' = f_x + f_y \frac{dy}{dx}$$

$$= f_x + f_y f$$

Next differentiation gives

$$\frac{d^3 y}{dx^3} = y''' = \frac{d}{dx}(f_x + f_y f)$$

$$= \frac{\partial}{\partial x}(f_x + f_y f) + \frac{\partial}{\partial y}(f_x + f_y f)\frac{dy}{dx}$$

$$= f_{xx} + f_{xy} f + f_y f_x + (f_{xy} + f_{yy} f + f_y f_y) f$$

$$= f_{xx} + f_{xy} f + f_y f_x + f_{xy} f + f_{yy} f^2 + f_y^2 f$$

$$= f_{xx} + 2 f_{xy} f + f^2 f_{yy} + f f_y^2 + f_y f_x$$

Next successive differentiation will give y'''', y''''', ... etc. By substituting $x = x_0$, $y = y_0$, we get

$$(y')_0, (y'')_0, (y''')_0, ...$$

Taylor's expansion of $f(x)$ about $x = x_0$ is given by

$$f(x) = f(x_0) + \frac{1}{1!}(x - x_0)\, f'(x_0) + \frac{1}{2!}(x - x_0)^2 f''(x_0) + ...$$

$\Rightarrow$

$$y(x) = y_0 + \frac{1}{1!}(x - x_0)(y')_0 + \frac{1}{2!}(x - x_0)^2 (y'')_0 + ...$$

This relation gives the value of y corresponding to every value of x for which it is convergent.

10.4 EULER'S METHOD

To find the numerical solution of a first order and first degree differential equation

$$\frac{dy}{dx} = f(x, y), \, y(x_0) = y_0 \tag{10.8}$$

at a point x_n, we divide the interval $[x_0, x_n]$ into n subintervals of equal width h so that

$$h = \frac{x_n - x_0}{n}, \text{ for } n = 0, 1, 1, 2, ..., n \tag{10.9}$$

The equation of the tangent line to the curve $y = y(x)$ at (x_0, y_0) is given by

$$y - y_0 = \left(\frac{dy}{dx}\right)(x - x_0)$$

In subinterval $[x_0, x_1]$, the slope is $\left(\dfrac{dy}{dx}\right)_{x_0} = f(x_0, y_0)$, so we have

$$y - y_0 = f(x_0, y_0)(x - x_0) \tag{10.10}$$

Now suppose the ordinate corresponding to x_1 meets this tangent line in Eq. (10.10) in (x_1, y_1), then

$$y_1 - y_0 = f(x_0, y_0)(x_1 - x_0)$$
$$\Rightarrow \qquad y_1 = y_0 + hf(x_0, y_0) \tag{10.11}$$

In the subinterval $[x_1, x_2]$, the slope is $\left(\dfrac{dy}{dx}\right)_{x_1} = f(x_1, y_1)$. So we have the tangent line

$$y - y_1 = f(x_1, y_1)(x - x_1) \tag{10.12}$$

If the ordinate corresponding to x_2 meets this tangent line in Eq. (10.12) in interval $[x_2, y_2]$, then

$$y_2 - y_1 = f(x_1, y_1)(x_2 - x_1)$$
$$\Rightarrow \qquad y_2 = y_1 + hf(x_1, y_1) \tag{10.13}$$

Continuing this process n times we get

$$y_{n+1} = y_n + hf(x_n, y_n) \tag{10.14}$$

This is known as Euler's method of finding the approximate solution of our initial value problem in Eq. (10.8). If the slope of the tangent line changes rapidly over an interval, then its value at the beginning of the interval gives a poor approximation in comparison with its average value over the interval. This error accumulates in succeeding intervals and as such the final value of y has very much error. Hence, this method is completely useless.

10.4.1 Error Estimates for the Euler Method

Let the exact solution at $x = x_n$ of the differential equation be $y(x_n)$ and the approximate solution obtained by Eq. (10.14) is y_n. Supposing the existence of the higher order derivatives, the Taylor's series for $y(x_{n+1})$ about $x = x_n$ is given by

$$y(x_{n+1}) = y(x_n) + hy'(x_n) + \frac{h^2}{2}y''(\rho_n), \, x_n \le \rho_n \le x_{n+1} \tag{10.15}$$

The local error is the error in y_{n+1} which is the result of the equation $y_{n+1} = y_n + hy'_n$ using exact values for y_n and y'_n and if this is denoted by E_{n+1}, then

$$E_{n+1} = -\frac{1}{2}h^2 y''(\rho_n)$$

The total solution error is given by

$$e_n = y_n - y(x_n) \tag{10.16}$$

Since y_0 is exact, it follows that $e_0 = 0$.

Besides local error E_{n+1}, there will be another error due to rounding off say R_{n+1} in computing y_{n+1} using Eq. (10.14). Thus we have

$$y_{n+1} = y_n + hf(x_n, y_n) + R_{n+1} \tag{10.17}$$

Thus total error is

$$\begin{aligned} e_{n+1} &= e_n + h[f(x_n, y_n) - y'(x_n)] + R_{n+1} + E_{n+1} \\ &= e_n + h[f(x_n, y_n) - f(x_n, y(x_n)] + R_{n+1} + E_{n+1} \end{aligned}$$

If $\dfrac{\partial f}{\partial y}$ is continuous, then by using mean value theorem, the expression can be written as

$$\begin{aligned} f(x_n, y_n) - f(x_n, y(x_n)) &= (y_n - y_{x_n}) f_y(x_n, \xi_n) \\ &= e_n f_y(x_n, \xi_n) \end{aligned}$$

where $\xi_n \in [y(x_n), y_n]$. Thus

$$\begin{aligned} e_{n+1} &= e_n + h e_n f_y(x_n, \xi_n) + R_{n+1} + E_{n+1} \\ &= e_n [1 + hf_y(x_n, \xi_n) + R_{n+1} + E_{n+1} \end{aligned}$$

Here the first term is the propagated error due to the error in the previous approximation y_n. Thus we obtain

$$e_0 = 0$$
$$e_1 = R_1 + E_1$$
$$e_2 = [1 + hf_y(x_1, \xi_1)] (R_1 + E_1) + R_2 + L_2$$

and so on.

10.5 MODIFIED EULER'S METHOD

From Euler's formula, we have

$$y_{n+1} = y_n + hf(x_n, y_n) \tag{10.18}$$

Let $y(x_n) = y_n$, denote the initial value using Eq. (10.18) an appropriate value of $y_n^{(0)}$ can be calculated by

$$y_n^{(0)} = y_{n-1} + \int_{x_{n-1}}^{x_n} f(x, y) \, dx$$

$$\Rightarrow \qquad y_n = y_{n-1} + hf(x_{n-1}, y_{n-1}) \tag{10.19}$$

Replacing $f(x, y)$ by $f(x_{n-1}, y_{n-1})$ in $x_{n-1} \le x < x_n$ and using Trapezoidal rule in $[x_{n-1}, x_n]$ we can write

$$y_n^{(0)} = y_{n-1} + \frac{h}{2}[f(x_{n-1}, y_{n-1}) + f(x_n, y_n)] \tag{10.20}$$

Replacing $f(x_n, y_n)$ by its approximate value $f(x_n, y_n^{(0)})$ at the end point of the interval $[x_{n-1}, x_n]$, we have

$$y_n^{(1)} = y_{n-1} + \frac{h}{2}[f(x_{n-1}, y_{n-1}) + f(x_n, y_n^{(0)})] \tag{10.21}$$

Proceeding as above, we get the iteration formula

$$y_n^{(r)} = y_{n-1} + \frac{h}{2}[f(x_{n-1}, y_{n-1}) + f(x_n, y_n^{(r-1)})] \tag{10.22}$$

where $y_n^{(r)}$ denote the rth approximation to y_n. Therefore, the Euler's modified formula is

$$y_n \approx y_n^{(r)} = y_{n-1} + \frac{h}{2}[f(x_{n-1}, y_{n-1}) + f(x_n, y_n^{(r-1)})] \tag{10.23}$$

Example 1: Using Picard's method of successive approximation, obtain a solution of the differential equation

$$\frac{dy}{dx} = x^4 y + x, y(0) = 3$$

Solution: Given initial value problem is

$$\frac{dy}{dx} = x^4 y + x, y(0) = 3$$

Here $x_0 = 0, y_0 = 3, f(x, y) = x^4 y + x$.

Equivalent integral equation for this problem is

$$y = y_0 + \int_{x_0}^{x} f(x, y) \, dx = 3 + \int_{0}^{x}(x^4 y + x) \, dx$$

The first approximation y_1 is given by

$$y_1 = 3 + \int_{0}^{x}(x^4 y_0 + x) \, dx$$

$$= 3 + \left[\frac{3}{5}x^5 + \frac{1}{2}x^2\right]_0^x = 3 + \frac{1}{2}x^2 + \frac{3}{5}x^5$$

The second approximation y_2 is given by

$$y_2 = 3 + \int_{0}^{x}(x^4 y_1 + x) \, dx$$

$$= 3 + \int_{0}^{x}\left[x + 3x^4 + \frac{1}{2}x^6 + \frac{3}{5}x^9\right] dx$$

$$= 3 + \left[\frac{1}{2}x^2 + \frac{3}{5}x^5 + \frac{1}{14}x^7 + \frac{3}{50}x^{10}\right]_0^x$$

$$= 3 + \frac{1}{2}x^2 + \frac{3}{5}x^5 + \frac{1}{14}x^7 + \frac{3}{50}x^{10}$$

The third approximation y_3 is given by

$$y_3 = 3 + \int_0^x (x^4 y_2 + x) \, dx$$

$$= 3 + \int_0^x \left[x + 3x^4 + \frac{1}{2}x^6 + \frac{3}{5}x^9 + \frac{1}{14}x^{11} + \frac{3}{50}x^{14} \right] dx$$

$$= 3 + \frac{1}{2}x^2 + \frac{3}{5}x^5 + \frac{1}{14}x^7 + \frac{3}{50}x^{10} + \frac{1}{168}x^{12} + \frac{1}{250}x^{15}$$

Example 2: Using Picard's method to find the value of y at $x = 0.25$, 0.5 and 1.0 upto three decimal places as a solution of the differential equation

$$\frac{dy}{dx} = \frac{x^2}{y^2 + 1}, \, y(0) = 0$$

Solution: Equivalent integral equation is

$$y = \int_0^x \frac{x^2}{y^2 + 1} \, dx$$

where $x_0 = 0, y_0 = 0, f(x, y) = \frac{x^2}{y^2 + 1}$.

The first approximation y_1 is given by

$$y_1 = \int_0^x \frac{x^2}{0+1} dx = \frac{x^3}{3}$$

The second approximation y_2 is given by

$$y_2 = \int_0^x \frac{x^2}{\left(\frac{x^3}{3}\right)^2 + 1} \, dx = \tan^{-1}\left(\frac{x^3}{3}\right)$$

$$= \frac{x^3}{3} - \frac{1}{81}x^9 + \dots$$

Clearly y_1 and y_2 agree to the first term. The term $\frac{x^3}{3}$ alone will give the result correct to three places for the value of x for which

$$\frac{1}{81}x^9 \le 0.0005$$

$\Rightarrow$ $$x \le 0.7$$

Hence $$y(0.25) = \frac{1}{3}(0.25)^3 = 0.005$$

$$y(0.5) = \frac{1}{3}(0.5)^3 = 0.042$$

Also beyond the range, we have

$$y(1.0) = \frac{1}{3}(1)^3 - \frac{1}{81}(1)^9 = \frac{1}{3} - \frac{1}{81} = 0.321$$

Example 3: Apply Taylor's method to obtain approximate value of y at $x = 0.2$ as a solution of differential equation

$$\frac{dy}{dx} = 2y + 3e^x, y(0) = 0.$$

Compare it with exact solution.

Solution: Given that $\dfrac{dy}{dx} = 2y + 3e^x$ with initial conditions $y(0) = 0$. Hence $x_0 = 0$, $y_0 = 0$, $f(x, y) = 2y + 3e^x$.

Taylor's series of $y(x)$ about $x = 0$ is

$$y(x) = y_0 + \frac{x}{1!}(y')_0 + \frac{x^2}{2!}(y'')_0 + \frac{x^3}{3!}(y''')_0 + \dots \qquad \text{(i)}$$

Here $\qquad y' = 2y + 3e^x$ gives $(y')_0 = 2y_0 + 3e^0 = 3$

Again differentiating and using values, we have

$$y'' = 2y' + 3e^x \Rightarrow (y'')_0 = 2(y')_0 + 3e^0 = 2 \times 3 + 3 = 9$$

$$y''' = 2y'' + 3e^x \Rightarrow (y''')_0 = 2(y'')_0 + 3e^0 = 2 \times 9 + 3 = 21$$

$$y'''' = 2y''' + 3e^x \Rightarrow (y'''')_0 = 2(y''')_0 + 3e^0 = 2 \times 21 + 3 = 45$$

$$y''''' = 2y'''' + 3e^x \Rightarrow (y''''')_0 = 2(y'''')_0 + 3e^0 = 2 \times 45 + 3 = 93$$

Putting these values in Eq. (i), we get

$$y(x) = 0 + 3x + \frac{x^2}{2!}9 + \frac{x^3}{3!}21 + \frac{x^4}{4!}45 + \frac{x^5}{5!}93 + \dots$$

$$= 3x + \frac{9x^2}{2} + \frac{7x^3}{2} + \frac{15}{8}x^4 + \frac{93}{120}x^5 + \dots$$

Taking $x = 0.2$, we get

$$y(0.2) = 3(0.2) + \frac{9}{2}(0.2)^2 + \frac{7}{2}(0.2)^3 + \frac{15}{8}(0.2)^4 + \frac{93}{120}(0.2)^5 + \dots$$

$$= 0.8112648$$

Exact solution: Given differential equation is a linear differential equation whose integrating factor is e^{-2x}. Thus solution is

$$ye^{-2x} = \int e^{-2x} 3e^x dx + c$$

$$= -3e^{-x} + c$$

$$= -3e^{-x} + 3 \text{ as } y(0) = 0 \text{ gives } c = 3$$

$$\Rightarrow \qquad y = 3e^{2x} - 3e^x$$

Putting $x = 0.2$, we have

$$y(0.2) = 3e^{2(0.2)} - 3e^{(0.2)}$$

$$= 4.475474 - 3.6642082$$

$$= 0.8112658$$

which is same as approximated value upto five decimal places.

Example 4: Using Taylor's series method, solve the differential equation $\dfrac{dy}{dx} = xy - 1$, $y(1) = 2$ at the point $x = 1.02$

Solution: Given that $\dfrac{dy}{dx} = xy - 1$ with initial condition $y(1) = 2$.

Here $x_0 = 1$, $y_0 = 2$, $f(x, y) = xy - 1$.

Taylor's series of $y(x)$ about $x = 1$ is

$$y(x) = y_0 + \frac{1}{1!}(x-1)(y')_0 + \frac{(x-1)^2}{2!}(y'')_0 + \frac{(x-1)^3}{3!}(y''')_0 + \dots \quad (i)$$

From the given differential equation

$$y' = xy - 1 \Rightarrow (y')_0 = x_0 y_0 - 1 = 1 \times 2 - 1 = 2 - 1 = 1$$
$$y'' = xy' + y \Rightarrow (y'')_0 = x_0(y')_0 + (y)_0 = 1 \times 1 + 2 = 3$$
$$y''' = xy'' + 2y' \Rightarrow (y''')_0 = x_0(y'')_0 + 2(y')_0 = 1 \times 3 + 2 \times 1 = 5$$
$$y'''' = xy''' + 3y'' \Rightarrow (y'''')_0 = x_0(y''')_0 + 3(y'')_0 = 1 \times 5 + 3 \times 3 = 14$$
$$y''''' = xy'''' + 4y''' \Rightarrow (y''''')_0 = x_0(y'''')_0 + 4(y''')_0 = 1 \times 14 + 4 \times 5 = 34$$
$$y'''''' = xy''''' + 4y'''' \Rightarrow (y'''''')_0 = x_0(y''''')_0 + 5(y''')_0 = 1 \times 34 + 5 \times 14 = 104$$

Putting all these values in Eq. (i), we get

$$y(x) = 2 + (x-1) \times 1 + \frac{1}{2}(x-1)^2 \times 3 + \frac{1}{6}(x-1)^3 \times 5 + \frac{1}{24}(x-1)^4 \times 14$$

$$+ \frac{1}{120}(x-1)^5 \times 34 + \frac{1}{720}(x-1)^6 \times 104 + \dots$$

$$= 2 + (x-1) + \frac{3}{2}(x-1)^2 + \frac{5}{6}(x-1)^3 + \frac{7}{12}(x-1)^4 + \frac{17}{60}(x-1)^5 + \frac{13}{90}(x-1)^6 + \dots$$

Putting $x = 1.02$ in this relation, we have

$$y(1.02) = 2 + (1.02 - 1) + \frac{3}{2}(1.02 - 1)^2 + \frac{5}{6}(1.02 - 1)^3 + \frac{7}{12}(1.02 - 1)^4$$

$$+ \frac{17}{60}(1.02 - 1)^5 + \frac{13}{90}(1.02 - 1)^6 + \dots$$

$$= 2.0206066$$

Example 5: Using Euler's method, find y at $x = 0.1$, for the given differential equation

$$\frac{dy}{dx} = \frac{y - x}{y + x}, \quad y(0) = 1$$

Solution: Divide the interval $[0, 0.1]$ into five subintervals of width $h = \dfrac{0.1 - 0}{5} = 0.02$.

Hence $x_1 = 0.02$, $x_2 = 0.04$, $x_3 = 0.06$, $x_4 = 0.08$, $x_5 = 0.1$.

Also $x_0 = 0$, $y_0 = 1$ and $f(x, y) = \dfrac{y - x}{y + x}$.

By Euler's method, we have

$$y_{n+1} = y_n + hf(x_n, y_n) \text{ for } n = 0, 1, 2, 3, \dots$$

Thus $\quad y_1 = y_0 + hf(x_0, y_0) = 1 + (0.02)\dfrac{1-0}{1+0} = 1.02$

$$y_2 = y_1 + hf(x_1, y_1) = 1.02 + (0.02)\left(\dfrac{1.02 - 0.02}{1.02 + 0.02}\right) = 1.0392307$$

$$y_3 = y_2 + hf(x_2, y_2) = 1.0392307 + (0.02)\left(\dfrac{1.0392307 - 0.04}{1.0392307 + 0.06}\right) = 1.05774810$$

$$y_4 = y_3 + hf(x_3, y_3) = 1.05774810 + (0.02)\left(\dfrac{1.05774810 - 0.06}{1.05774810 + 0.06}\right) = 1.07560093$$

$$y_5 = y_4 + hf(x_4, y_4) = 1.07560093 + (0.02)\left(\dfrac{1.07560093 - 0.08}{1.07560093 + 0.08}\right) = 1.09283181$$

Hence the value of y at $x = 1$ is 1.09283181.

Example 6: Given $\dfrac{dy}{dx} = x^3 + y,\ y(0) = 1$. Find $y(0.3)$ by using Euler's method using $h = 0.1$.

Solution: Given $x_0 = 0,\ y_0 = 1$ and $f(x, y) = x^3 + y$.

Here $h = 0.1$, so $x_1 = 0.1,\ x_2 = 0.2,\ x_3 = 0.3$.

By Euler's method, we have

$$y_{n+1} = y_n + hf(x_n, y_n) \text{ for } n = 0, 1, 2, 3, \dots$$

Thus
$$y_1 = y_0 + hf(x_0, y_0) = 1 + (0.1)\,(x_0^3 + y_0)$$
$$= 1 + (0.1)\,(0 + 1) = 1.1$$
$$y_2 = y_1 + hf(x_1, y_1) = y_1 + h(x_1^3 + y_1) = 1.1 + (0.1)\,[(0.1)^3 + 1.1]$$
$$= 1.1 + (0.1)\,(0.001 + 1.1) = 1.1 + (0.1)\,(1.101) = 1.2101$$
$$y_3 = y_2 + hf(x_2, y_2) = y_2 + h(x_2^3 + y_2)$$
$$= 1.2101 + (0.1)\,[(0.2)^3 + 1.2101)]$$
$$= 1.2101 + (0.1)\,(0.008 + 1.2101)$$
$$= 1.2101 + (0.1)\,(1.2181)$$
$$= 1.2101 + 0.12181$$
$$= 1.33191$$

Hence the value of $y(0.3) = 1.33191$.

Example 7: By using Euler's modified method, find the value of y at $x = 0.2$ and $x = 0.5$ for the solution of differential equation

$$\dfrac{dy}{dx} = \log_e(x + y),\ y(0) = 1.$$

Solution: Given that $\dfrac{dy}{dx} = \log_e(x + y),\ y(0) = 1$.

Here $x_0 = 0,\ y_0 = 1,\ f(x, y) = \log_e(x + y)$.

Let $x_1 = 0.2$ and $x_2 = 0.5$, we have to compute y_1 and y_2.

Now $\quad f(x_0, y_0) = \dfrac{dy}{dx} = \log_e(0 + 1) = 0$

Thus
$$y_1^{(1)} = y_0 + hf(x_0, y_0) = 1 + (0.2) \times 0 = 1$$

$$\left(\frac{dy}{dx}\right)_1^{(1)} = f(x_1, y_1^{(1)}) = \log_e(x_1 + y_1^{(1)}) = \log_e(0.2 + 1)$$
$$= \log_e(1.2) = 0.18232156$$

Hence
$$y_1^{(2)} = y_0 + \frac{h}{2}\left[\left(\frac{dy}{dx}\right)_0 + \left(\frac{dy}{dx}\right)_1^{(1)}\right]$$
$$= 1 + \frac{0.2}{2}[0 + \log_e(1.2)])$$
$$= 1 + (0.1) \ (0 + 0.18232156) = 1.018232156$$

Again
$$\left(\frac{dy}{dx}\right)_1^{(2)} = f(x_1, y_1^{(2)}) = \log_e(x_1 + y_1^{(2)})$$
$$= \log_e(0.2 + 1.018232156)$$
$$= \log_e(1.218232156) = 0.19740076$$

This gives
$$y_1^{(3)} = y_0 + \frac{h}{2}\left[\left(\frac{dy}{dx}\right)_0 + \left(\frac{dy}{dx}\right)_1^{(2)}\right]$$
$$= 1 + \frac{0.2}{2}[0 + 0.19740076]$$
$$= 1.019740076$$

Now
$$\left(\frac{dy}{dx}\right)_1^{(3)} = f(x_1, y_1^{(3)}) = \log_e(x_1 + y_1^{(3)}) = \log(0.2 + 1.019740076)$$
$$= \log_e(1.219740076) = 0.19863778$$

This gives
$$y_1^{(4)} = y_0 + \frac{h}{2}\left[\left(\frac{dy}{dx}\right)_0 + \left(\frac{dy}{dx}\right)_1^{(3)}\right]$$
$$= 1 + \frac{0.2}{2}[0 + 0.19863778]$$
$$= 1.019863778$$

Now
$$\left(\frac{dy}{dx}\right)_1^{(4)} = f(x_1, x_1^{(4)}) = \log_e(x_1 + y_1^{(4)})$$
$$= \log_e(0.2 + 1.019863778)$$
$$= \log_e(1.219863778) = 0.1987392$$

$$y_1^{(5)} = y_0 + \frac{h}{2}\left[\left(\frac{dy}{dx}\right)_0 + \left(\frac{dy}{dx}\right)_1^{(4)}\right]$$
$$= 1 + \frac{0.2}{2}[0 + 0.1987392]$$
$$= 1.01987392$$

$$\left(\frac{dy}{dx}\right)_1^{(5)} = f(x_1, y_1^{(5)}) = \log(0.2 + 1.01987392)$$

$$= \log_e(1.21987392) = 0.19874751$$

$$y_1^{(6)} = y_0 + \frac{h}{2}\left[\left(\frac{dy}{dx}\right)_0 + \left(\frac{dy}{dx}\right)_1^{(5)}\right]$$

$$= 1 + \frac{0.2}{2}[0 + 0.19874751]$$

$$= 1.019874751$$

$$\left(\frac{dy}{dx}\right)_1^{(6)} = f(x_1, y_1^{(6)}) = \log_e(0.2 + 1.019874751)$$

$$= \log_e(1.219874751) = 0.19874819$$

$$y_1^{(7)} = y_0 + \frac{0.2}{2}\left[\left(\frac{dy}{dx}\right)_0 + \left(\frac{dy}{dx}\right)_1^{(6)}\right]$$

$$= 1 + (0.1)(0 + 0.19874819)$$

$$= 1.019874819$$

Here $y_1^{(6)}$ and $y_1^{(7)}$ are same upto six decimal places, so

$$y_1 = y(0.2) = 1.019874$$

Now we obtain $y_2 = y(0.5)$. Here $h = 0.5 - 0.2 = 0.3$.

We have $y_2^{(1)} = y_1 + hf(x_1, y_1)$

$$= 1.019874 + (0.3)\log_e(0.2 + 1.019874)$$

$$= 1.019874 + (0.3)(0.1987476)$$

$$= 1.019874 + 0.05962427$$

$$= 1.07949827$$

$$y_2^{(2)} = y_1 + \frac{h}{2}\left[\left(\frac{dy}{dx}\right)_1 + \left(\frac{dy}{dx}\right)_2^{(1)}\right]$$

$$= 1.019874 + \frac{0.3}{2}[\log_e(x_1 + y_1) + \log_e(x_2 + y_2^{(1)})]$$

$$= 1.019874 + (0.15)[\log_e(0.2 + 1.019874) + \log_e(0.5 + 1.07949827)]$$

$$= 1.019874 + (0.15)[0.1987476 + 0.45710725] = 1.11825222$$

$$y_2^{(3)} = y_1 + \frac{h}{2}\left[\left(\frac{dy}{dx}\right)_1 + \left(\frac{dy}{dx}\right)_2^{(2)}\right]$$

$$= y_1 + \frac{h}{2}[\log_e(x_1 + y_1) + \log_e(x_2 + y_2^{(2)})]$$

$$= 1.019874 + \frac{0.3}{2}[\log_e(0.2 + 1.019874) + \log_e(0.5 + 1.11825222)]$$

$$= 1.019874 + (0.15)[0.1987476 + 0.48134669]$$

$$= 1.019874 + 0.10201414 = 1.12188814$$

$$y_2^{(4)} = y_1 + \frac{h}{2}\left[\left(\frac{dy}{dx}\right)_1 + \left(\frac{dy}{dx}\right)_2^{(3)}\right]$$

$$= y_1 + \frac{h}{2}[\log_e(x_1 + y_1) + \log_e(x_2 + y_2^{(3)})]$$

$$= 1.019874 + \frac{0.3}{2}[\log_e(0.2 + 1.019874) + \log_e(0.5 + 1.12188814)]$$

$$= 1.019874 + (0.15)[0.1987476 + 0.48359099]$$

$$= 1.019874 + 0.10235078 = 1.12222478$$

$$y_2^{(5)} = y_1 + \frac{h}{2}[\log_e(x_1 + y_1) + \log_e(x_2 + y_2^{(4)})]$$

$$= 1.019874 + \frac{0.3}{2}[\log_e(0.2 + 1.019874) + \log_e(0.5 + 1.12222478)]$$

$$= 1.019874 + (0.15)[0.1987476 + 0.48379853]$$

$$= 1.019874 + 0.10238192 = 1.12225592$$

$$y_2^{(6)} = y_1 + \frac{h}{2}[\log_e(x_1 + y_1) + \log_e(x_2 + y_2^{(5)})]$$

$$= 1.019874 + \frac{0.3}{2}[\log_e(0.2 + 1.019874 + \log_e(0.5 + 1.12225592)]$$

$$= 1.019874 + (0.15)[0.19874757 + 0.08381772]$$

$$= 1.019874 + 0.1238479 = 1.12225879$$

$$y_2^{(7)} = y_1 + \frac{h}{2}[\log_e(x_1 + y_1) + \log_e(x_2 + y_2^{(6)})]$$

$$= 1.019874 + (0.15)[\log_e(0.2 + 1.019874) + \log_e(0.5 + 1.12225879)]$$

$$= 1.019874 + (0.15)[0.1987476 + 0.48381949]$$

$$= 1.019874 + 0.10238506 = 1.12225906$$

Here $y_2^{(6)}$ and $y_2^{(7)}$ are same up to five decimal places, so $y_2 = y(0.5) = 1.122259$.

Example 8: Using Taylor's series method, solve $\dfrac{dy}{dx} = x - y^2$, $y(0) = 1$ at $x = 0.1$ correct to four decimal places.

Solution: Given $\qquad \dfrac{dy}{dx} = x - y^2$

$$y(0) = 1$$

$\therefore x_0 = 0$, $y_0 = 1$, $h = 0.1$, $x_1 = x_0 + h = 0 + 0.1 = 0.1$

Now
$$y'(x) = x - y^2 \qquad\qquad \therefore y_0' = -1$$
$$y''(x) = 1 - 2yy' \qquad\qquad \therefore y_0'' = 3$$
$$y'''(x) = -2yy'' - 2y'^2 \qquad\qquad \therefore y_0''' = -8$$
$$y''''(x) = -2yy''' - 6y'y'' \qquad\qquad \therefore y_0'''' = 34$$
$$y'''''(x) = -2yy'''' - 8y'y''' - 6y''^2 \qquad\qquad \therefore y_0''''' = -186$$

By Taylor's series

$$y(x) = 1 + xy_0' + \frac{x^2}{2!} y_0'' + \frac{x^3}{6} y_0''' + \frac{x^4}{24} y_0'''' + \frac{x^5}{120} y_0''''' + \dots$$

$$= 1 - x + \frac{3}{2}x^2 - \frac{4}{3}x^3 + \frac{17}{12}x^4 - \frac{31}{20}x^5 + \dots$$

$$y(0.1) = 1 - 0.1 + \frac{3}{2}(0.1)^2 - \frac{4}{3}(0.1)^3 + \frac{17}{12}(0.1)^4 - \frac{31}{20}(0.1)^5 + \dots$$

Neglecting 6th and higher order terms of above expression, we get

$$y(0.1) = 0.9138$$

Example 9: Given the differential equation $y'' - xy' - y = 0$ with the conditions $y(0) = 1$ and $y'(0) = 0$, use Taylor's series method to determine the value of $y(0.1)$.

Solution: We have

$$y'' - xy' - y = 0 \tag{i}$$

With conditions $y(0) = 1$ and $y'(0) = 0$, the Eq. (i) can be written as

$$y''(x) = xy'(x) + y(x) \tag{ii}$$

On differentiation, we get

$$y''' = xy''(x) + y'(x) + y'(x) = xy''(x) + 2y'(x) \tag{iii}$$

$$y'''' = xy'''(x) + y''(x) + 2y''(x) = xy'''(x) + 3y''(x) \tag{iv}$$

$$y''''' = xy''''(x) + y'''(x) + 3y'''(x) = xy''''(x) + 4y'''(x) \tag{v}$$

$$y'''''' = xy'''''(x) + y''''(x) + 4y''''(x) = xy'''''(x) + 5y''''(x) \tag{vi}$$

Putting $x = 0$ in Eqs (ii)–(vi), we obtain

$$y_{(0)}'' = 0 \times 0 + 1 = 1 \qquad\qquad \Rightarrow y''(0) = 1$$

$$y_{(0)}''' = 0\,(1) + 2 \times 0 = 0 \qquad\qquad \Rightarrow y'''(0) = 0$$

$$y_{(0)}'''' = 0 \times (0) + 3\,(1) = 0 + 3 \qquad \Rightarrow y''''(0) = 3$$

$$y_{(0)}''''' = 0\,(0 + 3) + 4\,(0) = 0 + 70 \qquad \Rightarrow y'''''(0) = 0$$

$$y_{(0)}'''''' = 0\,(0 + 70) + 5\,(0 + 3) = 0 + 120 + 15 \qquad \Rightarrow y''''''(0) = 15$$

By Taylor's series, we get

$$y(x) = y(0) + xy'(0) + \frac{x^2}{2} y''(0) + \frac{x^3}{6} y'''(0) + \frac{x^4}{24} y''''(0)$$

$$+ \frac{x^5}{120} y'''''(0) + \frac{x^6}{720} y''''''(0) + \dots$$

Hence

$$y(0.1) = 1 + \frac{(0.1)^2}{2} + \frac{(0.1)^4}{24}(3) + \frac{(0.1)^6}{720}(15) + \dots$$

$$= 1 + 0.005 + 0.0000125, \text{ neglecting the last term}$$

$$= 1.0050125$$

correct to seven decimal places.

EXERCISE 10.1

1. Evaluate by means of Taylor's series expansion, the following problem at $x = 0.1$, 0.2 to four significant figures $y'' - x(y')^2 + y^2 = 0; y(0) = 1, y(0) = 0$.

[Ans. 0.8112648]

2. Solve $\dfrac{dy}{dx} = x + y$, $y(1) = 0$ numerically up to $x = 1.2$ with $h = 0.1$. Compare the found result with the value of the explicit solution.

[Ans. 0.9950, 0.9801]

3. For the differential equation $\dfrac{dy}{dx} = -xy^2$, $y(0) = 2$. Calculate $y(0.2)$ by Taylor's series method retaining four nonzero terms only.

[Ans. 1.923072]

4. Using Picard's method, solve the differential equation $\dfrac{dy}{dx} = x + y$, $y(0) = 1$ at $x = 0.1, 0.2$.

[Ans. 1.1133415, 1.24280039]

5. Use Picard's method to approximate y when $x = 0.2$ given that $\dfrac{dy}{dx} = x - y$, $y(0) = 1$.

[Ans. 0.837]

6. Using Taylor's series method, compute $y(4.1)$ and $y(4.2)$ as solution of differential equation $\dfrac{dy}{dx} = \dfrac{1}{x^2 + y}$, $y(4) = 4$.

[Ans. 4.005, 4.0098]

7. Solve $y' = y^2 + x$; $y(0) = 1$, using Taylor's series method and compute $y(0.1)$ and $y(0.2)$

[Ans. 1.1164, 1.2725]

8. Use Euler's modified method to find the solution of $\dfrac{dy}{dx} = x + |\sqrt{y}|$, $y(0) = 1$ for the range $0 \le x \le 0.6$ for the step size 0.2

[Ans. $y(0.2) = 1.2309479$, $y(0.4) = 1.5254031$, $y(0.6) = 1.8862$]

9. Use Euler's method to find the solution of $\dfrac{dy}{dx} = x^2 + y^2$, $y(0) = 0$ in the range $0 \le x \le 0.5$ for the step size 0.1.

[Ans. 0, 0.001, 0.0050001, 0.0140026, 0.0300222]

10. Using Euler's method, solve for y at $x = 0.1$ the differential equation $\dfrac{dy}{dx} = x + y + xy$, $y(0) = 1$ taking step size $h = 0.025$.

[Ans. 1.1448]

11. Solve the differential equation $\dfrac{dy}{dx} = 2 + \sqrt{xy}$, $y(1) = 1$ for y at $x = 2$, taking step size 0.2 using Euler's modified method.

[Ans. 5.051]

12. Using modified Euler's method, find y at $x = 0.2$ taking step size 0.1 as solution of differential equation, $\dfrac{dy}{dx} = -xy^2$, $y(0) = 2$.

[1.9227]

10.6 RUNGE–KUTTA METHODS

While the Taylor's series method for solving differential equation numerically is handicapped by the labour involved in determining the higher order derivatives, Euler's method is slow and process of finding successive approximation too laborious. But, in Runge–Kutta methods, the derivatives of higher order are not required and we require only the given function values at different points. The derivation of second, third and fourth order Runge–Kutta method is tedious. So, without bothering to give the details of derivation, we give here only the working procedure to solve the initial value problem

$$\frac{dy}{dx} = f(x, y), \ y(x_0) = y_0 \tag{10.24}$$

10.6.1 Runge–Kutta First Order Method

Consider the differential equation

$$\frac{dy}{dx} = f(x, y) \tag{10.25}$$

subject to the initial condition

$$y(x_0) = y_0 \tag{10.26}$$

In accordance with Euler's method, we have

$$y_1 = y_0 + hf(x_0, y_0)$$
$$= y_0 + hy_0' \qquad [\because y' = f(x, y)]$$

Now $y_1 = y(x_0 + h)$. Expanding it by Taylor's series, we get

$$y_1 = y_0 + hy_0' + \frac{h^2}{2!} y_0'' + \cdots$$

$\Rightarrow$ Euler's method agrees with Taylor's series solution upto the term h^1. Hence Euler's method is first order Runge-Kutta method.

10.6.2 Runge–Kutta Second Order Method

To solve the first order differential equation

$$\frac{dy}{dx} = f(x, y) \tag{10.27}$$

subject to the initial condition

$$y(x_0) = y_0 \tag{10.28}$$

If h is the step-size for equidistant values of x and initial values are (x_0, y_0), we compute next values y_1 at $x_1 = x_0 + h$ from the following formula

$$k_1 = hf(x_0, y_0)$$
$$k_2 = hf(x_0 + h, y_0 + k_1)$$
$$\Delta y = \frac{1}{2}(k_1 + k_2)$$

Then
$$x_1 = x_0 + h$$
$$y_1 = y_0 + \Delta y = y_0 + \frac{1}{2}(k_1 + k_2)$$

In a similar manner, the increment in y for the second interval is computed by means of the formula

$$k_1 = hf(x_1, y_1)$$
$$k_2 = hf(x_1 + h, y_1 + k_1)$$
$$\Delta y = \frac{1}{2}(k_1 + k_2)$$

and so on. The Runge–Kutta method of second order is nothing but the modified Euler's method.

10.6.3 Runge–Kutta Third Order Method

The working procedure of third order Runge–Kutta method is given below to solve $\frac{dy}{dx} = f(x, y)$, $y(x_0) = y_0$.

$$k_1 = hf(x_0, y_0)$$

$$k_2 = hf\left(x_0 + \frac{h}{2}, y_0 + \frac{k_1}{2}\right)$$

$$k_3 = hf(x_0 + h, y_0 + 2k_2 - k_1)$$

and
$$\Delta y = \frac{1}{6}(k_1 + 4k_2 + k_3)$$

then
$$x_1 = x_0 + h$$

and
$$y_1 = y_0 + \Delta y = y_0 + \frac{1}{6}(k_1 + 4k_2 + k_3)$$

In a similar manner, we can calculate next approximation.

10.6.4 Runge–Kutta Fourth Order Method

The most commonly used Runge–Kutta formula is Runge–Kutta fourth order formula. The method is particularly suitable in cases when computation of higher derivative is complicated. Consider

$$\frac{dy}{dx} = f(x, y)$$

and
$$y(x_0) = y_0$$

then fourth order Runge–Kutta method algorithm is given below

$$k_1 = hf(x_0, y_0)$$

$$k_2 = hf\left(x_0 + \frac{h}{2}, y_0 + \frac{k_1}{2}\right)$$

$$k_3 = hf\left(x_0 + \frac{h}{2}, y_0 + \frac{k_2}{2}\right)$$

$$k_4 = hf(x_0 + h, y_0 + k_3)$$

where h is the step-size for equidistant values of x. Then

$$x_1 = x_0 + h$$

and
$$y_1 = y_0 + \Delta y = y_0 + \frac{1}{6}(k_1 + 2k_2 + 2k_3 + k_4)$$

In a similar manner, the increment in y for the second interval is computed by means of the formula

$$k_1 = hf(x_1, y_1)$$

$$k_2 = hf\left(x_1 + \frac{h}{2}, y_1 + \frac{k_1}{2}\right)$$

$$k_3 = hf\left(x_1 + \frac{h}{2}, y_1 + \frac{k_2}{2}\right)$$

$$k_4 = hf(x_1 + h, y_1 + k_3)$$

and
$$\Delta y = \frac{1}{6}(k_1 + 2k_2 + 2k_3 + k_4)$$

Then
$$x_2 = x_0 + 2h, \ y_2 = y_1 + \Delta y$$

and so on.

Note: Runge–Kutta methods are referred to single step methods.

Example 10: Obtain the value of y at $x = 0.1, 0.2$ using Runge–Kutta method of (i) second order (ii) third order for the differential equation $\dfrac{dy}{dx} = -2y$, given $y(0) = 1$.

Solution: Here $f(x, y) = \dfrac{dy}{dx} = -2y, \ x_0 = 0, \ y_0 = 1, \ h = 1$.

i. **Second order:** $k_1 = hf(x_0, y_0) = (0.1)(-2y_0) = (0.1)(-2 \times 1) = -0.2$

$k_2 = hf(x_0 + h, y_0 + k_1) = 0.1f(0.1, 0.8)$

$= 0.1 \times -2 \times 0.8 = -0.16$

$$\Delta y = \frac{1}{2}(k_1 + k_2) = \frac{1}{2}(-0.2 - 0.16) = -0.18$$

$$y_1 = y_0 + \Delta y = 1 - 0.18 = 0.82$$

$$x_1 = x_0 + h = 0 + 0.1 = 0.1$$

$$y_1 = y(x_1) = y(0.1) = 0.82$$

Again, for the second interval, we have

$k_1 = hf(x_1, y_1) = (0.1)(-2 \times 0.82) = -0.164$

$k_2 = hf(x_1 + h, y_1 + k_1) = 0.1f(0.2, 0.656)$

$= 0.1 \times (-2 \times 0.656) = -0.1312$

$$\Delta y = \frac{1}{2}(k_1 + k_2) = \frac{1}{2}(-0.164 - 0.1312) = -0.1476$$

$$y_2 = y_1 + \Delta y = 0.82 - 0.1476 = 0.6724$$

$$x_2 = x_0 + 2h = 0 + 2 \times 0.1 = 0.2$$

$$y(0.2) = 0.6724$$

ii. **Third order:** $k_1 = hf(x_0, y_0) = (0.1)(-2y_0) = (0.1)(-2 \times 1) = -0.2$

$$k_2 = hf\left(x_0 + \frac{h}{2}, y_0 + \frac{k_1}{2}\right) = 0.1\left[-2\left(y_0 + \frac{k_1}{2}\right)\right]$$

$$= 0.1\left[-2\left(1 - \frac{0.02}{2}\right)\right] = 0.1[-2(1 - 0.01)] = -0.198$$

$k_3 = hf(x_0 + h, y_0 + 2k_2 - k_1)$

$= 0.1[-2(y_0 + 2k_2 - k_1)] = 0.1[-2(1 + 2 \times -0.198 - (-0.2)]$

$= 0.1[-2(0.804)] = -0.1608$

$$\Delta y = \frac{1}{6}(k_1 + 4k_2 + k_3) = \frac{1}{6}(-0.2 + 4 \times -0.198) + (-0.1608) = -0.192$$

then
$$x_1 = x_0 + h = 0 + 0.1 = 0.1$$
$$y_1 = y(0.1) = y_0 + \Delta y = 1 - 0.192 = 0.808$$

Again, taking (x_1, y_1) in place of (x_0, y_0), repeat the process

∴
$$k_1 = hf(x_1, y_1) = (0.1)(-2y_1) = 0.1[-2 \times 0.808] = -0.1616$$

$$k_2 = hf\left(x_1 + \frac{h}{2}, y_1 + \frac{k_1}{2}\right)$$

$$= 0.1\left[-2\left(y_1 + \frac{k_1}{2}\right)\right] = 0.1\left[-2\left(0.808 - \frac{0.2}{2}\right)\right] = -0.1416$$

$$k_3 = hf(x_0 + h, y_1 + 2k_2 - k_1) = 0.1[-2(y_1 + 2k_2 - k_1)]$$
$$= 0.1[-2(0.808 + 2 \times -0.1416) - (-0.1616)]$$
$$= 0.1[-2(0.6864)] = -0.13728$$

$$\Delta y = \frac{1}{2}(k_1 + 4k_2 + k_3) = \frac{1}{6}[-0.1616 + 4 \times (-0.1416) + (-0.13728)]$$
$$= -0.1442$$

then
$$x_2 = x_1 + h = 0.1 + 0.1 = 0.2$$
$$y_2 = y(0.2) = y_1 + \Delta y = 0.808 - 0.1442 = 0.6638$$
$$y(0.2) = 0.6638$$

Example 11: Using Runge–Kutta method of order 4, find the numerical solution at $x = 1.2$ and $x = 1.4$ for $\dfrac{dy}{dx} = \dfrac{x^2 + y^2}{xy}$, with $y(1) = 3$.

Solution: Here $\qquad f(x, y) = \dfrac{dy}{dx} = \dfrac{x^2 + y^2}{xy}$

Let $x_0 = 1$, $y_0 = 3$, $h = 0.2$, we have

$$k_1 = hf(x_0, y_0) = 0.1\left[\frac{x_0^2 + y_0^2}{x_0 y_0}\right] = 0.2\left[\frac{1^2 + 3^2}{1 \times 3}\right] = 0.6666$$

$$k_2 = hf\left(x_0 + \frac{h}{2}, y_0 + \frac{k_1}{2}\right)$$

$$= 0.2\left[\frac{\left(x_0 + \frac{h}{2}\right)^2 + \left(y_0 + \frac{k_1}{2}\right)^2}{\left(x_0 + \frac{h}{2}\right)\left(y_0 + \frac{k_1}{2}\right)}\right]$$

$$= 0.2\left[\frac{\left(1 + \frac{0.2}{2}\right)^2 + \left(3 + \frac{0.6666}{2}\right)^2}{\left(1 + \frac{0.2}{2}\right)\left(3 + \frac{0.6666}{2}\right)}\right] = 0.2\left[\frac{(1.1)^2 + (3.3333)^2}{1.1 \times 3.3333}\right]$$

$$= 0.2 \left[\frac{12.3208}{3.6666} \right] = 0.6720$$

$$k_3 = hf\left(x_0 + \frac{h}{2}, y_0 + \frac{k_2}{2}\right) = 0.2 \left[\frac{\left(x_0 + \frac{h}{2}\right)^2 + \left(y_0 + \frac{k_2}{2}\right)^2}{\left(x_0 + \frac{h}{2}\right)\left(y_0 + \frac{k_2}{2}\right)} \right]$$

$$= 0.2 \left[\frac{\left(1 + \frac{0.2}{2}\right)^2 + \left(3 + \frac{0.6720}{2}\right)^2}{\left(1 + \frac{0.2}{2}\right)\left(3 + \frac{0.6720}{2}\right)} \right]$$

$$= 0.2 \left[\frac{(1.1)^2 + (3.336)^2}{(1.1)(3.336)} \right] = 0.8485$$

$$k_4 = hf(x_0 + h, y_0 + k_3) = 0.2 \left[\frac{(x_0 + h)^2 + (y_0 + k_3)^2}{(x_0 + h)(y_0 + k_3)} \right]$$

$$= 0.2 \left[\frac{(1 + 0.2)^2 + (3 + 0.8485)^2}{(1 + 0.2)(3 + 0.8485)} \right]$$

$$= 0.2 \left[\frac{16.2509}{4.6182} \right] = 0.7037$$

$$\Delta y = \frac{1}{6}(k_1 + 2k_2 + 2k_3 + k_4)$$

$$= \frac{1}{6}(0.6666 + 2 \times 0.6720 + 2 \times 0.8485 + 0.7037) = 0.7352$$

Thus $\quad\quad x_1 = x_0 + h = 1 + 0.2 = 1.2$

$$y(1.2) = y_1 = y_0 + \Delta y = 3 + 0.7352 = 3.7352$$

Again $\quad\quad k_1 = hf(x_1, y_1) = 0.2 \left[\frac{x_1^2 + y_1^2}{x_1 y_1} \right] = 0.2 \left[\frac{(1.2)^2 + (3.7352)^2}{(1.2)(3.7352)} \right]$

$$= 0.2 \left[\frac{15.3917}{4.4822} \right] = 0.6867$$

$$k_2 = hf\left(x_1 + \frac{h}{2}, y_1 + \frac{k_1}{2}\right) = 0.2 \left[\frac{\left(x_1 + \frac{h}{2}\right)^2 + \left(y_1 + \frac{k_1}{2}\right)^2}{\left(x_1 + \frac{h}{2}\right)\left(y_1 + \frac{k_1}{2}\right)} \right]$$

$$= 0.2 \left[\frac{\left(1.2 + \frac{0.2}{2}\right)^2 + \left(3.7352 + \frac{0.6867}{2}\right)^2}{\left(1.2 + \frac{0.2}{2}\right)\left(3.7352 + \frac{0.6867}{2}\right)} \right]$$

$$= 0.2\left[\frac{1.69 + 16.6345}{1.3 \times 4.07855}\right] = \frac{3.6649}{5.3021} = 0.6912$$

$$k_3 = hf\left(x_1 + \frac{h}{2}, y_1 + \frac{k_2}{2}\right) = 0.2\left[\frac{\left(x_1 + \frac{h}{2}\right)^2 + \left(y_1 + \frac{k_2}{2}\right)^2}{\left(x_1 + \frac{h}{2}\right)\left(y_1 + \frac{k_2}{2}\right)}\right]$$

$$= 0.2\left[\frac{(1.3)^2 + (4.0808)^2}{1.3 \times 4.0808}\right] = 0.6915$$

$$k_4 = hf(x_1 + h, y_1 + k_3) = 0.2\left[\frac{(x_1 + h)^2 + (y_1 + k_3)^2}{(x_1 + h)(y_1 + k_3)}\right]$$

$$= 0.2\left[\frac{(1.2 + 0.2)^2 + (3.7352 + 0.6915)^2}{1.4 \times 4.4267}\right] = 0.6956$$

$$\Delta y = \frac{1}{6}(k_1 + 2k_2 + 2k_3 + k_4)$$

$$= \frac{1}{6}[0.6867 + 2 \times 0.6912 + 2 \times 0.6915 + 0.6956] = 0.6912$$

Thus
$$x_2 = x_1 + h = 1.2 + 0.2 = 1.4$$
$$y(1.4) = y_2 = y_1 + \Delta y = 3.7352 + 0.6912 = 4.4264$$

Example 12: Apply Runge–Kutta fourth order method to find $y(0.1)$ for the initial value problem, $\dfrac{dy}{dx} = y - x$. Given $y(0) = 2$ [APJAKTU 2015]

Solution: Here $f(x, y) = \dfrac{dy}{dx} = y - x, x_0 = 0, y_0 = 2, h = 0.1$

$$k_1 = hf(x_0, y_0) = 0.1(y_0 - x_0) = 0.1(2 - 0) = 0.2$$

$$k_2 = hf\left(x_0 + \frac{h}{2}, y_0 + \frac{k_1}{2}\right) = 0.1\left[\left(y_0 + \frac{k_1}{2}\right) - \left(x_0 + \frac{h}{2}\right)\right]$$

$$= 0.1\left[\left(2 + \frac{0.2}{2}\right) - \left(0 + \frac{0.1}{2}\right)\right] = 0.205$$

$$k_3 = hf\left(x_0 + \frac{h}{2}, y_0 + \frac{k_2}{2}\right) = 0.1\left[\left(y_0 + \frac{k_2}{2}\right) - \left(x_0 + \frac{h}{2}\right)\right]$$

$$= 0.1\left[\left(2 + \frac{0.205}{2}\right) - \left(0 + \frac{0.1}{2}\right)\right] = 0.20525$$

$$k_4 = hf(x_0 + h, y_0 + k_3) = 0.1[(y_0 + k_3) - (x_0 + h)]$$
$$= 0.1[(2 + 0.20525) - (0 + 0.1)] = 0.210525$$

$$\Delta y = \frac{1}{6}(k_1 + 2k_2 + 2k_3 + k_4)$$

$$= \frac{1}{6}[0.2 + 2 \times 0.205 + 2 \times 0.20525 + 0.210525] = 0.2052$$

Thus
$$x_1 = x_0 + h = 0 + 0.1 = 0.1$$
$$y_1 = y(0.1) = y_0 + \Delta y = 2 + 0.2052 = 2.2052$$

Example 13: Compute $y(0.2)$ given $\dfrac{dy}{dx} + y + xy^2 = 0$, $y(0) = 1$ by taking $h = 0.1$ using Runge–Kutta method of fourth order.

Solution: Here $\dfrac{dy}{dx} = f(x, y) = -(xy^2 + y)$, $x_0 = 0$, $y_0 = 1$, $h = 0.1$.

$$k_1 = hf(x_0, y_0) = (0.1)[-(x_0 y_0^2 + y_0)] = -0.1$$

$$k_2 = hf\left(x_0 + \frac{h}{2}, y_0 + \frac{k_1}{2}\right) = 0.1 f(0.05, 0.95)$$
$$= -0.1[(0.05)(0.95)^2 + 0.95] = -0.0995$$

$$k_3 = hf\left(x_0 + \frac{h}{2}, y_0 + \frac{k_2}{2}\right) = 0.1 f(0.05, 0.95025)$$
$$= 0.1 \times -[(0.05) \times (0.95025)^2 + 0.95025] = -0.0995$$
$$k_4 = hf(x_0 + h, y_0 + k_3) = 0.1 f(0.1, 0.9005)$$
$$= 0.1 \times -[(0.1)(0.9005)^2 + 0.9005] = -0.0982$$

$$\Delta y = \frac{1}{6}(k_1 + 2k_2 + 2k_3 + k_4)$$

$$= \frac{1}{6}[-0.1 + 2 \times -0.0995 + 2 \times -0.0995 + (-0.0982)]$$

$$= -0.09936$$

Thus
$$x_1 = x_0 + h = 0 + 0.1 = 0.1$$
$$y_1 = y(0.1) = y_0 + \Delta y = 1 - 0.09936 = 0.9006$$

Again
$$k_1 = hf(x_1, y_1) = 0.1 f(0.1, 0.9006) = -0.0982$$

$$k_2 = hf\left(x_1 + \frac{h}{2}, y_1 + \frac{k_1}{2}\right) = 0.1 f(0.15, 0.8515) = -0.0960$$

$$k_3 = hf\left(x_1 + \frac{h}{2}, y_1 + \frac{k_2}{2}\right) = 0.1 f(0.15, 0.8526) = -0.0962$$

$$k_4 = hf(x_1 + h, y_1 + k_3) = 0.1 f(0.2, 0.8044) = -0.0934$$

$$\Delta y = \frac{1}{6}(k_1 + 2k_2 + 2k_3 + k_4)$$

$$= \frac{1}{6}[-0.0982 + 2 \times -0.0960 + 2 \times -0.0962 - 0.0934]$$

$$= -0.096$$

Thus
$$x_2 = x_1 + h = 0.1 + 0.1 = 0.2$$
$$y_2 = y(0.2) = y_1 + \Delta y = 0.9006 - 0.096 = 0.8046$$

Example 14: Solve the initial value problem $u' = -2tu^2$, $u(0) = 1$ with $h = 0.2$ on the interval $0 \le t \le 0.4$. Use Runge–Kutta fourth order method and compare your result with exact solution.

Solution: Here $f(t, u) = -2tu^2$, $u_0 = 1$, $t_0 = 0$, $h = 0.2$.

We have
$$k_1 = hf(t_0, u_0) = 0$$

$$k_2 = hf\left(t_0 + \frac{h}{2}, u_0 + \frac{k_1}{2}\right) = 0.2f(0.1, 1) = -0.04$$

$$k_3 = hf\left(t_0 + \frac{h}{2}, u_0 + \frac{k_2}{2}\right) = 0.2f(0.1, 0.98) = -0.038416$$

$$k_4 = hf(t_0 + h, u_0 + k_3) = 0.2f(0.2, 0.961584) = -0.0739715$$

Hence
$$\Delta u = \frac{1}{6}(k_1 + 2k_2 + 2k_3 + k_4) = -0.03846725$$

and
$$u_1 = u(0.2) = u_0 + \Delta u = 1 - 0.03846725 = 0.9615328$$

Again
$$k_1 = hf(t_1, u_1) = -0.0739636$$

$$k_2 = hf\left(t_1 + \frac{h}{2}, u_1 + \frac{k_1}{2}\right) = -0.1025753$$

$$k_3 = hf\left(t_1 + \frac{h}{2}, u_1 + \frac{k_2}{2}\right) = -0.0994255$$

$$k_4 = hf(t_1 + h, u_1 + k_3) = -0.1189166$$

Hence
$$\Delta u = \frac{1}{6}(k_1 + 2k_2 + 2k_3 + k_4) = -0.0994803$$

and
$$u_2 = u(0.4) = u_1 + \Delta u = 0.9615328 - 0.0994803 = 0.8620525$$

Absolute error in numerical solution are
$$e(0.2) = |0.961539 - 0.961533| = 0.000006$$
$$e(0.4) = |0.862069 - 0.862053| = 0.000016$$

Example 15: Given $y' = x^2 - y$, $y(0) = 1$. Find $y(0.1)$, $y(0.2)$ using Runge-Kutta methods of (i) second order (ii) third order.

Solution: Given
$$y' = \frac{dy}{dx} = f(x, y)$$
$$y' = x^2 - y, x_0 = 0, y_0 = 1$$

$\therefore$
$$f(x_0, y_0) = x_0^2 - y_0 = 0 - 1 = -1$$

Let
$$h = 0.1$$

Runge-Kutta Second Order Method
$$k_1 = hf(x_0, y_0) = (0.1)(-1) = -0.1$$
$$k_2 = hf(x_0 + h, y_0 + k_1) = 0.1f(0.1, 0.9)$$
$$= (0.1)[(0.1)^2 - 0.9] = -0.089$$

$$\Delta y = \frac{1}{2}(k_1 + k_2) = \frac{1}{2}[(-0.1) + (-0.089)] = -0.0945$$

$$y_1 = y(0.1) = y_0 + \Delta y = 1 - 0.0945 = 0.9055$$

Again taking $x_1 = 0.1$, $y_1 = 0.9055$

$$k_1 = hf(x_1, y_1) = h(x_1^2 - y_1) = (0.1)[(0.1)^2 - 0.9055] = -0.08955$$

$$k_2 = hf(x_1 + h, y_1 + k_1) = hf(0.2, 0.81595)$$

$$= (0.1)[(0.2)^2 - 0.81595] = -0.077595$$

$\therefore$

$$\Delta y = \frac{1}{2}(k_1 + k_2)$$

$$= \frac{1}{2}[(-0.08955) + (-0.077595)] = \frac{1}{2} \times (-0.16745)$$

$$= 0.0835725$$

$$y_2 = y(0.2) = y_1 + \Delta y = 0.9055 - 0.0835725$$

$$= 0.8219275$$

Runge-Kutta Third Order Method

$$k_1 = hf(x_0, y_0) = -0.1$$

$$k_2 = hf\left(x_0 + \frac{h}{2}, y_0 + \frac{k_1}{2}\right)$$

$$= 0.1f\left[0 + \frac{0.1}{2}, 1 - \frac{0.1}{2}\right]$$

$$= 0.1f(0.05, 0.95) = 0.1[(0.05)^2 - 0.95]$$

$$= -0.09475$$

$$k_3 = hf[x_0 + h, y_0 - k_1 + 2k_2]$$

$$= 0.1f[0.1, 1 - (-0.1) + 2 \times -0.09475]$$

$$= 0.1f[0.1, 1 + 0.1 - 0.1895] = 0.1f(0.1, 0.9105)$$

$$= 0.1[(0.1)^2 - 0.9105] = -0.09005$$

$$\Delta y = \frac{1}{6}(k_1 + 4k_2 + k_3)$$

$$= \frac{1}{6}[-0.1 + 4 \times -0.09475 - 0.09005]$$

$$= \frac{1}{6}[-0.1 - 0.379 - 0.09005]$$

$$= -0.09484$$

$$y_1 = y(0.1) = y_0 + \Delta y = 1 - 0.09484 = 0.90516$$

Again taking $x_1 = 0.1$, $y_1 = 0.90516$, $h = 0.1$

$$k_1 = hf(x_1, y_1) = (0.1)[x_1^2 - y_1) = 0.1[(0.1)^2 - 0.90516]$$

$$= -0.089516$$

$$k_2 = hf\left[x_1 + \frac{h}{2}, y_1 + \frac{k_1}{2}\right]$$

$$= 0.1f\left[0.1 + \frac{0.1}{2}, 0.90516 - \frac{0.089516}{2}\right]$$

$$= 0.1f(0.15, 0.860402)$$

$$= 0.1[(0.15)^2 - 0.860402] = 0.1[0.0225 - 0.860402]$$

$$= -0.0837902$$

$$k_3 = hf[x_1 + h, y_1 - k_1 + 2k_2]$$

$$= 0.1f[0.1 + 0.1, 0.90516 - (-0.089516) + 2 \times -0.0837902]$$

$$= 0.1f[0.2, 0.8270956]$$

$$= 0.1[(0.2)^2 - 0.8270956] = -0.0787096$$

$$\Delta y = \frac{1}{6}(k_1 + 4k_2 + k_3)$$

$$= \frac{1}{6}[-0.089516 + 4 \times -0.0837902 - 0.0787096]$$

$$= -0.0839051$$

$$y_2 = y(0.2) = y_1 + \Delta y = 0.90516 - 0.0839051$$

$$= 0.8212549$$

10.6.5 Runge–Kutta Method for Simultaneous First Order Differential Equations

Consider the simultaneous differential equation of the form

$$\frac{dy}{dx} = f_1(x, y, z)$$

and

$$\frac{dz}{dx} = f_2(x, y, z)$$

with initial conditions $y(x_0) = y_0$, $z(x_0) = z_0$.

Now starting from (x_0, y_0, z_0), the increments Δy and Δz in y and z respectively are given by formulae

$$k_1 = hf_1(x_0, y_0, z_0)$$

$$k_2 = hf_1\left[x_0 + \frac{h}{2}, y_0 + \frac{k_1}{2}, z_0 + \frac{l_1}{2}\right]$$

$$k_3 = hf_1\left[x_0 + \frac{h}{2}, y_0 + \frac{k_2}{2}, z_0 + \frac{l_2}{2}\right]$$

$$k_4 = hf_1[x_0 + h, y_0 + k_3, z_0 + l_3]$$

$$\Delta y = \frac{1}{6}(k_1 + 2k_2 + 2k_3 + k_4)$$

$$l_1 = hf_2(x_0, y_0, z_0)$$

$$l_2 = hf_2\left[x_0 + \frac{h}{2}, y_0 + \frac{k_1}{2}, z_0 + \frac{l_1}{2}\right]$$

$$l_3 = hf_2 \left[x_0 + \frac{h}{2}, y_0 + \frac{k_2}{2}, z_0 + \frac{l_2}{2} \right]$$

$$l_4 = hf_2 [x_0 + h, y_0 + k_3, z_0 + l_3]$$

$$y_1 = y_0 + \Delta y \text{ and } z_1 = z_0 + \Delta z$$

In order to find y_2 and z_2, we replace x_0, y_0, z_0 by x_1, y_1, z_1 in above formula.

10.6.6 Runge–Kutta Method for Second Order Differential Equation

Consider a second order differential equation

$$\frac{d^2y}{dx^2} = f\left(x, y, \frac{dy}{dx}\right)$$

with initial conditions $y(x_0) = y_0$ and $y'(x_0) = y_0'$.

Let $$\frac{dy}{dx} = z$$

So that $$\frac{d^2y}{dx^2} = \frac{dz}{dx}$$

Hence, differential equation reduces to

$$\frac{dz}{dx} = f(x, y, z)$$

with initial conditions $y(x_0) = y_0, z(x_0) = z_0$.

Thus, the problem is converted into simultaneous differential equations of the form

$$\frac{dy}{dx} = z = f_1(x, y, z)$$

and $$\frac{dz}{dx} = f_2(x, y, z)$$

with initial conditions $y(x_0) = y_0$ and $z(x_0) = z_0$.

These equations can now be solved by the method discussed previously.

Example 16: Find $y(0.1)$ and $z(0.1)$ from the system of equations, $\dfrac{dy}{dx} = x + z, \dfrac{dz}{dx} = x - y^2$ given $y(0) = 2, z(0) = 1$ usin/g Runge–Kutta method of fourth order.

Solution: Here $f_1(x, y, z) = x + z$

and $f_2(x, y, z) = x - y^2$

$$h = 0.1, x_0 = 0, y_0 = 2, z_0 = 1$$

We use
$$k_1 = hf_1(x_0, y_0, z_0) = 0.1f_1(0, 2, 1) = 0.1$$
$$l_1 = hf_2(x_0, y_0, z_0) = 0.1f_2(0, 2, 1) = -0.4$$

$$k_2 = hf_1\left(x_0 + \frac{h}{2}, y_0 + \frac{k_1}{2}, z_0 + \frac{l_1}{2}\right)$$

$$= 0.1f_1(0.05, 2.05, 0.8) = 0.085$$

$$l_2 = hf_2\left(x_0 + \frac{h}{2}, y_0 + \frac{k_1}{2}, z_0 + \frac{l_1}{2}\right)$$

$$= 0.1f_2(0.05, 2.05, 0.8) = -0.41525$$

$$k_3 = hf_1\left(x_0 + \frac{h}{2}, y_0 + \frac{k_2}{2}, z_0 + \frac{l_2}{2}\right)$$

$$= 0.1f_1(0.05, 2.0425, 0.79238) = 0.084238$$

$$l_3 = hf_2\left(x_0 + \frac{h}{2}, y_0 + \frac{k_2}{2}, z_0 + \frac{l_2}{2}\right)$$

$$= 0.1f_2(0.05, 2.0425, 0.79238) = -0.4122$$

$$k_4 = hf_1(x_0 + h, y_0 + k_3, z_0 + l_3)$$

$$= 0.1f_1(0.1, 2.084238, 0.5878) = 0.06878$$

$$l_4 = hf_2(x_0 + h, y_0 + k_3, z_0 + l_3)$$

$$= 0.1f_2(0.1, 2.084238, 0.5878) = -0.4244$$

$$\Delta y = \frac{1}{6}(k_1 + 2k_2 + 2k_3 + k_4)$$

$$= \frac{1}{6}(0.1 + 2 \times 0.085 + 2 \times 0.084238 + 0.06878) = 0.0845$$

$$y_1 = y(0.1) = 2.0845$$

$$\Delta z = \frac{1}{6}(l_1 + 2l_2 + 2l_3 + l_4)$$

$$= \frac{1}{6}(-0.4 + 2 \times (-0.41525) + 2 \times (-0.4122) + (-0.4244)] = -0.4132$$

$$z(0.1) = z_1 = z_0 + \Delta z = 1 - 0.4132 = 0.5868$$

Example 17: Find $y(0.1)$ given $y'' + 2xy' - 4y = 0$, $y(0) = 0.2$, $y'(0) = 0.5$.

Solution: Here $\dfrac{d^2y}{dx^2} = -2xy' + 4y$, $y(0) = 0.2$, $y'(0) = 0.5$, $h = 0.1$, $x_0 = 0$.

Let $$\frac{dy}{dx} = z$$

Then given equation becomes

$$\frac{dz}{dx} = -2xz + 4y$$

Thus the system of equations is

$$\frac{dy}{dx} = f_1(x, y, z) = z$$

$$\frac{dz}{dx} = f_2(x, y, z) = -2xz + 4y$$

Given $y_0 = 0.2$, $z_0 = y'(0) = 0.5$

$$k_1 = hf_1(x_0, y_0, z_0) = 0.1f_1(0, 0.2, 0.5) = 0.05$$

$$l_1 = hf_2(x_0, y_0, z_0) = 0.1f_2(0, 0.2, 0.5) = 0.08$$

$$k_2 = hf_1\left(x_0 + \frac{h}{2}, y_0 + \frac{k_1}{2}, z_0 + \frac{l_1}{2}\right) = 0.1f_1(0.05, 0.225, 0.54) = 0.054$$

$$l_2 = hf_2\left(x_0 + \frac{h}{2}, y_0 + \frac{k_1}{2}, z_0 + \frac{l_1}{2}\right) = 0.1f_2\,(0.05, 0.225, 0.54) = 0.0846$$

$$k_3 = hf_1\left(x_0 + \frac{h}{2}, y_0 + \frac{k_2}{2}, z_0 + \frac{l_2}{2}\right) = 0.1f_1\,(0.05, 0.227, 0.5423) = 0.05423$$

$$l_3 = hf_2\left(x_0 + \frac{h}{2}, y_0 + \frac{k_2}{2}, z_0 + \frac{l_2}{2}\right) = 0.1f_2\,(0.05, 0.227, 0.5423) = 0.085377$$

$$k_4 = hf_1\left(x_0 + h, y_0 + k_3, z_0 + l_3\right) = 0.1f_1\,(0.1, 0.25423, 0.585377) = 0.0585377$$

$$l_4 = hf_2\left(x_0 + h, y_0 + k_3, z_0 + l_3\right) = 0.1f_2\,(0.1, 0.25423, 0.585377) = 0.089984$$

$$\Delta y = \frac{1}{6}(k_1 + 2k_2 + 2k_3 + k_4)$$

$$= \frac{1}{6}(0.05 + 2 \times 0.054 + 2 \times 0.05423 + 0.0585377) = 0.05416$$

$$\Delta z = \frac{1}{6}(l_1 + 2l_2 + 2l_3 + l_4)$$

$$= \frac{1}{6}(0.08 + 2 \times 0.0846 + 2 \times 0.085377 + 0.089984) = 0.084989$$

$$y_1 = y(0.1) = y_0 + \Delta y = 0.25416$$

$$z_1 = z(0.1) = z_0 + \Delta z = 0.584989$$

Example 18: Solve $\dfrac{d^2y}{dx^2} = x\dfrac{dy}{dx} - y$; $y(0) = 3$, $y'(0) = 0$ to approximate $y(0.1)$.

Solution: Here $\dfrac{d^2y}{dx^2} = x\dfrac{dy}{dx} - y$; $y(0) = 3$, $y'(0) = 0$, $x_0 = 0$, $h = 0.1$.

Consider $\qquad\qquad \dfrac{dy}{dx} = z$

then $\qquad\qquad \dfrac{dz}{dx} = \dfrac{d^2y}{dx^2}$

Then given equation becomes

$$\frac{dz}{dx} = xz - y$$

Thus system of equations is

$$\frac{dy}{dx} = f_1(x, y, z) = z$$

$$\frac{dz}{dx} = f_2(x, y, z) = xz - y$$

given $y(0) = 3$, $z(0) = 0$

$$k_1 = hf_1(x_0, y_0, z_0) = 0.1f_1(0, 3, 0) = 0$$

$$l_1 = hf_2(x_0, y_0, z_0) = 0.1f_2(0, 3, 0) = -0.3$$

$$k_2 = hf_1\left(x_0 + \frac{h}{2}, y_0 + \frac{k_1}{2}, z_0 + \frac{l_1}{2}\right) = 0.1f_1\,(0.05, 3, -0.15) = -0.015$$

$$l_2 = hf_2\left(x_0 + \frac{h}{2}, y_0 + \frac{k_2}{2}, z_0 + \frac{l_2}{2}\right) = 0.1f_2(0.05, 3, -0.15) = -0.30075$$

$$k_3 = hf_1\left(x_0 + \frac{h}{2}, y_0 + \frac{k_2}{2}, z_0 + \frac{l_2}{2}\right) = 0.1f_1(0.05, 2.9925, -0.150375) = -0.0150375$$

$$l_3 = hf_2\left(x_0 + \frac{h}{2}, y_0 + \frac{k_2}{2}, z_0 + \frac{l_2}{2}\right) = 0.1f_2(0.05, 2.9925, -0.150375) = -0.3000018$$

$$k_4 = hf_1(x_0 + h, y_0 + k_3, z_0 + l_3) = 0.1f_1(0.1, 2.9849625, -0.3000018) = -0.03000018$$

$$l_4 = hf_2(x_0 + h, y_0 + k_3, z_0 + l_3) = 0.1f_2(0.1, 2.9849625, -0.3000018) = -0.3014962$$

$$\Delta y = \frac{1}{6}(k_1 + 2k_2 + 2k_3 + k_4)$$

$$= \frac{1}{6}[0 + 2 \times -0.015 + 2 \times -0.0150375 + (-0.03000018)] = -0.01501253$$

$$\Delta z = \frac{1}{6}(l_1 + 2l_2 + 2l_3 + l_4)$$

$$= \frac{1}{6}[-0.3 + 2 \times (-0.30075) + 2 \times (-0.3000018) - 0.3014962] = -0.3004999$$

$$y_1 = y(0.1) = y_0 + \Delta y = 2.984987$$
$$z_1 = z(0.1) = z_0 + \Delta z = -0.3004999$$

EXERCISE 10.2

1. Solve $\dfrac{dy}{dx} = y - x$, by second order Runge–Kutta method given that $y = 2$ when $x = 0$. Also find $y(0.1)$ and $y(0.2)$ and the corresponding error.

   [Ans. $y(0.1) = 2.2050, y(0.2) = 2.4210$, error $= 0.001, 0.004$]

2. Given $\dfrac{dy}{dx} = x^2 - y, y(0) = 1$, find $y(0.1), y(0.2)$ using Runge–Kutta methods of (i) second order, (ii) third order.

   [Ans. (i) $y(0.1) = 0.9055, y(0.2) = 0.8219275$; (ii) $y(0.1) = 0.90515, y(0.2) = 0.8212449$

3. Using second order Runge–Kutta method, find y at $x = 0.1$.

   Given $\dfrac{dy}{dx} = \dfrac{(1+x)y^2}{2}; y(0) = 1$ [Ans. $y(0.1) = 1.0552$]

4. Using Runge–Kutta method of fourth order, solve $\dfrac{dy}{dx} = \dfrac{y^2 - x^2}{y^2 + x^2}$ given $y(0) = 1$ at $x = 0.2, 0.4$.   [Ans. $y(0.2) = 1.19598, y(0.4) = 1.3751$

5. Find the value of $y(1.1)$ using Runge–Kutta method of fourth order, given that $y' = y^2 + xy, y(1) = 1.0$, take $h = 0.05$ [Ans. $y(1.05) = 1.109496, y(1.1) = 1.241506$]

6. Apply Runge–Kutta fourth order method to solve $10y' = x^2 + y^2; y(0) = 1$ for $0 < x \leq 0.4$ and $h = 0.1$.

   [Ans. $y(0.1) = 1.0101, y(0.2) = 1.0206, y(0.3) = 1.0317, y(0.4) = 1.0437$]

7. Evaluate $y(1.2)$ and $y(1.4)$; given $y' = \dfrac{2xy + e^x}{x^2 + xe^x}$, $y(1) = 0$.

[Ans. $y(1.2) = 0.1402$, $y(1.4) = 0.2705$]

8. Solve $\dfrac{dy}{dx} + \dfrac{y}{x} = \dfrac{1}{x^2}$, $y(1) = 1$ for $y(1.1)$. [Ans. $y(1.1) = 0.9958$]

9. Use Runge–Kutta fourth order method with $h = 0.1$ to find $x(0.1)$ and $x(0.2)$ where

$\dfrac{dx}{dt} = t - x$ and $x(0) = 0$. [Ans. $x(0.1) = 0.0048375$, $x(0.2) = 0.0187305$]

10. Find by Runge–Kutta method an approximate value of y for $x = 0.8$ given that $y = 0.41$ when $x = 0.4$ and $y' = \sqrt{x + y}$. [Ans. 0.848]

11. Solve $\dfrac{dy}{dx} = yz + x$, $\dfrac{dz}{dx} = xz + y$, given that $y(0) = 1$, $z(0) = -1$ for $y(0.1)$, $z(0.1)$.

[Ans. $y(0.1) = 0.9139363$, $z(0.1) = -0.9092176$]

12. Evaluate $y(1.1)$, $z(1.1)$ given $\dfrac{dy}{dx} = xyz$, $\dfrac{dz}{dx} = \dfrac{xy}{z}$, $y(1) = \dfrac{1}{3}$, $z(1) = 1$.

[Ans. $y(1.1) = 0.3707$, $z(1.1) = 1.03615$]

13. Obtain the value of $x(0.1)$ given $\dfrac{d^2x}{dt^2} = t\dfrac{dx}{dt} - 4x$, $x(0) = 3$, $x'(0) = 0$.

[Ans. $x(0.1) = 2.9399$]

14. Solve $\dfrac{d^2y}{dx^2} - x^2\dfrac{dy}{dx} - 2xy = 0$ given that $y(0) = 1$, $y'(0) = 0$ for $y(0.1)$ using Runge–Kutta method. [Ans. 1.005334]

15. Use Runge-Kutta method to approximate y, when $x = 0.1$ and $x = 0.2$, given that

$y(0) = 1$ and $\dfrac{dy}{dx} = x + y$. [Ans. 1.1103, 1.2428]

10.7 PREDICTOR-CORRECTOR METHODS

Runge–Kutta method is called single step method because this method use only the information from the last step computed and no attention is paid for the nature of the solution at the earlier points.

Now we shall take up the methods which require values x_0, x_{-1}, x_{-2}, ... to compute value of y at x_1. A predictor formula predicts the value of y at $x = x_1$ and then corrector formula is used to improve the predicted value of y at $x = x_1$.

In this section, we will obtain two important predictor–corrector methods, namely, Milne's method and Adam–Bashforth (or Adam's) predictor–corrector method.

10.8 MILNE'S METHOD

Suppose our aim is to solve the following differential equation

$$\frac{dy}{dx} = f(x, y),\ y(x_0) = y_0 \tag{10.29}$$

Starting from $y_0 = y(x_0)$, we have to estimate successively $y_1 = y(x_1)$, $y_2 = y(x_2)$, $y_3 = y(x_3)$ where $x_1 = x_0 + h$, $x_2 = x_0 + 2h$ and $x_3 = x_0 + 3h$.

By Newton's forward interpolation formula, we have

$$y = y_0 + u\Delta y_0 + \frac{u(u-1)}{2!}\Delta^2 y_0 + \frac{u(u-1)(u-2)}{3!}\Delta^3 y_0 + \ldots$$

where $u = \dfrac{x-x_0}{h} \Rightarrow x = x_0 + uh$ and $dx = hdu$

Changing y to y', we get

$$y' = y_0' + u\Delta y_0' + \frac{u(u-1)}{2!}\Delta^2 y_0' + \frac{u(u-1)(u-2)}{3!}\Delta^3 y_0' + \ldots \quad (10.30)$$

Now integrating Eq. (10.30) from x_0 to $x_0 + 4h$, we have

$$\int_{x_0}^{x_4=x_0+4h} y'dx = \int_{x_0}^{x_0+4h}\left[y_0' + u\Delta y_0' + \frac{u(u-1)}{2!}\Delta^2 y_0' + \frac{u(u-1)(u-2)}{3!}\Delta^3 y_0' + \ldots\right]dx$$

$$\Rightarrow \quad [y]_{x_0}^{x_4} = h\left[y_0' u + \frac{u^2}{2}\Delta y_0' + \frac{\left(\dfrac{u^3}{3} - \dfrac{u^2}{2}\right)}{2}\Delta^2 y_0' + \frac{\left(\dfrac{u^4}{4} - u^3 + u^2\right)}{6}\Delta^3 y_0' + \ldots\right]_0^4$$

$$\Rightarrow \quad y_4 - y_0 = h\left[4y_0' + 8\Delta y_0' + \frac{20}{3}\Delta^2 y_0' + \frac{8}{3}\Delta^3 y_0' + \frac{14}{45}\Delta^4 y_0' + \ldots\right]$$

$$= \frac{4h}{3}\left[3y_0' + 6\Delta y_0' + 5\Delta^2 y_0' + 2\Delta^3 y_0' + \frac{7}{30}\Delta^4 y_0' + \ldots\right]$$

$$= \frac{4h}{3}\left[3y_0' + 6(y_1' - y_0') + 5(y_2' - 2y_1' + y_0') + \right.$$

$$\left. 2(y_3' - 3y_2' + 3y_1' - y_0') + \frac{7}{30}\Delta^4 y_0' + \ldots\right]$$

$$= \frac{4h}{3}(2y_1' - y_2' + 2y_3') + \frac{14h}{45}\Delta^4 y_0' + \ldots \quad (10.31)$$

On neglecting fourth and higher order differences, we have

$$y_4 - y_0 = \frac{4h}{3}(2y_1' - y_2' + 2y_3')$$

$$\Rightarrow \quad y_4 = y_0 + \frac{4h}{3}(2y_1' - y_2' + 2y_3')$$

This is Milne's predictor formula. It will predict the value of y_4 when y_0, y_1, y_2 and y_3 are known.

This formula can also be written in the following manner

$$y_4 = y_0 + \frac{4h}{3}(2f_1 - f_2 + 2f_3) \quad (10.32)$$

To get Milne's corrector formula, integrate Eq. (10.30) between the limits x_0 to $x_0 + 2h$.

Therefore

$$\int_{x_0}^{x_0+2h} y'dx = \int_{x_0}^{x_0+2h}\left[y_0' + u\Delta y_0' + \frac{u(u-1)}{2!}\Delta^2 y_0' + ...\right]dx$$

$$y_2 - y_0 = h\int_0^2\left[y_0' + u\Delta y_0' + \frac{(u^2-u)}{2}\Delta^2 y_0' + ...\right]du$$

$$= h\left[uy_0' + \frac{u^2}{2}\Delta y_0' + \frac{\left(\frac{u^3}{3}-\frac{u^2}{2}\right)}{2}\Delta^2 y_0' + ...\right]_0^2$$

$$= h\left[2y_0' + 2\Delta y_0' + \frac{1}{3}\Delta^2 y_0' + ...\right]$$

Neglecting third and higher order differences, we have

$$y_2 - y_0 = h\left[2y_0' + 2(y_1'-y_0') + \frac{1}{3}(y_2'-2y_1'+y_0')\right]$$

$$= \frac{h}{3}[y_0' + 4y_1' + y_2']$$

$$\Rightarrow \qquad y_2 = y_0 + \frac{h}{3}[y_0' + 4y_1' + y_2']$$

This is Milne's corrector formula. It can also be written as

$$y_2 = y_0 + \frac{h}{3}[f_0 + 4f_1 + f_2]$$

In general, the predictor formula is given by

$$y_{n+1} = y_{n-3} + \frac{4h}{3}[2f_{n-2} - f_{n-1} + 2f_n]$$

and the corrector formula is given by

$$y_{n+1} = y_{n-1} + \frac{h}{3}[f_{n-1} - 4f_n + f_{n+1}]$$

Example 19: Find $y(2)$ if $y(x)$ is the solution of $\frac{dy}{dx} = \frac{x+y}{2}$ where $y(0) = 2$, $y(0.5) = 2.636$, $y(1) = 3.595$ and $y(1.5) = 4.968$ using Milne's predictor–corrector formula.

[APJAKTU 2014]

Solution: Here $\qquad \frac{dy}{dx} = f(x,y) = \frac{1}{2}(x+y) \qquad\qquad$ (i)

$x_0 = 0, x_1 = 0.5, x_2 = 1.0, x_3 = 1.5, h = 0.5, y_0 = 2, y_1 = 2.636, y_2 = 3.595, y_3 = 4.968$.
By Milne's predictor formula

$$y_4 = y_0 + \frac{4h}{3}(2f_1 - f_2 + 2f_3) \qquad\qquad (ii)$$

$$f_1 = f(x_1, y_1) = \frac{1}{2}(x_1 + y_1) = \frac{1}{2}(0.5 + 2.636) = 1.5680$$

$$f_2 = f(x_2, y_2) = \frac{1}{2}(x_2 + y_2) = \frac{1}{2}(1.0 + 3.595) = 2.2975$$

$$f_3 = f(x_3, y_3) = \frac{1}{2}(x_3 + y_3) = \frac{1}{2}(1.5 + 4.968) = 3.2340$$

From Eq. (ii) $y_4 = 2 + \frac{4 \times 0.5}{3}[2 \times 1.5680 - 2.2975 + 2 \times 3.2340] = 6.8710$

By Milne's corrector formula

$$y_4 = y_2 + \frac{h}{3}(f_2 + 4f_3 + f_4) \qquad \text{(iii)}$$

But $$f_4 = f(x_4, y_4) = \frac{1}{2}(x_4 + y_4)$$

$$= \frac{1}{2}(2 + 6.8710) = 4.4355$$

By Eq. (iii) $y_4 = 3.595 + \frac{0.5}{3}(2.2975 + 4 \times 3.2340 + 4.4355) = 6.8732$

$\therefore$ Corrected value of y at $x = 2.0$ is 6.8732.

Example 20: Apply Milne's predictor–corrector method, find $y(0.8)$ if $y(x)$ is the solution of $\frac{dy}{dx} = 1 + y^2$. Given $y(0) = 0$, $y(0.2) = 0.2027$, $y(0.4) = 0.4228$ and $y(0.6) = 0.6841$.

Solution: Here $\frac{dy}{dx} = f(x, y) = 1 + y^2$

Given $x_0 = 0, x_1 = 0.2, x_2 = 0.4, x_3 = 0.6; y_0 = 0, y_1 = 0.2027, y_2 = 0.4228, y_3 = 0.6841$ and $h = 0.2$

$$f_0 = f(x_0, y_0) = 1 + y_0^2 = 1 + 0 = 1$$
$$f_1 = f(x_1, y_1) = 1 + y_1^2 = 1.04109$$
$$f_2 = f(x_2, y_2) = 1 + y_2^2 = 1.17876$$
$$f_3 = f(x_3, y_3) = 1 + y_3^2 = 1.46799$$

By Milne's predictor formula

$$y_4 = y_0 + \frac{4h}{3}(2f_1 - f_2 + 2f_3)$$

$$y_4 = y(0.8) = 0 + \frac{4 \times 0.2}{3}[2 \times 1.04109 - 1.17876 + 2 \times 1.46799] = 1.02384$$

Corresponding to this predicted value, we have

$$f_4 = f(x_4, y_4) = 1 + y_4^2 = 1 + (1.02384)^2 = 2.04825$$

By Milne's corrector formula, we have

$$y_4 = y_2 + \frac{h}{3}(f_2 + 4f_3 + f_4)$$

$$= 0.4228 + \frac{0.2}{3}[1.17876 + 4 \times 1.46799 + 2.04825]$$

$$y_4 = y(0.8) = 1.02940$$

Example 21: Apply Milne's method to find a solution of the differential equation $\frac{dy}{dx} = x - y^2$ in the range $0 \le x \le 1$ for the boundary condition $y = 0$ at $x = 0$.

Solution: Here $\qquad \frac{dy}{dx} = f(x, y) = x - y^2$ $\qquad\qquad$ (i)

Given $x_0 = 0$, $y_0 = 0$.

In Milne's method, we need three known functional values to predict the fourth one. To obtain y_1, y_2 and y_3, we use Picard's method.

By Picard's method, we have

$$y = y_0 + \int_0^x f(x, y)\, dx$$

For first approximation y_1, we put y_0 in place of y

$$y_1 = y_0 + \int_0^x f(x, y_0)\, dx = 0 + \int_0^x f(x, 0)\, dx = \int_0^x x\, dx = \frac{x^2}{2}$$

Similarly, $\qquad y_2 = y_0 + \int_0^x f(x, y_1)\, dx = 0 + \int_0^x f\left(x, \frac{x^2}{2}\right) dx$

$$= \int_0^x \left[x - \left(\frac{x^2}{2}\right)^2 \right] dx = \frac{x^2}{2} - \frac{x^5}{20}$$

$$y_3 = y_0 + \int_0^x f(x, y_2)\, dx = 0 + \int_0^x f\left(x, \frac{x^2}{2} - \frac{x5}{20}\right) dx$$

$$= \int_0^x \left[x - \left(\frac{x^2}{2} - \frac{x^5}{20}\right)^2 \right] dx = \int_0^x \left(x - \frac{x^4}{4} + \frac{x^7}{20} - \frac{x^{10}}{400} \right) dx$$

$$= \frac{x^2}{2} - \frac{x^5}{20} + \frac{x^8}{160} - \frac{x^{11}}{4400} \qquad\qquad \text{(ii)}$$

Taking $h = 0.2$, we have $x_0 = 0$, $x_1 = 0.2$, $x_2 = 0.4$, $x_3 = 0.6$. Putting these values of x in Eq. (ii), we get

$$y_1 = \frac{(0.2)^2}{2} - \frac{(0.2)^5}{20} + \frac{(0.2)^8}{160} - \frac{(0.2)^{11}}{4400} = 0.019984$$

$$y_2 = \frac{(0.4)^2}{2} - \frac{(0.4)^5}{20} + \frac{(0.4)^8}{160} - \frac{(0.4)^{11}}{4400} = 0.079492$$

$$y_3 = \frac{(0.6)^2}{2} - \frac{(0.6)^5}{20} + \frac{(0.6)^8}{160} - \frac{(0.6)^{11}}{4400} = 0.1762169$$

and
$$f_0 = f(x_0, y_0) = 0$$
$$f_1 = f(x_1, y_1) = x_1 - y_1^2 = 0.2 - (0.019984)^2 = 0.1996006$$
$$f_2 = f(x_2, y_2) = x_2 - y_2^2 = 0.4 - (0.079492)^2 = 0.3936810$$
$$f_3 = f(x_3, y_3) = x_3 - y_3^2 = 0.6 - (0.1762169)^2 = 0.5689476$$

By Milne's predicted formula, we have

$$y_4 = y_0 + \frac{4h}{3}(2f_1 - f_2 + 2f_3)$$

$$= 0 + \frac{4 \times 0.2}{3}[2 \times 0.1996006 - 0.3936810 + 2 \times 0.5689476]$$

$$= 0.3049107$$

Corresponding to this predicted value of y_4, we have
$$f_4 = f(x_4, y_4) = x_4 - y_4^2 = 0.8 - (0.3049107)^2 = 0.7070294$$

Using Milne's corrector formula, we find

$$y_4 = y_2 + \frac{h}{3}(f_2 + 4f_3 + f_4)$$

$$= 0.079492 + \frac{0.2}{3}[0.3936810 + 4 \times 0.5689476 + 0.7070294]$$

$$= 0.304592$$

which is the improved value of y_4.

Corresponding to this corrected value of y_4, we have
$$f_4 = f(x_4, y_4) = x_4 - y_4^2 = 0.8 - (0.304592)^2 = 0.7072236$$

Next, we shall predict the value of y_5, we have

$$y_5 = y_1 + \frac{4h}{3}(2f_2 - f_3 + 2f_4)$$

$$= 0.019984 + \frac{4 \times 0.2}{3}[2 \times 0.3936810 - 0.5689476 + 2 \times 0.7072236]$$

$$= 0.4554138$$

Corresponding to this predicted value of y_5, we have
$$f_5 = f(x_5, y_5) = x_5 - y_5^2 = 1.0 - (0.4554138)^2 = 0.7925982$$

By using Milne's corrected formula

$$y_5 = y_3 + \frac{h}{3}(f_3 + 4f_4 + f_5)$$

$$= 0.1762169 + \frac{0.2}{3}[0.5689476 + 4 \times 0.7072236 + 0.7925982]$$

$$y_5 = y(1.0) = 0.4555795$$

which is the corrected value of $y(1.0)$.

Example 22: Using Runge–Kutta method of order four, find y for $x = 0.3, 0.6, 0.9$ given

that $\dfrac{dy}{dx} = -y; y(0) = 1$. Continue the solution at $x = 1.2$ using Milne's method.

Solution: Here $f(x, y) = \dfrac{dy}{dx} = -y$

Given $x_0 = 0$, $y_0 = 1$, $h = 0.3$

$$k_1 = hf(x_0, y_0) = 0.3f(0, 1) = 0.3 \times -1 = -0.3$$

$$k_2 = hf\left(x_0 + \frac{h}{2}, y_0 + \frac{k_1}{2}\right) = 0.3f(0.15, 0.85) = -0.255$$

$$k_3 = hf\left(x_0 + \frac{h}{2}, y_0 + \frac{k_2}{2}\right) = 0.3f(0.15, 0.8875) = -0.26625$$

$$k_4 = hf(x_0 + h, y_0 + k_3) = 0.3f(0.3, 0.73375) = -0.220125$$

$$\Delta y = \frac{1}{6}(k_1 + 2k_2 + 2k_3 + k_4)$$

$$= \frac{1}{6}[-0.3 + 2 \times -0.255 + 2 \times -0.26625 - 0.220125]$$

$$= -0.2604375$$

$$y_1 = y(0.3) = y_0 + \Delta y = 1 - 0.2604375 = 0.7395625$$

Again, we have $x_1 = 0.3$, $y_1 = 0.7395625$

$$k_1 = hf(x_1, y_1) = 0.3f(0.3, 0.7395625) = -0.2218688$$

$$k_2 = hf\left(x_1 + \frac{h}{2}, y_1 + \frac{k_1}{2}\right)$$

$$= 0.3f(0.45, 0.6286281) = -0.1885884$$

$$k_3 = hf\left(x_1 + \frac{h}{2}, y_1 + \frac{k_2}{2}\right) = 0.3f(0.45, 0.6452683) = -0.1935805$$

$$k_4 = hf(x_1 + h, y_1 + k_3) = 0.3f(0.6, 0.545982) = -0.1637946$$

$$\Delta y = \frac{1}{6}(k_1 + 2k_2 + 2k_3 + k_4)$$

$$= \frac{1}{6}[-0.2218688 + 2 \times -0.1885884 + 2 \times -0.1935805 - 0.1637946]$$

$$= -0.1916669$$

$$y_2 = y(0.6) = y_1 + \Delta y = 0.5478956$$

Again $x_2 = 0.6$, $y_2 = 0.5478956$

$$k_1 = hf(x_2, y_2) = 0.3f(0.6, 0.5478956) = -0.1643687$$

$$k_2 = hf\left(x_2 + \frac{h}{2}, y_2 + \frac{k_1}{2}\right) = 0.3f(0.75, 0.4657112) = -0.1397134$$

$$k_3 = hf\left(x_2 + \frac{h}{2}, y_2 + \frac{k_2}{2}\right) = 0.3f(0.75, 0.4780389) = -0.1434117$$

$$k_4 = hf(x_2 + h, y_2 + k_3) = 0.3f(0.9, 0.4044839) = -0.1213452$$

$$\Delta y = \frac{1}{6}(k_1 + 2k_2 + 2k_3 + k_4)$$

$$= \frac{1}{6}[-0.1643687 + 2 \times -0.1397134 + 2 \times -0.1434117 - 0.1213452]$$

$$= -0.1419940$$

$$y(0.9) = y_3 = y_2 + \Delta y = 0.5478956 - 0.1419940 = 0.4059016$$

Since $y(0.3) = 0.7395625$, $y(0.6) = 0.5478956$, $y(0.9) = 0.4059016$

Now
$$f_1 = f(x_1, y_1) = -y_1 = -0.7395625$$
$$f_2 = f(x_2, y_2) = -y_2 = -0.5478956$$
$$f_3 = f(x_3, y_3) = -y_3 = -0.4059016$$

By Milne's predicted formula, we have

$$y_4 = y_0 + \frac{4h}{3}(2f_1 - f_2 + 2f_3)$$

$$= 1 + \frac{4 \times 0.3}{3}[2 \times (-0.7395625) - (-0.5478956) + 2 \times (-0.4059016)]$$

$$= 0.3027870$$

Corresponding to this predicted value of y_4, we have
$$f_4 = f(x_4, y_4) = -y_4 = -0.3027870$$

Using Milne's corrected formula, we have

$$y_4 = y_2 + \frac{h}{3}(f_2 + 4f_3 + f_4)$$

$$= 0.5478956 + \frac{0.3}{3}[-0.5478956 + 4 \times -0.4059016 + (-0.3027870)]$$

$$= 0.3004667$$

Thus $y_4 = y(1.2) = 0.3004667$.

10.9 ADAM–BASHFORTH (OR ADAM'S) PREDICTOR–CORRECTOR METHOD

Let $\frac{dy}{dx} = f(x, y)$, $y(x_0) = y_0$ numerically making use of either Taylor's series or Runge–Kutta method, we first calculate $y_{-1} = y(x_{-1})$, $y_{-2} = y(x_{-2})$ and $y_{-3} = y(x_{-3})$ where $x_{-1} = x_0 - h$, $x_{-2} = x_0 - 2h$, $x_{-3} = x_0 - 3h$. By Newton backward formula for interpolation, we have

$$y = y_0 + u\nabla y_0 + \frac{u(u+1)}{2}\nabla^2 y_0 + \frac{u(u+1)(u+2)}{6}\nabla^3 y_0 + \dots \quad (10.33)$$

where $x = x_0 + uh$.

Changing y to y', we get

$$y' = y_0' + u\nabla y_0' + \frac{u(u+1)}{2}\nabla^2 y_0' + \frac{u(u+1)(u+2)}{6}\nabla^3 y_0' + \dots \quad (10.34)$$

Now integrating Eq. (10.34) from x_0 to $x_0 + h$, we have

$$\int_{x_0}^{x_1 = x_0 + h} y'dx = \int_{x_0}^{x_0 + h} \left[y_0' + u\nabla y_0' + \frac{u(u+1)}{2}\nabla^2 y_0' + \frac{u(u+1)(u+2)}{6}\nabla^3 y_0' + \dots \right] dx$$

$$\Rightarrow \qquad y_1 - y_0 = h\left[uy_0' + \frac{u^2}{2}\nabla y_0' + \frac{\left(\dfrac{u^3}{3} + \dfrac{u^2}{2}\right)}{2}\nabla^2 y_0' + \frac{\left(\dfrac{u^4}{4} + \dfrac{3u^3}{3} + \dfrac{2u^2}{2}\right)}{6}\nabla^3 y_0' + ... \right]_0^1$$

$$= h\left[y_0' + \frac{\nabla y_0'}{2} + \frac{5}{12}\nabla^2 y_0' + \frac{3}{8}\nabla^3 y_0' + ... \right] \qquad (10.35)$$

Neglecting the fourth and higher order derivatives and using

$$\nabla y_0' = y_0' - y_{-1}' = f_0 - f_{-1}$$
$$\nabla^2 y_0' = y_0' - 2y_{-1}' + y_{-2}' = f_0 - 2f_{-1} + f_{-2}$$
and
$$\nabla^3 y_0' = y_0' - 3y_{-1}' + 3y_{-2}' - y_{-3}' = f_0 - 3f_{-1} + 3f_{-2} - f_{-3}$$

We put the above values in Eq. (10.35), we get

$$y_1 = y_0 + h\left[f_0 + \frac{1}{2}(f_0 - f_{-1}) + \frac{5}{12}(f_0 - 2f_{-1} + f_{-2}) \right.$$

$$\left. + \frac{3}{8}(f_0 - 3f_{-1} + 3f_{-2} - f_{-3}) \right]$$

$$= y_0 + \frac{h}{24}(55f_0 - 59f_{-1} + 37f_{-2} - 9f_{-3}) \qquad (10.36)$$

which is known as Adams–Bashforth predictor formula.

Just to improve the predicted value of y_1 from this predictor formula, we first find f_1 using the predicted value y_1. Again we start with Newton's backward formula for interpolation to derive the corrector formula at y_1.

$$y' = y_1' + u\nabla y_1' + \frac{u(u+1)}{2}\nabla^2 y_1' + \frac{u(u+1)(u+2)}{6}\nabla^3 y_1' + ...$$

On integrating from x_0 to $x_0 + h$ (corresponding limits for u will be -1 to 0 as $x = x_1 + uh$) followed by simplification will give

$$y_1 = y_0 + h\left[f_1 - \frac{1}{2}\nabla f_1 - \frac{1}{12}\nabla^2 f_1 - \frac{1}{24}\nabla^3 f_1 - ... \right]$$

Neglecting fourth and higher order derivative terms, we get

$$y_1 = y_0 + h\left[f_1 - \frac{1}{2}\nabla f_1 - \frac{1}{2}\nabla^2 f_1 - \frac{1}{24}\nabla^3 f_1 \right] \qquad (10.37)$$

Now, $\qquad \nabla f_1 = f_1 - f_0, \nabla^2 f_1 = f_1 - 2f_0 - f_{-1}, \nabla^3 f_1 = f_1 - 3f_0 + 3f_{-1} - f_{-2}$

Putting the above values in Eq. (10.37), we get

$$y_1 = y_0 + h\left[f_1 - \frac{1}{2}(f_1 - f_0) - \frac{1}{12}(f_1 - 2f_0 - f_{-1}) - \frac{1}{24}(f_1 - 3f_0 + 3f_{-1} - f_{-2}) \right]$$

$$= y_0 + \frac{h}{24}(9f_1 + 19f_0 - 5f_{-1} + f_{-2})$$

which is known as Adams–Bashforth corrector formula. Adams–Bashforth predictor-corrector formula are denoted generally as

$$y_{n+1} = y_n + \frac{h}{24}[55f_n - 59f_{n-1} + 37f_{n-2} - 9f_{n-3}] \text{ (Predicted formula)}$$

$$= y_n + \frac{h}{24}[9f_{n+1} + 19f_n - 5f_{n-1} + f_{n-2}] \text{ (Corrected formula)}$$

Example 23: Using Runge–Kutta method of order 4, find y for $x = 0.1, 0.2, 0.3$ given that $\dfrac{dy}{dx} = xy + y^2$, $y(0) = 1$. Continue the solution at $x = 0.4$ using any predictor–corrector method.

Solution: To solve this problem we will use Adams–Bashforth predictor–corrector method.

Let $f(x, y) = \dfrac{dy}{dx} = xy + y^2$, $x_0 = 0$, $x_1 = 0.1$, $x_2 = 0.2$, $x_3 = 0.4$, $x_4 = 0.4$, $y_0 = 1$

$$k_1 = hf(x_0, y_0) = (0.1)\, f(0, 1) = (0.1)1 = 0.1$$

$$k_2 = hf\left(x_0 + \frac{h}{2}, y_0 + \frac{k_1}{2}\right) = 0.1f(0.05, 1.05)$$

$$= (0.1)[(0.05)(1.05) + (1.05)^2] = 0.1155$$

$$k_3 = hf\left(x_0 + \frac{h}{2}, y_0 + \frac{k_2}{2}\right) = 0.1f(0.05, 1.0578)$$

$$= 0.1[(0.05)(1.0578) + (1.0578)^2] = 0.1172$$

$$k_4 = hf(x_0 + h, y_0 + k_3) = 0.1f(0.1, 1.1172)$$

$$= 0.1[(0.1)(1.1172) + (1.1172)^2] = 0.13598$$

$$\Delta y = \frac{1}{6}(k_1 + 2k_2 + 2k_3 + k_4)$$

$$= \frac{1}{6}(0.1 + 2 \times 0.1155 + 2 \times 0.1172 + 0.13598] = 0.1169$$

$$y_1 = y_0 + \Delta y = 1 + 0.1169 = 1.1169$$

Again $$k_1 = hf(x_1, y_1) = 0.1f(0.1, 1.1169) = 0.1359$$

$$k_2 = hf\left(x_1 + \frac{h}{2}, y_1 + \frac{k_1}{2}\right) = 0.1f(0.15, 1.1849) = 0.1582$$

$$k_3 = hf\left(x_1 + \frac{h}{2}, y_1 + \frac{k_2}{2}\right) = 0.1f(0.15, 1.196) = 0.16098$$

$$k_4 = hf(x_1 + h, y_1 + k_3) = 0.1f(0.2, 1.2779) = 0.1889$$

$$\Delta y = \frac{1}{6}(k_1 + 2k_2 + 2k_3 + k_4)$$

$$= \frac{1}{6}[0.1359 + 2 \times 0.1582 + 2 \times 0.16098 + 0.1889] = 0.1605$$

$$y_2 = y_1 + \Delta y = 1.2774$$

Again $$k_1 = hf(x_2, y_2) = 0.1f(0.2, 1.2774) = 0.1887$$

$$k_2 = hf\left(x_2 + \frac{h}{2}, y_2 + \frac{k_1}{2}\right) = 0.1f(0.25, 1.3718) = 0.2225$$

$$k_3 = hf\left(x_2 + \frac{h}{2}, y_2 + \frac{k_2}{2}\right) = 0.1f(0.25, 1.3887) = 0.2274$$

$$k_4 = hf(x_2 + h, y_2 + k_3) = 0.1f(0.3, 1.5048) = 0.2716$$

$$\Delta y = \frac{1}{6}(k_1 + 2k_2 + 2k_3 + k_4)$$

$$= \frac{1}{6}[0.1887 + 2 \times 0.2225 + 2 \times 0.2274 + 0.2716] = 0.22668$$

$$y_3 = y_2 + \Delta y = 1.2774 + 0.22668 = 1.5041$$

By Adams–Bashforth predicted formula, we have

$$y_4 = y_3 + \frac{h}{24}[55f_3 - 59f_2 + 37f_1 - 9f_0]$$

Now

$$f_0 = x_0 y_0 + y_0^2 = 0 \times 1 + (1)^2 = 1$$

$$f_1 = x_1 y_1 + y_1^2 = (0.1)(1.1169) + (1.1169)^2 = 1.3592$$

$$f_2 = x_2 y_2 + y_2^2 = (0.2)(1.2774) + (1.2774)^2 = 1.8872$$

$$f_3 = x_3 y_3 + y_3^2 = (0.3)(1.5041) + (1.5041)^2 = 2.7135$$

So,

$$y_4 = 1.5041 + \frac{0.1}{24}[55 \times 2.7135 - 59 \times 1.8872 + 37 \times 1.3592 - 9 \times 1]$$

$$= 1.8341$$

Corresponding to this predicted value of y_4, we have

$$f_4 = x_4 y_4 + y_4^2 = (0.4)(1.8341) + (1.8341)^2 = 4.0976$$

Using Adams–Bashforth corrector formula, we have

$$y_4 = y_3 + \frac{h}{24}[9f_4 + 19f_3 - 5f_2 + f_1]$$

$$= 1.5041 + \frac{0.1}{24}[9 \times 4.0976 + 19 \times 2.7135 - 5 \times 1.8872 + 1.3592]$$

$$= 1.8389$$

Hence

$$y(0.4) = 1.8389$$

Example 24: Determine $y(0.4)$ given the equation $\dfrac{dy}{dx} = \dfrac{1}{2}xy$ using Adams–Bashforth method, given that $y(0) = 1$, $y(0.1) = 1.0025$, $y(0.2) = 1.0101$, $y(0.3) = 1.0228$.

Solution: Here $f(x, y) = \dfrac{1}{2}xy$, $x_0 = 0$, $x_1 = 0.1$, $x_2 = 0.2$, $x_3 = 0.3$, $y_0 = 1$, $y_1 = 1.0025$, $y_2 = 1.0101$, $y_3 = 1.0228$.

Now,

$$f_0 = \frac{1}{2}x_0 y_0 = \frac{1}{2} \times 0 \times 1 = 0$$

$$f_1 = \frac{1}{2}x_1 y_1 = \frac{1}{2} \times 0.1 \times 1.0025 = 0.050125$$

$$f_2 = \frac{1}{2}x_2 y_2 = \frac{1}{2} \times 0.2 \times 1.0101 = 0.10101$$

$$f_3 = \frac{1}{2}x_3 y_3 = \frac{1}{2} \times 0.3 \times 1.0228 = 0.15342$$

Now, Adams–Bashforth predictor formula is

$$y_4 = y_3 + \frac{h}{24}(55f_3 - 59f_2 + 37f_1 - 9f_0)$$

$$= 1.0228 + \frac{0.1}{24}[55 \times 0.15342 - 59 \times 0.10101 + 37 \times 0.050125 - 9 \times 0]$$

$$= 1.04085$$

Corresponding to this predicted value of y_4, we have

$$f_4 = \frac{1}{2}x_4y_4 = \frac{1}{2} \times 0.4 \times 1.04085 = 0.20817$$

Using Adams–Bashforth corrector formula, we have

$$y_4 = y_3 + \frac{h}{24}[9f_4 + 19f_3 - 5f_2 + f_1]$$

$$= 1.0228 + \frac{0.1}{24}[9 \times 0.20817 + 19 \times 0.15342 - 5 \times 0.10101 + 0.050125]$$

$$= 1.04085$$

Example 25: Using Adams–Bashforth predictor–corrector method, find $y(1.4)$ given that $x^2y' + xy = 1$; $y(1) = 1$, $y(1.1) = 0.996$, $y(1.2) = 0.986$, $y(1.3) = 0.972$.

Solution: Here $x^2y' + xy = 1$

or
$$x^2\frac{dy}{dx} = 1 - xy$$

or
$$\frac{dy}{dx} = f(x,y) = \frac{1-xy}{x^2}$$

Let $x_0 = 1$, $x_1 = 1.1$, $x_2 = 1.2$, $x_3 = 1.3$.
Given $y_0 = 1$, $y_1 = 0.996$, $y_2 = 0.986$, $y_3 = 0.972$.

Now
$$f_0 = \frac{1-x_0y_0}{x_0^2} = \frac{1-1\times1}{1^2} = 0$$

$$f_1 = \frac{1-x_1y_1}{x_1^2} = \frac{1-(1.1)\times(0.996)}{(1.1)^2} = -0.079008$$

$$f_2 = \frac{1-x_2y_2}{x_2^2} = \frac{1-(1.2)\times(0.986)}{(1.2)^2} = -0.127222$$

$$f_3 = \frac{1-x_3y_3}{x_3^2} = \frac{1-(1.3)\times(0.972)}{(1.3)^2} = -0.155976$$

Now, Adams–Bashforth predictor formula, we have

$$y_4 = y_3 + \frac{h}{24}[55f_3 - 59f_2 + 37f_1 - 9f_0]$$

$$= 0.972 + \frac{0.1}{24}[55 \times (-0.155976) - 59 \times (-0.127222) + 37 \times (-0.079008) - 9 \times 0]$$

$$= 0.955350$$

Corresponding to this predictor value of y_4, we have

$$f_4 = \frac{1 - x_4 y_4}{x_4^2}$$

$$= \frac{1 - (1.4) \times 0.955350}{(1.4)^2} = -0.172189$$

Using Adams–Bashforth corrector formula, we have

$$y_4 = y_3 + \frac{h}{24}[9f_4 + 19f_3 - 5f_2 + f_1]$$

$$= 0.972 + \frac{0.1}{24}[9 \times -0.172189 + 19 \times -0.155976 - 5 \times -0.127222 + (-0.079008)]$$

$$y_4 = y(1.4) = 0.955516$$

EXERCISE 10.3

1. Using Adams–Bashforth method to find $y(1.4)$ given:

   $\dfrac{dy}{dx} = x^2(1+y)$, $y(1) = 1$, $y(1.1) = 1.233$, $y(1.2) = 1.548$ and $y(1.3) = 1.979$.

   [Ans. $y(1.4) = 2.5749$]

2. Find $y(0.1)$, $y(0.2)$, $y(0.3)$ from $\dfrac{dy}{dx} = x^2 - y$; $y(0) = 1$ by using Taylor's series method and hence obtain $y(0.4)$ using Adams–Bashforth method.

   [Ans. $y(0.4) = 0.6896522$]

3. Using Adams–Bashforth formula, find $y(0.4)$ and $y(0.5)$ if y satisfies the differential equation $\dfrac{dy}{dx} = 3e^x + 2y$ with $y(0) = 0$. Compute y at $x = 0.1, 0.2, 0.3$ by means of Runge–Kutta method. [Ans. $y(0.4) = 2.2089$, $y(0.5) = 3.20798$]

4. Find $y(0.4)$ given $y' = y - \dfrac{2x}{y}$, $y(0) = 1$, $y(0.1) = 1.0959$, $y(0.2) = 1.1841$, $y(0.3) = 1.2662$ using Adam's method. [Ans. $y(0.4) = 1.3431$]

5. Given $y' = x^2 + \dfrac{1}{2}y$, $y(1) = 2$, $y(1.1) = 2.2156$, $y(1.2) = 2.4649$, $y(1.3) = 2.7514$, use Adam's method to estimate $y(1.4)$ correct to four decimal places.

   [Ans. $y(1.4) = 3.0794$]

6. Given $\dfrac{dy}{dx} = x^2 + y$, $y(0) = 1$, find $y(0.1)$ by Picard's method, $y(0.2)$ by Picard method, $y(0.2)$ by modified Euler method, $y(0.3)$ by Runge–Kutta method and $y(0.4)$ by Adam's method.

7. Solve initial value problem $\dfrac{dy}{dx} = 1 + xy^2$, $y(0) = 1$ for $x = 0.4, 0.5$ by using Milne's predictor–corrector method when it is given that $y(0.1) = 1.105$, $y(0.2) = 1.223$, $y(0.3) = 1.355$. [Ans. $1.5381, 1.7473$]

8. Find the solution of differential equation $\dfrac{dy}{dx} = x + y$ using Milne's method at $x = 0.5$ given that $y(0) = 1$.

[Ans. 1.7970324]

9. Solve the differential equation $\dfrac{dy}{dx} = 2e^x - y$ at $x = 0.4, 0.5$ using Milne's predictor–corrector method. Given that $y(0) = 2$, $y(0.1) = 2.010$, $y(0.2) = 2.040$, $y(0.3) = 2.090$.

[Ans. 2.162, 2.256]

10. Use Milne's method to obtain the solution of the following differential equation at the required points.

(i) At $x = 0.8$, $\dfrac{dy}{dx} = x - y^2$. Given that $y(0) = 0$, $y(0.2) = 0.02$, $y(0.4) = 0.0795$, $y(0.6) = 0.1762$.

(ii) At $x = 0.4$, $2\dfrac{dy}{dx} = (1 + x^2)y^2$. Given that $y(0) = 1$, $y(0.1) = 1.06$, $y(0.2) = 1.12$, $y(0.3) = 1.21$.

[Ans. (i) 0.3049, (ii) 1.2797]

11 Algebraic Eigen Values and Eigen Vectors of Matrices

11.1 INTRODUCTION

The importance of eigen values can be seen in many physical problems such as stability of an aircraft, the natural frequency of the vibrations of a beam etc. The eigen values are part of the Jordan cononical forms and similar forms which are the convenient and revealing way of representing matrices, so eigen values occurs naturally in the analysis of many mathematical problems. Due to this reason, any system of first order linear differential equations with constant coefficients can be solved in terms of eigen values of its coefficient matrix. Due to these and other reasons, the study of eigen values and eigen vectors is necessary.

11.2 EIGEN VALUES AND EIGEN VECTORS

Let $A = [a_{ij}]$ be any square matrix of order n and there exist a number λ and nonzero vector x such that

$$Ax = \lambda x \qquad (11.1)$$

then λ is called eigen value or latent root and x is called the corresponding eigen vector of the matrix A. Equation (11.1) gives

$$(A - \lambda I)x = 0.$$

$$\Rightarrow \begin{bmatrix} a_{11} - \lambda & a_{12} & a_{13} & \cdots & a_{1n} \\ a_{21} & a_{22} - \lambda & a_{23} & \cdots & a_{2n} \\ a_{31} & a_{32} & a_{33} - \lambda & \cdots & a_{3n} \\ \cdots & \cdots & \cdots & \cdots & \cdots \\ \cdots & \cdots & \cdots & \cdots & \cdots \\ a_{n1} & a_{n2} & a_{n3} & \cdots & a_{nn} - \lambda \end{bmatrix} x = 0$$

It represents a homogeneous system of n linear equations

$$(a_{11} - \lambda) x_1 + a_{12}x_2 + a_{13}x_3 + \ldots + a_{1n}x_n = 0$$
$$a_{21}x_1 + (a_{22} - \lambda) x_2 + a_{23}x_3 + \ldots + a_{2n}x_n = 0$$
$$a_{31}x_1 + a_{32}x_2 + (a_{33} - \lambda) x_3 \ldots + a_{3n}x_n = 0$$
$$\cdots\cdots\cdots\cdots\cdots\cdots\cdots\cdots\cdots\cdots\cdots \cdots \cdots$$
$$\cdots\cdots\cdots\cdots\cdots\cdots\cdots\cdots\cdots\cdots\cdots \cdots \cdots$$
$$a_{n1}x_1 + a_{n2}x_2 + a_{n3}x_3 \ldots + (a_{nn} - \lambda)x_n = 0$$

which will have a nontrivial solution if and only if

$$|A - \lambda I| = 0$$

$$\Rightarrow \quad \begin{bmatrix} a_{11} - \lambda & a_{12} & a_{13} & \cdots & a_{1n} \\ a_{21} & a_{22} - \lambda & a_{23} & \cdots & a_{2n} \\ a_{31} & a_{32} & a_{33} - \lambda & \cdots & a_{3n} \\ \cdots & \cdots & \cdots & \cdots & \cdots \\ \cdots & \cdots & \cdots & \cdots & \cdots \\ a_{n1} & a_{n2} & a_{n3} & \cdots & a_{nn} - \lambda \end{bmatrix} = 0$$

On expanding it, we get a polynomial of degree n in λ

$$\lambda^n - (a_{11} + a_{22} + a_{33} + \dots + a_{nn})\lambda^{n-1} + \dots + (-1)^n |A| = 0 \qquad (11.2)$$

Equation (11.2) is called characteristic equation of matrix A and has n roots say λ_1, λ_2, λ_3, ..., λ_n. These roots are called eigen values of the matrix A. Corresponding to these values of λ, Eq. (11.1) has non-trivial solution. Corresponding to each eigen value λ, Eq. (11.1) will have a nonzero solution $x = [x_1, x_2, ..., x_n]^T$, which is called the eigen vector corresponding to the eigen value λ. If x_1, x_2, x_3, ..., x_n be the eigen vectors corresponding to the eigen values λ_1, λ_2, ..., λ_n, then we have

$$Ax_1 = \lambda_1 x_1, \ Ax_2 = \lambda_2 x_2, \ Ax_3 = \lambda_3 x_3, \ ..., \ Ax_n = \lambda_n x_n$$

Out of these n eigen values, some or all of them may be coincide. From Eq. (11.2), we have

$$\text{Sum of roots of Eq. (11.2)} = \sum_{i=1}^{n} \lambda_i = \sum_{i=1}^{n} a_{ii}$$

$$\text{Product of roots of Eq. (11.2)} = \prod_{i=1}^{n} \lambda_i = |A|.$$

If Eq. (11.1) is premultiplied by matrix A, we obtain
$$A^2 x = \lambda A x = \lambda^2 x$$

On repeating premultiplication by matrix A, we get
$$A^m x = \lambda^m x$$

This implies that if A has the eigen values λ_1, λ_2, λ_3, ..., λ_n, then A^m has the eigen values $\lambda_1^m, \lambda_2^m, \lambda_3^m, ..., \lambda_n^m$ while the eigen vectors remain same. Some important properties of eigen values are as follows:

1. If $B = P^{-1}AP$, then B is similar to matrix A and eigen values of A and B are the same where P is nth order non-singular matrix, it means any similarity transformation applied to a matrix leaves its eigen values unchanged.

2. Matrices A and A^T (transpose of A) have the same eigen values but different eigen vectors.

3. For any scalar k, the eigen values of the matrix kA are k times the eigen values of matrix A.

4. If λ is the eigen value of matrix A, then $(\lambda + k)$ is the eigen value of the matrix $(A + kI)$ where I is a unit matrix of same order as A.

5. The diagonal elements of a triangular matrix are the eigen values of that triangular matrix.

6. For a non-singular matrix A, the eigen values of A^{-1} are the reciprocal of the eigen values of A.

7. If λ is an eigen value of a nonsingular matrix A, then $\dfrac{|A|}{\lambda}$ is an eigen value of matrix adj (A).

8. The eigen values of a Hermitian and real symmetric matrix are all real.

9. The eigen values of a skew-Hermitian matrix are either zero or purely imaginary number.

10. If the eigen value λ_p is repeated k times, then there exist k orthonormal eigen vectors corresponding to λ_p.

11. Corresponding to any eigen vector, there exists only one eigen value, but if x is an eigen vector corresponding to the eigen value λ of the matrix A, then kx (k being any scalar) is also the eigen vector corresponding to the same eigen value.

12. Let $\lambda_1, \lambda_2, \lambda_3, ..., \lambda_n$ are the eigen values, the matrix A and $f(A)$ is a rational function of A, then the eigen values of $f(A)$ are $f(\lambda_1), f(\lambda_2), ..., f(\lambda_n)$ and

$$f(A) = |f(\lambda_1) f(\lambda_2) \ f(\lambda_3) ... f(\lambda_n)|$$

Methods of finding eigen values and eigen vectors can be studied from the books of matrices and linear algebra. Here we study bounds of eigen values.

Theorem: If λ be an eigen value of the matrix A, then for some $k \, (1 \leq k \leq n)$

$$|\lambda - a_{kk}| \leq |a_{k1}| + |a_{k2}| + ... + a_{kk-1}| + |a_{kk+1}| + ... + |a_{kn}| = \rho_k \text{ (say)}$$

i.e. all the eigen values of A lies in the union of the circles with centres a_{kk} and radii ρ_k.

Proof: Let λ is the eigen value of the $n \times n$ square matrix A and x be the corresponding eigen vector, then we have

$$Ax = \lambda x$$

This implies that

$$a_{11}x_1 + a_{12}x_2 + a_{13}x_3 + ... + a_{1n}x_n = \lambda x_1$$
$$a_{21}x_1 + a_{22}x_2 + a_{23}x_3 + ... + a_{2n}x_n = \lambda x_2$$
$$\text{.. ... .}$$
$$a_{k1}x_1 + a_{k2}x_2 + a_{k3}x_3 + ... + a_{kn}x_n = \lambda x_k$$
$$\text{.. ... .}$$
$$a_{n1}x_1 + a_{n2}x_2 + a_{n3}x_3 + ... + a_{nn}x_n = \lambda x_n$$

Let x_k be the largest component of x, then we have

$$\left| \frac{x_m}{x_k} \right| \leq 1 \ \forall \ m = 1, 2, ..., n$$

Dividing the kth equation by x_k, we get

$$a_{k1}\frac{x_1}{x_k} + a_{k2}\frac{x_2}{x_k} + a_{k3}\frac{x_3}{x_k} + ... + a_{kk} + ... + a_{kn}\frac{x_n}{x_k} = \lambda$$

$\Rightarrow$
$$a_{k1}\frac{x_1}{x_k} + a_{k2}\frac{x_2}{x_k} + a_{k3}\frac{x_3}{x_k} + ... + a_{kn}\frac{x_n}{x_k} = \lambda - a_{kk}$$

$\Rightarrow$
$$|\lambda - a_{kk}| \leq |a_{k1}|\left|\frac{x_1}{x_k}\right| + |a_{k2}|\left|\frac{x_2}{x_k}\right| + ... + |a_{kn}|\left|\frac{x_n}{x_k}\right|$$

$$\leq |a_{k1}| + |a_{k2}| + |a_{k3}| + |a_{kk-1}| + |a_{kk+1}| + |a_{kn}|, \text{ since } \left|\frac{x_n}{x_k}\right| \leq 1$$

$$\leq |a_{k1}| + |a_{k2}| + |a_{k3}| + ... + |a_{kk-1}| + |a_{kk+1}| + ... + |a_{kn}| = \rho_k \text{ (say)}$$

which shows that all the eigen values of A lie within or on the union of the circles with centres at a_{kk} and radii ρ_k.

These circles are called Gerschgorin circles. The bounds thus obtained are all independent. All the eigen values of A must lie in the intersection of these bounds. Such bounds are called Gerschgorin bound.

Remark: If any of the Gerschgorin circles is isolated, then it has exactly one eigen value.

11.3 GERSCHGORIN CIRCLE THEOREM*

If A is an $n \times n$ matrix and R_i denotes the circle in the complex plane

$$|z - a_{ii}| \le \sum_{\substack{j=1, \\ j \ne i}}^{n} |a_{ij}|$$

Then the eigen values of the matrix A are contained within

$$R = \bigcup_{i=1}^{n} R_i$$

Further, the union of any k of these circles that do not intersect the remaining $(n - k)$ must contain precisely k (counting multiplicities) of the eigen values.

For example, consider the matrix

$$A = \begin{bmatrix} 4 & 1 & 1 \\ 0 & 2 & 1 \\ -2 & 0 & 9 \end{bmatrix}$$

For this matrix Gerschgorin circles are

$$|z - a_{11}| \le |a_{12}| + |a_{13}| \Rightarrow |z - 4| \le 2 \ (G_1)$$
$$|z - a_{22}| \le |a_{21}| + |a_{23}| \Rightarrow |z - 2| \le 1 \ (G_2)$$
$$|z - a_{33}| \le |a_{31}| + |a_{32}| \Rightarrow |z - 9| \le 2 \ (G_3)$$

In the complex plane, these circles are shown in Fig. 11.1.

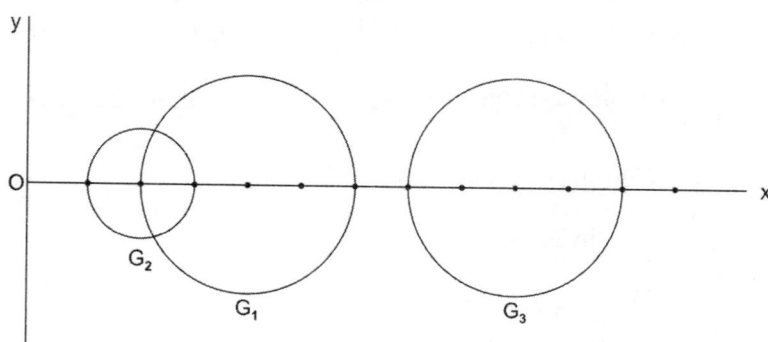

Fig. 11.1: Gerschgorin circles

As G_1, G_2 are disjoint from G_3, there must be precisely two eigen values in $R_1 \cup R_2$ and one eigen value within G_3.

Now, we observe numerical methods to find eigen values.

* Only statement

10.4 JACOBI'S METHOD

We know that if A is a real symmetric matrix, then all its eigen values are real and there exists a real orthogonal matrix B such that $B^{-1}AB$ is a diagonal matrix whose diagonal elements are the eigen values of A. Due to these properties in the method, we diagonalise a real symmetric matrix A by applying a series of orthogonal transformations $B_1, B_2, ..., B_r$ in such a way that the product $B = B_1 B_2 B_3 ... B_r$ satisfies the relation

$$B^{-1}AB = D, \text{ a diagonal matrix}$$

For this, we choose the numerically largest non-diagonal elements a_{ij} and form a submatrix like

$$A_1 = \begin{bmatrix} a_{ii} & a_{ij} \\ a_{ji} & a_{jj} \end{bmatrix}.$$

where $a_{ij} = a_{ji}$ which can be easily diagonalised. Consider the orthogonal matrix

$$B_1 = \begin{bmatrix} \cos\theta & -\sin\theta \\ \sin\theta & \cos\theta \end{bmatrix}$$

Now we have

$$B_1^{-1}A_1B_1 = \begin{bmatrix} \cos\theta & \sin\theta \\ -\sin\theta & \cos\theta \end{bmatrix} \begin{bmatrix} a_{ii} & a_{ij} \\ a_{ji} & a_{jj} \end{bmatrix} \begin{bmatrix} \cos\theta & -\sin\theta \\ \sin\theta & \cos\theta \end{bmatrix}$$

$$= \begin{bmatrix} a_{ii}\cos\theta + a_{ji}\sin\theta & a_{ij}\cos\theta + a_{jj}\sin\theta \\ -a_{ii}\sin\theta + a_{ji}\cos\theta & -a_{ij}\sin\theta + a_{jj}\cos\theta \end{bmatrix} \begin{bmatrix} \cos\theta & -\sin\theta \\ \sin\theta & \cos\theta \end{bmatrix}$$

$$= \begin{bmatrix} a_{ii}\cos^2\theta + a_{ji}\sin\theta\cos\theta & -a_{ii}\cos\theta\sin\theta - a_{ji}\sin^2\theta \\ + a_{ij}\cos\theta\sin\theta + a_{jj}\sin^2\theta & + a_{ij}\cos^2\theta + a_{jj}\cos\theta\sin\theta \\ -a_{ii}\cos\theta\sin\theta + a_{ji}\cos^2\theta & a_{ii}\sin^2\theta - a_{ji}\cos\theta\sin\theta \\ -a_{ij}\sin^2\theta + a_{jj}\sin\theta\cos\theta & -a_{ij}\sin\theta\cos\theta + a_{jj}\cos^2\theta \end{bmatrix}$$

$$= \begin{bmatrix} a_{ii}\cos^2\theta + a_{ij}\sin 2\theta + a_{jj}\sin^2\theta & (-a_{ii} + a_{jj})\dfrac{1}{2}\sin 2\theta + a_{ji}\cos 2\theta \\ (-a_{ii} + a_{jj})\dfrac{1}{2}\sin 2\theta + a_{ji}\cos 2\theta & a_{ii}\sin^2\theta - a_{ji}\sin 2\theta + a_{jj}\cos^2\theta \end{bmatrix}$$

This matrix reduces to diagonal form if

$$a_{ij}\cos 2\theta + \frac{1}{2}(a_{jj} - a_{ii})\sin 2\theta = 0$$

$$\Rightarrow \qquad \tan 2\theta = \frac{2a_{ij}}{(a_{ii} - a_{jj})}$$

This equation gives four values of θ, but to get the least possible rotation we choose the value of θ in the range $-\dfrac{\pi}{4} \le \theta \le \dfrac{\pi}{4}$.

Hence $B_1^{-1}A_1B_1$ is reduced to diagonal form.

In the next step, the largest non-diagonal element (in magnitude) in the new rotated matrix is found and above procedure is repeated using the diagonal matrix B_2. In order to anihilate the non-diagonal elements, we have to perform a series of such transformations. After r such transformations, we obtain

$$B_r^{-1} B_{r-1}^1 \dots B_3^{-1} B_2^{-1} B_1^{-1} A_1 B_1 B_2 B_3 \dots B_r = B^{-1}AB$$

As $r \to \infty$, $B^{-1}AB$ approaches a diagonal matrix whose diagonal elements are the eigen values of A and the corresponding columns of

$$B = B_1 B_2 B_3 \dots B_r$$

are the eigen vectors of A.

In Jacobi's method, the element anihilated by a transformation, may not remain zero after the subsequent transformations. To overcome this drawback Given's and House-Holder's preferred to reduce the given symmetric matrix to tri-diagonal matrix. Given's and House-Holder's methods are modification of the Jacobi's method. These methods produce zero off-diagonal elements in each iteration which remain zero in subsequent iterations.

11.5 GIVEN'S METHOD

For a real symmetric matrix A, Given's method consists of the following steps.

Step 1: First reduce the given matrix A to a tridiagonal matrix. Consider the matrix A_1 as

$$A_1 = \begin{bmatrix} a_{11} & a_{12} & a_{13} \\ a_{21} & a_{22} & a_{23} \\ a_{31} & a_{32} & a_{33} \end{bmatrix}$$

and O_1 is the orthogonal rotation matrix in the plane $(2, 3)$ as

$$O_1 = \begin{bmatrix} 1 & 0 & 0 \\ 0 & \cos\theta & -\sin\theta \\ 0 & \sin\theta & \cos\theta \end{bmatrix}$$

These matrices gives

$$O_1^{-1} A_1 O_1 = \begin{bmatrix} 1 & 0 & 0 \\ 0 & \cos\theta & \sin\theta \\ 0 & -\sin\theta & \cos\theta \end{bmatrix} \begin{bmatrix} a_{11} & a_{12} & a_{13} \\ a_{21} & a_{22} & a_{23} \\ a_{31} & a_{32} & a_{33} \end{bmatrix} \begin{bmatrix} 1 & 0 & 0 \\ 0 & \cos\theta & -\sin\theta \\ 0 & \sin\theta & \cos\theta \end{bmatrix}$$

In the matrix $O^{-1}A_1O_1$, the element $(1, 3)$ of first row and third column is $a_{13}\cos\theta - a_{12}\sin\theta$ which will be zero if $\tan\theta = \dfrac{a_{13}}{a_{12}}$.

Corresponding to this value of θ, the above transformation gives zeros in $(1, 3)$ and $(3, 1)$ positions. Again perform the rotation in the plane $(2, 4)$ and put the resulting element $(1, 4) = 0$. This would not disturb the zero already obtained at $(1, 3)$ and $(3, 1)$ positions. Then the transformations are applied to the matrix in turn so as to anihilate the elements $(1, 3)$; $(1, 4)$; $(1, 5)$; ... $(1, n)$; $(2, 4)$; $(2, 5)$; ..., $(2, n)$ in this order. We arrive at the triorthogonal matrix

$$P = \begin{bmatrix} p_1 & q_1 & 0 & 0 & 0 & \cdots & 0 & 0 \\ q_1 & p_2 & q_2 & 0 & 0 & \cdots & 0 & 0 \\ 0 & q_2 & p_3 & q_3 & 0 & \cdots & 0 & 0 \\ 0 & 0 & q_3 & p_4 & q_4 & \cdots & 0 & 0 \\ 0 & 0 & 0 & q_4 & p_5 & \cdots & 0 & 0 \\ \cdots & \cdots & \cdots & \cdots & \cdots & \cdots & \cdots & \cdots \\ 0 & 0 & 0 & 0 & 0 & \cdots & p_{n-1} & q_{n-1} \\ 0 & 0 & 0 & 0 & 0 & \cdots & q_{n-1} & p_n \end{bmatrix}$$

Step 2: Find the eigen values of tridiagonal matrix. Let the resulting tridiagonal matrix

is $\begin{bmatrix} \alpha_{11} & \alpha_{12} & 0 \\ \alpha_{12} & \alpha_{22} & \alpha_{23} \\ 0 & \alpha_{23} & \alpha_{33} \end{bmatrix}$. The eigen values of this matrix are same as that of original matrix.

To find eigen values, we take the characteristic equation as

$$\begin{bmatrix} \alpha_{11} - \lambda & \alpha_{12} & 0 \\ \alpha_{12} & \alpha_{22} - \lambda & \alpha_{23} \\ 0 & \alpha_{23} & \alpha_{33} - \lambda \end{bmatrix} = 0 = f_3(\lambda)$$

$\Rightarrow$

$$f_0(\lambda) = 1$$
$$f_1(\lambda) = \alpha_{11} - \lambda = \alpha_{11} - \lambda f_0(\lambda)$$
$$f_2(\lambda) = \begin{vmatrix} \alpha_{11} - \lambda & \alpha_{12} \\ \alpha_{12} & \alpha_{22} - \lambda \end{vmatrix} = (\alpha_{22} - \lambda) f_1(\lambda) - \alpha_{12}^2 f_0(\lambda)$$
$$f_3(\lambda) = (\alpha_{33} - \lambda) f_2(\lambda) - \alpha_{23}^2 f_1(\lambda)$$

In general the recurrence relation is

$$f_k(1) = (\alpha_{kk} - \lambda) f_{k-1}(\lambda) - (\alpha_{k-1k})^2 f_{k-2}(\lambda); \, 2 \leq k \leq n$$

The characteristic equation $f_3(\lambda) = 0$ can also be solved by any other method.

Step 3: Find the eigen vectors of tridiagonal matrix. If O_1, O_2, O_3, ..., O_j are orthogonal matrices employed in reducing the matrix A to tridiagonal matrix P and Y be the eigen vector of the tridiagonal matrix P, then the corresponding eigen vector of the matrix A is given by

$$X = O_1 O_2 O_3 ... O_j Y$$

Notes:

1. The sequence of functions $f_0(\lambda)$, $f_1(\lambda)$, $f_2(\lambda)$, ..., $f_k(\lambda)$ is called the strum sequence. For various values of λ, a table of sequence is prepared and the number of changes in sign of the strum sequence is calculated. The difference between the number of changes of sign for consecutive values of λ gives an approximate location of the eigen values. After knowing the location of the eigen values, their exact values can be found by any iterative method.

2. In the Given's method, the number of rotations required are equivalent to the number of non-tridiagonal elements of the matrix. For example, for the 3 order square matrix only one rotation is required and for the 4 order square matrix, only three rotations are required. The amount of computation goes on decreasing from one rotation to next as the order of the matrix for computation also starts reducing.

Example 1: Obtain the eigen values and eigen vectors of the symmetric matrix

$$A = \begin{bmatrix} 0 & 1 & 1 \\ 1 & 0 & 1 \\ 1 & 1 & 0 \end{bmatrix}$$

Solution: The characteristic equation of matrix A is

$$|A - \lambda I| = 0$$

$$\Rightarrow \qquad \begin{bmatrix} -\lambda & 1 & 1 \\ 1 & -\lambda & 1 \\ 1 & 1 & -\lambda \end{bmatrix} = 0 \Rightarrow (\lambda + 1)^2 (\lambda - 2) = 0$$

The eigen values are $-1, -1, 2$.

For $\lambda = 2$, the eigen vectors are given by

$$|A - 2I| \, X = 0$$

$$\begin{bmatrix} -2 & 1 & 1 \\ 1 & -2 & 1 \\ 1 & 1 & -2 \end{bmatrix} \begin{bmatrix} x_1 \\ x_2 \\ x_3 \end{bmatrix} = 0$$

$$\Rightarrow \qquad \begin{aligned} -2x_1 + x_2 + x_3 &= 0 \\ x_1 - 2x_2 + x_3 &= 0 \\ x_1 + x_2 - 2x_3 &= 0 \end{aligned}$$

On solving, we get $x_1 = x_2 = x_3 = k$ (say). Thus eigen vectors is $x = [x_1 \, x_2 \, x_3]^T = [k \, k \, k]^T$.

For $\lambda = -1, -1$, the eigen vectors are given by

$$(A + I) \, X = 0$$

$$\Rightarrow \qquad \begin{bmatrix} +1 & 1 & 1 \\ 1 & +1 & 1 \\ 1 & 1 & +1 \end{bmatrix} \begin{bmatrix} x_1 \\ x_2 \\ x_3 \end{bmatrix} = 0$$

$\Rightarrow x_1 + x_2 + x_3 = 0$, all three equations are same. The trial solutions are $1, 0, -1$ and $1 + k, -1, -k$.

Hence eigen vectors are $X_1 = \begin{bmatrix} 1 \\ 0 \\ -1 \end{bmatrix}$ and $X_2 = \begin{bmatrix} 1+k \\ -1 \\ -k \end{bmatrix}$.

These eigen vectors are orthogonal, so we have

$$\langle X_1, X_2 \rangle = 0 \Rightarrow 1(1 + k) + 0(-1) + (-1)(-k) = 0$$

$$\Rightarrow \qquad 1 + k + k = 0 \Rightarrow k = -\frac{1}{2}$$

Hence the two vectors are $X_1 = \begin{bmatrix} 1 \\ 0 \\ 1 \end{bmatrix}$ and $X_2 = \begin{bmatrix} \dfrac{1}{2} \\ -1 \\ \dfrac{1}{2} \end{bmatrix}$

Hence orthogonal eigen vectors are given by $X_1 = \begin{bmatrix} \dfrac{1}{\sqrt{2}} \\ 0 \\ \dfrac{1}{\sqrt{2}} \end{bmatrix}$ and $X_2 = \begin{bmatrix} \dfrac{1}{\sqrt{6}} \\ -\dfrac{2}{\sqrt{6}} \\ \dfrac{1}{\sqrt{6}} \end{bmatrix}$.

Example 2: Using Jacobi's method, find all the eigen values and the eigen vectors of

the matrix $\begin{bmatrix} 1 & \sqrt{2} & 2 \\ \sqrt{2} & 3 & \sqrt{2} \\ 2 & \sqrt{2} & 1 \end{bmatrix}$.

Solution: The largest non-diagonal element is $a_{13} = a_{31} = 2$. Here the corresponding elements $a_{11} = 1$ and $a_{33} = 1$.

Therefore $\tan 2\theta = \dfrac{2a_{13}}{a_{11} - a_{33}} = \dfrac{2 \times 2}{1 - 1} = \dfrac{4}{0} = \infty \Rightarrow \theta = \dfrac{\pi}{4}$

Here $B_1 = \begin{bmatrix} \cos\theta & 0 & -\sin\theta \\ 0 & 1 & 0 \\ \sin\theta & 0 & \cos\theta \end{bmatrix} = \begin{bmatrix} \cos\dfrac{\pi}{4} & 0 & -\sin\dfrac{\pi}{4} \\ 0 & 1 & 0 \\ \sin\dfrac{\pi}{4} & 0 & \cos\dfrac{\pi}{4} \end{bmatrix}$

$= \begin{bmatrix} \dfrac{1}{\sqrt{2}} & 0 & -\dfrac{1}{\sqrt{2}} \\ 0 & 1 & 0 \\ \dfrac{1}{\sqrt{2}} & 0 & \dfrac{1}{\sqrt{2}} \end{bmatrix}$.

and $B_1^{-1} = \begin{bmatrix} \dfrac{1}{\sqrt{2}} & 0 & \dfrac{1}{\sqrt{2}} \\ 0 & 1 & 0 \\ -\dfrac{1}{\sqrt{2}} & 0 & \dfrac{1}{\sqrt{2}} \end{bmatrix}$

Hence the first transformation gives

$D_1 = B_1^{-1}AB = \begin{bmatrix} \dfrac{1}{\sqrt{2}} & 0 & \dfrac{1}{\sqrt{2}} \\ 0 & 1 & 0 \\ -\dfrac{1}{\sqrt{2}} & 0 & \dfrac{1}{\sqrt{2}} \end{bmatrix} \begin{bmatrix} 1 & \sqrt{2} & 2 \\ \sqrt{2} & 3 & \sqrt{2} \\ 2 & \sqrt{2} & 1 \end{bmatrix} \begin{bmatrix} \dfrac{1}{\sqrt{2}} & 0 & -\dfrac{1}{\sqrt{2}} \\ 0 & 1 & 0 \\ \dfrac{1}{\sqrt{2}} & 0 & \dfrac{1}{\sqrt{2}} \end{bmatrix}$

$= \begin{bmatrix} 3 & 2 & 0 \\ 2 & 3 & 0 \\ 0 & 0 & -1 \end{bmatrix}$

Next nondiagonal element is $a_{12} = a_{21} = 2$ and corresponding elements are $a_{11} = 3$, $a_{22} = 3$.

Therefore $\qquad \tan 2\phi = \dfrac{2a_{12}}{a_{11} - a_{22}} = \dfrac{2 \times 2}{3 - 3} = \dfrac{4}{0} = \infty$

$\Rightarrow \qquad\qquad\qquad \phi = \dfrac{\pi}{4}$

Here $\qquad\qquad B_2 = \begin{bmatrix} \cos\phi & -\sin\phi & 0 \\ \sin\phi & \cos\phi & 0 \\ 0 & 0 & 1 \end{bmatrix}$

$$= \begin{bmatrix} \cos\dfrac{\pi}{4} & -\sin\dfrac{\pi}{4} & 0 \\ \sin\dfrac{\pi}{4} & \cos\dfrac{\pi}{4} & 0 \\ 0 & 0 & 1 \end{bmatrix} = \begin{bmatrix} \dfrac{1}{\sqrt{2}} & -\dfrac{1}{\sqrt{2}} & 0 \\ \dfrac{1}{\sqrt{2}} & \dfrac{1}{\sqrt{2}} & 0 \\ 0 & 0 & 1 \end{bmatrix}$$

and $\qquad\qquad B_2^{-1} = \begin{bmatrix} \dfrac{1}{\sqrt{2}} & \dfrac{1}{\sqrt{2}} & 0 \\ -\dfrac{1}{\sqrt{2}} & \dfrac{1}{\sqrt{2}} & 0 \\ 0 & 0 & 1 \end{bmatrix}$

Hence the second transformation gives

$$D_2 = B_2^{-1}D_1B_2 = \begin{bmatrix} \dfrac{1}{\sqrt{2}} & \dfrac{1}{\sqrt{2}} & 0 \\ -\dfrac{1}{\sqrt{2}} & \dfrac{1}{\sqrt{2}} & 0 \\ 0 & 0 & 1 \end{bmatrix}\begin{bmatrix} 3 & 2 & 0 \\ 2 & 3 & 0 \\ 0 & 0 & -1 \end{bmatrix}\begin{bmatrix} \dfrac{1}{\sqrt{2}} & -\dfrac{1}{\sqrt{2}} & 0 \\ \dfrac{1}{\sqrt{2}} & \dfrac{1}{\sqrt{2}} & 0 \\ 0 & 0 & 1 \end{bmatrix} = \begin{bmatrix} 5 & 0 & 0 \\ 0 & 1 & 0 \\ 0 & 0 & -1 \end{bmatrix}$$

Thus the eigen values of given matrix are $5, 1, -1$. The corresponding eigen vectors are the columns of matrix B, where

$$B = B_1B_2 = \begin{bmatrix} \dfrac{1}{\sqrt{2}} & 0 & -\dfrac{1}{\sqrt{2}} \\ 0 & 1 & 0 \\ \dfrac{1}{\sqrt{2}} & 0 & \dfrac{1}{\sqrt{2}} \end{bmatrix}\begin{bmatrix} \dfrac{1}{\sqrt{2}} & -\dfrac{1}{\sqrt{2}} & 0 \\ \dfrac{1}{\sqrt{2}} & \dfrac{1}{\sqrt{2}} & 0 \\ 0 & 0 & 1 \end{bmatrix}$$

$$= \begin{bmatrix} \dfrac{1}{2} & -\dfrac{1}{2} & -\dfrac{1}{\sqrt{2}} \\ \dfrac{1}{\sqrt{2}} & \dfrac{1}{\sqrt{2}} & 0 \\ \dfrac{1}{2} & -\dfrac{1}{2} & \dfrac{1}{\sqrt{2}} \end{bmatrix}$$

Example 3: Reduce the following matrix to the tridiagonal form by Given's method

$$A = \begin{bmatrix} 2 & 1 & 3 \\ 1 & 4 & 2 \\ 3 & 2 & 3 \end{bmatrix}$$

Solution: Here the only non-tridiagonal element is $a_{13} = 3$. In order to anihilate a_{13}, we define the orthogonal matrix O_1 in the plane $(2, 3)$ as

$$O_1 = \begin{bmatrix} 1 & 0 & 0 \\ 0 & \cos\theta & -\sin\theta \\ 0 & \sin\theta & \cos\theta \end{bmatrix}$$

where θ is given by $\quad \tan\theta = \dfrac{a_{13}}{a_{12}} = \dfrac{3}{1} = 3 \Rightarrow \sin\theta = \dfrac{3}{\sqrt{10}}, \cos\theta = \dfrac{1}{\sqrt{10}}$

Hence

$$O_1^{-1}AO_1 = \begin{bmatrix} 1 & 0 & 0 \\ 0 & \cos\theta & \sin\theta \\ 0 & -\sin\theta & \cos\theta \end{bmatrix} \begin{bmatrix} 2 & 1 & 3 \\ 1 & 4 & 2 \\ 3 & 2 & 3 \end{bmatrix} \begin{bmatrix} 1 & 0 & 0 \\ 0 & \cos\theta & -\sin\theta \\ 0 & \sin\theta & \cos\theta \end{bmatrix}$$

$$= \begin{bmatrix} 1 & 0 & 0 \\ 0 & \dfrac{1}{\sqrt{10}} & \dfrac{3}{\sqrt{10}} \\ 0 & -\dfrac{3}{\sqrt{10}} & \dfrac{1}{\sqrt{10}} \end{bmatrix} \begin{bmatrix} 2 & 1 & 3 \\ 1 & 4 & 2 \\ 3 & 2 & 3 \end{bmatrix} \begin{bmatrix} 1 & 0 & 0 \\ 0 & \dfrac{1}{\sqrt{10}} & -\dfrac{3}{\sqrt{10}} \\ 0 & \dfrac{3}{\sqrt{10}} & \dfrac{1}{\sqrt{10}} \end{bmatrix}$$

$$= \begin{bmatrix} 2 & 1 & 3 \\ \sqrt{10} & \sqrt{10} & \dfrac{11}{\sqrt{10}} \\ 0 & -\sqrt{10} & -\dfrac{3}{\sqrt{10}} \end{bmatrix} \begin{bmatrix} 1 & 0 & 0 \\ 0 & \dfrac{1}{\sqrt{10}} & -\dfrac{3}{\sqrt{10}} \\ 0 & \dfrac{3}{\sqrt{10}} & \dfrac{1}{\sqrt{10}} \end{bmatrix}$$

$$= \begin{bmatrix} 2 & \sqrt{10} & 0 \\ \sqrt{10} & 4.3 & -1.9 \\ 0 & -1.9 & 2.7 \end{bmatrix}$$

Example 4: Reduce the following matrix to the tridiagonal form by Given's method

$$A = \begin{bmatrix} 1 & 2 & 4 \\ 2 & 1 & 2 \\ 4 & 2 & 1 \end{bmatrix}$$

Solution: Here the only non-tridiagonal element is $a_{13} = 3$. In order to anihilate a_{13}, we define the orthogonal matrix O_1 in the plane $(2, 3)$ as

$$O_1 = \begin{bmatrix} 1 & 0 & 0 \\ 0 & \cos\theta & -\sin\theta \\ 0 & \sin\theta & \cos\theta \end{bmatrix}$$

where θ is given by $\quad \tan\theta = \dfrac{a_{13}}{a_{12}} = \dfrac{4}{2} = 2 \Rightarrow \cos\theta = \dfrac{1}{\sqrt{5}}, \sin\theta = \dfrac{2}{\sqrt{5}}$

Hence
$$O_1^{-1}AO_1 = \begin{bmatrix} 1 & 0 & 0 \\ 0 & \cos\theta & \sin\theta \\ 0 & -\sin\theta & \cos\theta \end{bmatrix} \begin{bmatrix} 1 & 2 & 4 \\ 2 & 1 & 2 \\ 4 & 2 & 1 \end{bmatrix} \begin{bmatrix} 1 & 0 & 0 \\ 0 & \cos\theta & -\sin\theta \\ 0 & \sin\theta & \cos\theta \end{bmatrix}$$

$$= \begin{bmatrix} 1 & 0 & 0 \\ 0 & \dfrac{1}{\sqrt{5}} & \dfrac{2}{\sqrt{5}} \\ 0 & -\dfrac{2}{\sqrt{5}} & \dfrac{1}{\sqrt{5}} \end{bmatrix} \begin{bmatrix} 1 & 2 & 4 \\ 2 & 1 & 2 \\ 4 & 2 & 1 \end{bmatrix} \begin{bmatrix} 1 & 0 & 0 \\ 0 & \dfrac{1}{\sqrt{5}} & -\dfrac{2}{\sqrt{5}} \\ 0 & \dfrac{2}{\sqrt{5}} & \dfrac{1}{\sqrt{5}} \end{bmatrix}$$

$$= \begin{bmatrix} 1 & 2 & 4 \\ \dfrac{2}{\sqrt{5}} & \sqrt{5} & \dfrac{4}{\sqrt{5}} \\ 0 & 0 & -\dfrac{3}{\sqrt{5}} \end{bmatrix} \begin{bmatrix} 1 & 0 & 0 \\ 0 & \dfrac{1}{\sqrt{5}} & -\dfrac{2}{\sqrt{5}} \\ 0 & \dfrac{2}{\sqrt{5}} & \dfrac{1}{\sqrt{5}} \end{bmatrix}$$

$$= \begin{bmatrix} 1 & 2\sqrt{5} & 0 \\ 2\sqrt{5} & 2.6 & -1.2 \\ 0 & -1.2 & -0.6 \end{bmatrix}$$

Example 5: Determine the characteristic roots and vectors of the matrix

$$A = \begin{bmatrix} 8 & -6 & 2 \\ -6 & 7 & -4 \\ 2 & -4 & 3 \end{bmatrix}$$

Solution: The characteristic equations of A is

$$|A - \lambda I| = 0$$

$\Rightarrow \quad \begin{vmatrix} 8-\lambda & -6 & 2 \\ -6 & 7-\lambda & -4 \\ 2 & -4 & 3-\lambda \end{vmatrix} = 0$

On solving it, we have

$(8-\lambda)(\lambda^2 - 10\lambda + 21 - 16) + 6(-18 + 6\lambda + 8) + 2(24 - 14 + 2\lambda) = 0$

$\Rightarrow \qquad\qquad \lambda(\lambda^2 - 18\lambda + 45) = 0 \Rightarrow \lambda(\lambda - 3)(\lambda - 15) = 0$

Therefore, the characteristic roots of A are $0, 3, 15$.

The characteristic vector corresponding to characteristic root 0 is the nonzero solution of the equation

$$(A - OI)X = 0$$

$\Rightarrow \qquad \begin{bmatrix} 8 & -6 & 2 \\ -6 & 7 & -4 \\ 2 & -4 & 3 \end{bmatrix} \begin{bmatrix} x_1 \\ x_2 \\ x_3 \end{bmatrix} = 0$

Applying $R_3 \rightarrow 3R_3 + R_2$, we have

$$\begin{bmatrix} 8 & -6 & 2 \\ -6 & 7 & -4 \\ 0 & -5 & 5 \end{bmatrix} \begin{bmatrix} x_1 \\ x_2 \\ x_3 \end{bmatrix} = 0$$

Applying $R_2 \rightarrow 8R_2 + 6R_2$, we have

$$\begin{bmatrix} 8 & -6 & 2 \\ 0 & 20 & -20 \\ 0 & 0 & 5 \end{bmatrix} \begin{bmatrix} x_1 \\ x_2 \\ x_3 \end{bmatrix} = 0$$

Applying $R_3 \rightarrow R_3 + \dfrac{1}{4}R_2$, we have

$$\begin{bmatrix} 8 & -6 & 2 \\ 0 & 20 & -20 \\ 0 & 0 & 0 \end{bmatrix} \begin{bmatrix} x_1 \\ x_2 \\ x_3 \end{bmatrix} = 0$$

The coefficient matrix of these equations is of rank 2. Therefore, the equations has linearly dependent solutions. These equations can be written as

$$8x_1 - 6x_2 + 2x_3 = 0$$
$$20x_2 - 20x_3 = 0$$

Last equation gives $x_2 = x_3$. Let us take $x_2 = x_3 = 1$, then first equation gives $x_1 = \dfrac{1}{2}$.

Therefore eigen vector is $X_1 = \begin{bmatrix} \dfrac{1}{2} \\ 1 \\ 1 \end{bmatrix}$.

The characteristic vector corresponding to characteristic root 3 is the nonzero solution of the equation

$$(A - 3I)X = 0$$

$$\Rightarrow \qquad \begin{bmatrix} 5 & -6 & 2 \\ -6 & 4 & -4 \\ 2 & -4 & 0 \end{bmatrix} \begin{bmatrix} x_1 \\ x_2 \\ x_3 \end{bmatrix} = 0$$

Applying $R_3 \rightarrow 3R_3 + R_2$, we have

$$\begin{bmatrix} 5 & -6 & 2 \\ -6 & 4 & -4 \\ 0 & -8 & -4 \end{bmatrix} \begin{bmatrix} x_1 \\ x_2 \\ x_3 \end{bmatrix} = 0$$

Applying $R_2 \rightarrow 5R_2 + 6R_3$, we have

$$\begin{bmatrix} 5 & -6 & 2 \\ 0 & -16 & -8 \\ 0 & -8 & -4 \end{bmatrix} \begin{bmatrix} x_1 \\ x_2 \\ x_3 \end{bmatrix} = 0$$

Applying $R_3 \to 2R_3 - R_2$, we have

$$\begin{bmatrix} 5 & -6 & 2 \\ 0 & -16 & -8 \\ 0 & 0 & 0 \end{bmatrix} \begin{bmatrix} x_1 \\ x_2 \\ x_3 \end{bmatrix} = 0$$

The coefficient matrix of these equations is of rank 2. Therefore, the equations has linearly dependent solutions. These equations can be written as

$$5x_1 - 6x_2 + 2x_3 = 0$$
$$-16x_2 - 8x_3 = 0$$

Last equation gives $2x_2 = -x_3$. Let us take $x_2 = 1$, $x_3 = -2$, then first equation gives $x_1 = 2$. Therefore eigen vector is $X_2 = \begin{bmatrix} 2 \\ 1 \\ -2 \end{bmatrix}$.

The characteristic vector corresponding to characteristic root 15 is the nonzero solution of the equation

$$(A - 15I) X = 0$$

$$\Rightarrow \quad \begin{bmatrix} -7 & -6 & 2 \\ -6 & -8 & -4 \\ 2 & -4 & -12 \end{bmatrix} \begin{bmatrix} x_1 \\ x_2 \\ x_3 \end{bmatrix} = 0$$

Applying $R_3 \to 3R_3 + R_2$, we have

$$\begin{bmatrix} -7 & -6 & 2 \\ -6 & -8 & -4 \\ 0 & -20 & -40 \end{bmatrix} \begin{bmatrix} x_1 \\ x_2 \\ x_3 \end{bmatrix} = 0$$

Applying $R_2 \to 7R_2 - 6R_1$, we have

$$\begin{bmatrix} -7 & -6 & 2 \\ 0 & -20 & -40 \\ 0 & -20 & -40 \end{bmatrix} \begin{bmatrix} x_1 \\ x_2 \\ x_3 \end{bmatrix} = 0$$

Applying $R_3 \to R_3 - R_2$, we have

$$\begin{bmatrix} -7 & -6 & 2 \\ 0 & -20 & -40 \\ 0 & 0 & 0 \end{bmatrix} \begin{bmatrix} x_1 \\ x_2 \\ x_3 \end{bmatrix} = 0$$

The coefficient matrix of these equations is of rank 2. Therefore, the equations has linearly dependent solutions. These equations can be written as

$$-7x_1 - 6x_2 + 2x_3 = 0$$
$$-20x_2 - 40x_3 = 0$$

Last equation gives $x_2 = -2x_3$. Let us take $x_3 = 1$, $x_2 = -2$, then first equation gives $x_1 = 2$. Therefore eigen vector is $X_3 = \begin{bmatrix} 2 \\ -2 \\ 1 \end{bmatrix}$.

Example 6: Determine the characteristic roots and vectors of the matrix

$$A = \begin{bmatrix} 6 & -2 & 2 \\ -2 & 3 & -1 \\ 2 & -1 & 3 \end{bmatrix}$$

Solution: The characteristic equations of A is

$$|A - \lambda I| = 0$$

$$\Rightarrow \quad \begin{vmatrix} 6-\lambda & -2 & 2 \\ -2 & 3-\lambda & -1 \\ 2 & -1 & 3-\lambda \end{vmatrix} = 0$$

On solving it, we have

$$(6-\lambda)(\lambda^2 - 6\lambda + 9 - 1) + 2(2\lambda - 6 + 2) + 2(2 - 6 + 2\lambda) = 0$$

$$\Rightarrow \qquad 6\lambda^2 - 36\lambda + 48 - \lambda^3 + 6\lambda^2 - 8\lambda + 4\lambda - 8 + 4\lambda - 8 = 0$$

$$\Rightarrow \qquad -\lambda^3 + 12\lambda^2 - 36\lambda + 32 = 0$$

$$\Rightarrow \qquad (2-\lambda)(\lambda^2 - 10\lambda + 16) = 0$$

$$\Rightarrow \qquad (2-\lambda)(\lambda - 2)(\lambda - 8) = 0$$

Therefore, the characteristic roots of A are 2, 2, 8.

The characteristic vector corresponding to characteristic root 2 is the nonzero solution of the equations

$$(A - 2I)X = 0$$

$$\Rightarrow \qquad \begin{bmatrix} 4 & -2 & 2 \\ -2 & 1 & -1 \\ 2 & -1 & 1 \end{bmatrix} \begin{bmatrix} x_1 \\ x_2 \\ x_3 \end{bmatrix} = 0$$

Applying $R_3 \to 2R_3 - R_1$ and $R_2 \to 2R_3 + R_1$, we have

$$\begin{bmatrix} 4 & -2 & 2 \\ 0 & 0 & 0 \\ 0 & 0 & 0 \end{bmatrix} \begin{bmatrix} x_1 \\ x_2 \\ x_3 \end{bmatrix} = 0$$

The coefficient matrix of these equations is of rank 1. Therefore, the equations has $3 - 1 = 2$ linearly independent solutions. These equations can be written as

$$4x_1 - 2x_2 + 2x_3 = 0$$

The two linearly independent solutions of this equation are $\begin{bmatrix} -1 \\ 0 \\ 2 \end{bmatrix}$ and $\begin{bmatrix} 1 \\ 2 \\ 0 \end{bmatrix}$. Hence

corresponding eigen vectors are $X_1 = \begin{bmatrix} -1 \\ 0 \\ 2 \end{bmatrix}$ and $X_2 = \begin{bmatrix} 1 \\ 2 \\ 0 \end{bmatrix}$.

The characteristic vector corresponding to characteristic root 8 is the nonzero solution of the equations

$$(A - 8I)X = 0$$

$$\Rightarrow \quad \begin{bmatrix} -2 & -2 & 2 \\ -2 & -5 & -1 \\ 2 & -1 & -5 \end{bmatrix} \begin{bmatrix} x_1 \\ x_2 \\ x_3 \end{bmatrix} = 0$$

Applying $R_2 \rightarrow R_2 - R_1$ and $R_3 \rightarrow R_3 + (2R_1 - R_2)$, we get

$$\begin{bmatrix} -2 & -2 & 2 \\ 0 & -3 & -3 \\ 0 & 0 & 0 \end{bmatrix} \begin{bmatrix} x_1 \\ x_2 \\ x_3 \end{bmatrix} = 0$$

The coefficient matrix of these equations is of rank 2. Therefore, the equations has $3 - 2 = 1$ linearly independent solution. These equations can be written as

$$- 2x_1 - 2x_2 + 2x_3 = 0$$
$$- 3x_2 - 3x_3 = 0$$

Last equation gives $x_2 = -x_3$. Let us take $x_2 = 1, x_3 = -1$, then the first equation gives $x_1 = -2$. Therefore, the eigen vectors is $X_3 = \begin{bmatrix} -2 \\ 1 \\ -1 \end{bmatrix}$ or $\begin{bmatrix} 2 \\ -1 \\ 1 \end{bmatrix}$.

11.6 HOUSE–HOLDER'S METHOD

This method is indeed more complicated but requires half as much computation as Given's method for reduction of a symmetric matrix to tridiagonal matrix. In this method, the matrix is reduced to tri-diagonal matrix using elementary orthogonal transformations. Let $A = [a_{ij}]$ be an nth order real symmetric matrix. This method consists of pre- and post-multiplication of A by a symmetric orthogonal matrix O such that OAO reduces to the tridiagonal form.

Let the matrix O can be written as

$$O = I - 2ww^T$$

where w is a column matrix such that

$$w_1^2 + w_2^2 + ... + w_n^2 = 1$$

Now
$$O = I - 2ww^T \Rightarrow O^T = [I - 2ww^T]^T = I - 2ww^T = O$$
$$\Rightarrow \quad O^T O = (I - 2ww^T)^T (I - 2ww^T)$$
$$= (I - 2ww^T)(I - 2ww^T)$$
$$= I - 4ww^T + 4ww^T = I$$

Thus O is a symmetric orthogonal matrix.

Next, we take w_k with first $(k - 1)$ zero components, so that

$$w_k^T = [0, 0, 0, ..., 0, x_k, x_{k+1}, ..., x_n]$$

But
$$w_k^T w_k = 1 \Rightarrow x_k^2 + x_{k+1}^2 + ... + x_n^2 = 1$$
$$\Rightarrow \quad O_k^{-1} AO_k = O_k^T AO_k = O_k AO_k$$

We then form $A_k = O_k A_{k-1} O_k, k = 2, 3, ... n.$

First we determine x's so that zeros are created in the positions $(1, 3), (1, 4), ... (1, n)$ and $(3, 1), (4, 1), ... (n, 1)$. Then on the second transformation, we find x's so that zeros are created in the position $(2, 4), (2, 5), ..., (2, n)$ and $(4, 2), (5, 2), ... (n, 2)$. Thus after $(n - 2)$ such transformations, we arrive at a tridiagonal matrix.

11.7 Dominant Latent Root (Rayleigh's Power Method)

If $\lambda_1, \lambda_2, ..., \lambda_n$ are the eigen values of matrix $A = [a_{ij}]_{n \times n}$ such that $|\lambda_1| \geq |\lambda_2| \geq |\lambda_3| \geq ... \geq |\lambda_n|$, then λ_1 is called the dominant latent root of the matrix A.

If $X_1, X_2, X_3, ..., X_n$ are the eigen vectors corresponding to the eigen values $\lambda_1, \lambda_2, \lambda_3, ..., \lambda_n$ respectively, then any column vector Y_0 can be expressed as the linear combination of eigen vectors $X_1, X_2, ..., X_n$. That is

$$Y_0 = a_1 X_1 + a_2 X_2 + ... + a_n X_n, \text{ where } a's \text{ are scalars}$$

$$\Rightarrow \quad AY_0 = A(a_1 X_1 + a_2 X_2 + ... + a_n X_n)$$

$$= a_1 A X_1 + a_2 A X_2 + a_3 A X_3 + ... + a_n A X_n$$

$$= a_1 \lambda_1 X_1 + a_2 \lambda_2 X_2 + a_3 \lambda_3 X_3 + ... + a_n \lambda_n X_n$$

Again $\quad A^2 Y_0 = A(AY_0) = a_1 \lambda_1 A X_1 + a_2 \lambda_2 A X_2 + a_3 \lambda_3 A X_3 + ... + a_n \lambda_n A X_n$

$$= a_1 \lambda_1^2 X_1 + a_2 \lambda_2^2 X_2 + a_3 \lambda_3^2 X_3 + ... + a_n \lambda_n^2 X_n$$

Continuing in this way, we get

$$A^r Y_0 = a_1 \lambda_1^r X_1 + a_2 \lambda_2^r X_2 + ... + a_n \lambda_n^r X_n$$

$$= a_1 \lambda_1^r X_1 + \sum_{i=2}^{n} a_i \left(\frac{\lambda_i}{\lambda_1}\right)^r x_r$$

Now, as $\lambda_1 > \lambda_i \; \forall \, i = 2, 3, ..., n$ so for $r \to \infty$, we have $\left(\dfrac{\lambda_i}{\lambda_1}\right)^r \to 0$.

The contribution of the term $a_1 \lambda_1^r X_1$ to the sum on the right increases with r and therefore, every time we multiply a column vector by A, it becomes nearer to the eigen vector X_1. Then we make the largest component of the resulting column vector unity to avoid the factor a_1.

This shows that $A^r Y_0$ is an eigen vector to the largest eigen value λ_1.

If we take $\qquad AY_0 = k_1 Z_1 \; \Rightarrow \; Z_1 = \left(\dfrac{1}{k_1}\right) AY_0$

where Z_1 is find by dividing each element of AY_0 by its numerically largest element. Repeating this process, we get

$$A^{r+1} Y_0 = AZ_r = k_{r+1} Z_{r+1}$$

where k_{r+1} is the eigen value and Z_{r+1} is corresponding eigen vector.

This method is also known as power series method. It should be noted that the eigen value of matrix A is the reciprocal of the dominant eigen value of A^{-1}.

11.8 QR METHOD

This method is used to calculate eigen values of a matrix A may be non-symmetric, which was introduced by J.G.F. Francis. In this method matrix A is described as

$$A = QR$$

where Q is orthogonal and R is upper triangular matrices. If A is real, then Q and R can be choosen real. Let $A_1 = A$, then in kth step, we have

$$A_k = Q_k R_k$$

and $(k+1)$th step $\qquad A_{k+1} = R_k Q_k$

This process is continued until A_{k+1} converges to an upper triangular matrix. As Q_k is orthogonal so $Q_k^T Q_k = I$, we have

$$R_k = Q_k^T A_k$$

Hence $\qquad A_{k+1} = Q_k^T A_k Q_k$

This implies that A_{k+1} is orthogonally similar to A_k. Hence by induction A_{k+1} is orthogonally similar to A_1.

The process of reduction of A to upper triangular matrix using Jacobi's method is quite lengthy. By converting matrix into Hessanberg matrix through similarity transformation, we can make this reduction process smooth and fast.

A matrix B is called Hessanberg if

$$b_{ij} = 0 \text{ for all } i > j + 1$$

i.e. A matrix with all the elements below subdiagonal are zero is called an upper Hessanberg matrix.

11.9 REDUCTION OF A MATRIX TO UPPER HESSANBERG MATRIX

To understand the method, we consider a matrix A of order 5 as

$$A = \begin{bmatrix} a_{11} & a_{12} & a_{13} & a_{14} & a_{15} \\ a_{21} & a_{22} & a_{23} & a_{24} & a_{25} \\ a_{31} & a_{32} & a_{33} & a_{34} & a_{35} \\ a_{41} & a_{42} & a_{43} & a_{44} & a_{45} \\ a_{51} & a_{52} & a_{53} & a_{54} & a_{55} \end{bmatrix}$$

For this, we use $\qquad A_1 = L_1 A L_1^{-1}$

where L_1 is a lower triangular matrix of order 5 defined as

$$L_1 = \begin{bmatrix} 1 & 0 & 0 & 0 & 0 \\ 0 & 1 & 0 & 0 & 0 \\ 0 & l_{32} & 1 & 0 & 0 \\ 0 & l_{42} & 0 & 1 & 0 \\ 0 & l_{52} & 0 & 0 & 1 \end{bmatrix}$$

with $\qquad l_{32} = -\dfrac{a_{31}}{a_{21}}, \; l_{42} = -\dfrac{a_{41}}{a_{21}}, \; l_{52} = -\dfrac{a_{51}}{a_{21}}.$

Then $\qquad L_1^{-1} = \begin{bmatrix} 1 & 0 & 0 & 0 & 0 \\ 0 & 1 & 0 & 0 & 0 \\ 0 & -l_{32} & 1 & 0 & 0 \\ 0 & -l_{42} & 0 & 1 & 0 \\ 0 & -l_{52} & 0 & 0 & 1 \end{bmatrix}$

Now $\qquad A_1 = L_1 A L_1^{-1}$

$$= \begin{bmatrix} 1 & 0 & 0 & 0 & 0 \\ 0 & 1 & 0 & 0 & 0 \\ 0 & l_{32} & 1 & 0 & 0 \\ 0 & l_{42} & 0 & 1 & 0 \\ 0 & l_{52} & 0 & 0 & 1 \end{bmatrix} \begin{bmatrix} a_{11} & a_{12} & a_{13} & a_{14} & a_{15} \\ a_{21} & a_{22} & a_{23} & a_{24} & a_{25} \\ a_{31} & a_{32} & a_{33} & a_{34} & a_{35} \\ a_{41} & a_{42} & a_{43} & a_{44} & a_{45} \\ a_{51} & a_{52} & a_{53} & a_{54} & a_{55} \end{bmatrix} \begin{bmatrix} 1 & 0 & 0 & 0 & 0 \\ 0 & 1 & 0 & 0 & 0 \\ 0 & -l_{32} & 1 & 0 & 0 \\ 0 & -l_{42} & 0 & 1 & 0 \\ 0 & -l_{52} & 0 & 0 & 1 \end{bmatrix}$$

$$
= \begin{bmatrix}
a_{11} & a_{12} & a_{13} & a_{14} & a_{15} \\
a_{21} & a_{22} & a_{23} & a_{24} & a_{25} \\
0 & a_{32}^{(1)} & a_{33}^{(1)} & a_{34}^{(1)} & a_{35}^{(1)} \\
0 & a_{42}^{(1)} & a_{43}^{(1)} & a_{44}^{(1)} & a_{45}^{(1)} \\
0 & a_{52}^{(1)} & a_{53}^{(1)} & a_{54}^{(1)} & a_{55}^{(1)}
\end{bmatrix}
$$

Now to find $A_2 = L_2 A_1 L_2^{-1}$, we select

$$
L_2 = \begin{bmatrix}
1 & 0 & 0 & 0 & 0 \\
0 & 1 & 0 & 0 & 0 \\
0 & 0 & 1 & 0 & 0 \\
0 & 0 & l_{43} & 1 & 0 \\
0 & 0 & l_{53} & 0 & 1
\end{bmatrix} \text{ with } l_{43} = -\frac{a_{42}^{(1)}}{a_{32}^{(1)}}, l_{53} = -\frac{a_{52}^{(1)}}{a_{32}^{(1)}}
$$

After calculation, we get

$$
A_2 = \begin{bmatrix}
a_{11} & a_{12} & a_{13} & a_{14} & a_{15} \\
a_{21} & a_{22} & a_{23} & a_{24} & a_{25} \\
0 & a_{32}^{(1)} & a_{33}^{(1)} & a_{34}^{(1)} & a_{35}^{(1)} \\
0 & 0 & a_{43}^{(2)} & a_{44}^{(2)} & a_{45}^{(2)} \\
0 & 0 & a_{53}^{(2)} & a_{54}^{(2)} & a_{55}^{(2)}
\end{bmatrix}
$$

Again to find $A_3 = L_3 A_2 L_3^{-1}$, we select

$$
L_3 = \begin{bmatrix}
1 & 0 & 0 & 0 & 0 \\
0 & 1 & 0 & 0 & 0 \\
0 & 0 & 1 & 0 & 0 \\
0 & 0 & 0 & 1 & 0 \\
0 & 0 & 0 & l_{54} & 1
\end{bmatrix} \text{ with } l_{54} = -\frac{a_{52}^{(2)}}{a_{43}^{(2)}}
$$

After calculation, we get

$$
A_3 = \begin{bmatrix}
a_{11} & a_{12} & a_{13} & a_{14} & a_{15} \\
a_{21} & a_{22} & a_{23} & a_{24} & a_{25} \\
0 & a_{32}^{(1)} & a_{33}^{(1)} & a_{34}^{(1)} & a_{35}^{(1)} \\
0 & 0 & a_{43}^{(2)} & a_{44}^{(2)} & a_{45}^{(2)} \\
0 & 0 & 0 & a_{54}^{(3)} & a_{55}^{(3)}
\end{bmatrix}
$$

Which is the required Hessanberg matrix. Now we apply Jacobi's method to reduce Hessanberg matrix to upper triangular matrix. One can also use House-Holder's method to transform the given matrix A to upper Hessenberg form.

Example 7: Using House-Holder's method reduce the following matrix $A = \begin{bmatrix} 1 & 4 & 3 \\ 4 & 1 & 2 \\ 3 & 2 & 1 \end{bmatrix}$

in tri-diagonal form.

Solution: Let $W_2^T = \begin{bmatrix} 0 & x_2 & x_3 \end{bmatrix}$

then
$$O = I - 2W_2 W_2^T = \begin{bmatrix} 1 & 0 & 0 \\ 0 & 1 & 0 \\ 0 & 0 & 1 \end{bmatrix} - 2 \begin{bmatrix} 0 \\ x_2 \\ x_3 \end{bmatrix} \begin{bmatrix} 0 & x_2 & x_3 \end{bmatrix}$$

$$= \begin{bmatrix} 1 & 0 & 0 \\ 0 & 1 & 0 \\ 0 & 0 & 1 \end{bmatrix} - 2 \begin{bmatrix} 0 & 0 & 0 \\ 0 & x_2^2 & x_2 x_3 \\ 0 & x_2 x_3 & x_3^2 \end{bmatrix}$$

$$= \begin{bmatrix} 1 & 0 & 0 \\ 0 & 1 - 2x_2^2 & -2x_2 x_3 \\ 0 & -2x_2 x_3 & 1 - 2x_3^2 \end{bmatrix}$$

The element $(1, 3)$ of OAO is zero only when the corresponding element of AO is zero.

Now
$$AO = \begin{bmatrix} 1 & 4 & 3 \\ 4 & 1 & 2 \\ 3 & 2 & 1 \end{bmatrix} \begin{bmatrix} 1 & 0 & 0 \\ 0 & 1 - 2x_2^2 & -2x_2 x_3 \\ 0 & -2x_2 x_3 & 1 - 2x_3^2 \end{bmatrix}$$

$$= \begin{bmatrix} 1 & 4(1 - 2x_2^2) - 6x_2 x_3 & -8x_2 x_3 + 3(1 - 2x_3^2) \\ 4 & 1 - 2x_2^2 - 4x_2 x_3 & -2x_2 x_3 + 2(1 - 2x_3^2) \\ 3 & 2(1 - 2x_2^2) - 2x_2 x_3 & -4x_2 x_3 + (1 - 2x_3^2) \end{bmatrix}$$

For the element $(1, 3)$ is zero, we have
$$-8x_2 x_3 + 3(1 - 2x_3^2) = 0$$
$$\Rightarrow \qquad -8x_2 x_3 + 3 - 6x_3^2 = 0$$
$$\Rightarrow \qquad 3 - 2x_3(4x_2 + 3x_3) = 0$$
$$\Rightarrow \qquad 3 - 2x_3 p_1 = 0 \tag{i}$$
where $\qquad 4x_2 + 3x_3 = p_1 \tag{ii}$

But the sum of the square of the first row remains invariant, so we have
$$1^2 + [4(1 - 2x_2^2) - 6x_2 x_3]^2 + 0 = 1^2 + 4^2 + 3^2$$
$$\Rightarrow \qquad 4 - 8x_2^2 - 6x_2 x_3 = \pm \sqrt{25}$$
$$\Rightarrow \qquad 4 - 2x_2(4x_2 + 3x_3) = \pm 5$$
$$\Rightarrow \qquad 4 - 2p_1 x_2 = \pm 5 \tag{iii}$$

Multiplying Eq. (i) by x_3 and Eq. (iii) by x_2 and then adding, we get
$$3x_3 + 4x_2 - 2x_3^2 p_1 - 2x_2^2 p_1 = \pm 5x_2$$
$$\Rightarrow \qquad p_1 - 2p_1(x_2^2 + x_3^2) = \pm 5x_2 \text{ by Eq. (ii) } 4x_2 + 3x_3 = p_1$$
$$\Rightarrow \qquad p_1 - 2p_1 = \pm 5x_2 \text{ since } O \text{ is orthogonal} \Rightarrow x_2^2 + x_3^2 = 1$$
$$\Rightarrow \qquad p_1 = \pm 5x_2 \tag{iv}$$

Using Eq. (iv) in Eq. (iii), we have
$$4 - 2(\pm 5x_2) x_2 = \pm 5$$
$$\Rightarrow \qquad 4 \pm 10x_2^2 = \pm 5$$
$$\Rightarrow \qquad 10x_2^2 = 1 \text{ or } 10x_2^2 = 9$$

$\Rightarrow$
$$x_2 = \frac{1}{\sqrt{10}} \text{ or } x_2 = \frac{3}{\sqrt{10}}$$

Now from Eq. (i) $x_3 = \dfrac{3}{2p_1} = \pm\dfrac{3}{10x_2}$

Since x_3 contains x_2 in the denominator, we obtain best accuracy if x_2 is large. Therefore we take

$$x_2 = \frac{3}{\sqrt{10}} \Rightarrow x_3 = \pm\frac{1}{\sqrt{10}}$$

Taking +ve sign, we have

$$x_2 = \frac{3}{10}, x_3 = \frac{1}{\sqrt{10}}$$

Thus
$$w_2^T = \left[0, \frac{3}{\sqrt{10}}, \frac{1}{\sqrt{10}}\right]$$

and
$$O = I - 2w_2 w_2^T = \begin{bmatrix} 1 & 0 & 0 \\ 0 & -\dfrac{4}{5} & -\dfrac{3}{5} \\ 0 & -\dfrac{3}{5} & \dfrac{4}{5} \end{bmatrix}$$

Also
$$OAO = \begin{bmatrix} 1 & 0 & 0 \\ 0 & -\dfrac{4}{5} & -\dfrac{3}{5} \\ 0 & -\dfrac{3}{5} & \dfrac{4}{5} \end{bmatrix} \begin{bmatrix} 1 & 4 & 3 \\ 4 & 1 & 2 \\ 3 & 2 & 1 \end{bmatrix} \begin{bmatrix} 1 & 0 & 0 \\ 0 & -\dfrac{4}{5} & -\dfrac{3}{5} \\ 0 & -\dfrac{3}{5} & \dfrac{4}{5} \end{bmatrix}$$

$$= \begin{bmatrix} 1 & 4 & 3 \\ -5 & -2 & -\dfrac{11}{5} \\ 0 & 1 & -\dfrac{2}{5} \end{bmatrix} \begin{bmatrix} 1 & 0 & 0 \\ 0 & -\dfrac{4}{5} & -\dfrac{3}{5} \\ 0 & -\dfrac{3}{5} & \dfrac{4}{5} \end{bmatrix}$$

$$= \begin{bmatrix} 1 & -5 & 0 \\ -5 & \dfrac{73}{25} & -\dfrac{14}{25} \\ 0 & -\dfrac{14}{25} & -\dfrac{23}{25} \end{bmatrix}$$

which is the required tridiagonal matrix.

Example 8: Using House-Holder's method reduce the matrix $A = \begin{bmatrix} 2 & 1 & 3 \\ 1 & 4 & 2 \\ 3 & 2 & 3 \end{bmatrix}$ in tri-diagonal form.

Solution: Let $W_2^T = \begin{bmatrix} 0 & x_2 & x_3 \end{bmatrix}$

then $\qquad O = I - 2W_2 W_2^T = \begin{bmatrix} 1 & 0 & 0 \\ 0 & 1 & 0 \\ 0 & 0 & 1 \end{bmatrix} - 2 \begin{bmatrix} 0 \\ x_2 \\ x_3 \end{bmatrix} \begin{bmatrix} 0 & x_2 & x_3 \end{bmatrix}$

$$= \begin{bmatrix} 1 & 0 & 0 \\ 0 & 1-2x_2^2 & -2x_2 x_3 \\ 0 & -2x_2 x_3 & 1-2x_3^2 \end{bmatrix}$$

The element (1, 3) of OAO is zero only when the corresponding element of AO is zero.

Now $\qquad AO = \begin{bmatrix} 2 & 1 & 3 \\ 1 & 4 & 2 \\ 3 & 2 & 3 \end{bmatrix} \begin{bmatrix} 1 & 0 & 0 \\ 0 & 1-2x_2^2 & -2x_2 x_3 \\ 0 & -2x_2 x_3 & 1-2x_3^2 \end{bmatrix}$

$$= \begin{bmatrix} 2 & (1-2x_2^2)-6x_2 x_3 & -2x_2 x_3 + 3(1-2x_3^2) \\ 1 & 4(1-2x_2^2)-4x_2 x_3 & -8x_2 x_3 + 2(1-2x_2^2) \\ 3 & 2(1-2x_2^2)-6x_2 x_3 & -4x_2 x_3 + 3(1-2x_3^2) \end{bmatrix}$$

For the element (1, 3) is zero, we have

$$-2x_2 x_3 + 3(1-2x_3^2) = 0$$

$\Rightarrow \qquad -2x_2 x_3 + 3 - 6x_3^2 = 0$

$\Rightarrow \qquad -2x_3(x_2 + 3x_3) + 3 = 0$

$\Rightarrow \qquad 3 - 2x_3 p_1 = 0 \qquad\qquad\qquad\qquad\qquad\qquad\qquad$ (i)

where $\qquad\qquad p_1 = x_2 + 3x_3 \qquad\qquad\qquad\qquad\qquad\qquad\qquad$ (ii)

But the sum of the square of the first row remains invariant, so we have

$$2^2 + [(1-2x_2^2)-6x_2 x_3]^2 + 0 = 2^2 + 1^2 + 3^2$$

$\Rightarrow \qquad\qquad 1-2x_2^2 - 6x_2 x_3 = \pm\sqrt{10}$

$\Rightarrow \qquad\qquad 1 - 2x_2(x_2 + 3x_3) = \pm\sqrt{10}$

$\Rightarrow \qquad\qquad 1 - 2x_2 p_1 = \pm\sqrt{10} \qquad\qquad\qquad\qquad\qquad\qquad$ (iii)

Multiplying Eq. (i) by x_3 and Eq. (iii) by x_2 and then adding, we get

$$x_2 + 3x_3 - 2x_3^2 p_1 - 2x_2^2 p_1 = \pm\sqrt{10}\, x_2$$

$\Rightarrow \qquad\qquad p_1 - 2p_1(x_2^2 + x_3^2) = \pm\sqrt{10}\, x_2$

$\Rightarrow \qquad\qquad p_1 - 2p_1 = \pm\sqrt{10}\, x_2$, as O is orthogonal $\Rightarrow x_2^2 + x_3^2 = 1$

$\Rightarrow \qquad\qquad p_1 = \pm\sqrt{10}\, x_2 \qquad\qquad\qquad\qquad\qquad\qquad$ (iv)

Using Eq. (iv) in Eq. (iii), we have

$$1 - 2x_2(\pm\sqrt{10}\, x_2) = \pm\sqrt{10}$$

$$\Rightarrow \qquad 1 \pm 2\sqrt{10}\, x_2^2 = \pm\sqrt{10}$$

$$\Rightarrow \qquad x_2^2 = \frac{\sqrt{10}-1}{2\sqrt{10}} \text{ or } x_2^2 = \frac{\sqrt{10}+1}{2\sqrt{10}}$$

$$\Rightarrow \qquad x_2 = 0.811242 \text{ or } x_2 = 0.584722$$

Now from Eq. (i) $\qquad x_3 = \dfrac{3}{2p_1} = \pm\dfrac{3}{2\sqrt{10}x_2}$

Since x_3 contains x_2 in the denominator, we obtain best accuracy if x_2 is large. Therefore we take $x^2 = 0.811242$

$$\Rightarrow \qquad x_3 = \frac{3}{2\sqrt{10}(0.811242)} = 0.584710$$

Thus $\qquad W_2^T = [0, 0.811242, 0.584710]$

and $\qquad O = I - 2W_2 W_2^T = \begin{bmatrix} 1 & 0 & 0 \\ 0 & -0.3162274 & -0.948683 \\ 0 & -0.948683 & 0.3162274 \end{bmatrix}$

Also $OAO = \begin{bmatrix} 1 & 0 & 0 \\ 0 & -0.3162274 & -0.948683 \\ 0 & -0.948683 & 0.3162274 \end{bmatrix}\begin{bmatrix} 2 & 1 & 3 \\ 1 & 4 & 2 \\ 3 & 2 & 3 \end{bmatrix}\begin{bmatrix} 1 & 0 & 0 \\ 0 & -0.3162274 & -0.948683 \\ 0 & -0.948683 & 0.3162274 \end{bmatrix}$

$$= \begin{bmatrix} 2 & -3.16227716 & 0 \\ -3.16227716 & 4.299997732 & 1.899998944 \\ 0 & 1.899998944 & 2.699999595 \end{bmatrix}$$

which is the required tridiagonal matrix.

Example 9: Determine the largest eigen value and corresponding eigen vector of the matrix

$$A = \begin{bmatrix} 2 & -1 & 0 \\ -1 & 2 & -1 \\ 0 & -1 & 2 \end{bmatrix}.$$

Solution: Let the initial eigen vector is $X\,[1\ 0\ 0]^T$. Then

$$AX = \begin{bmatrix} 2 & -1 & 0 \\ -1 & 2 & -1 \\ 0 & -1 & 2 \end{bmatrix}\begin{bmatrix} 1 \\ 0 \\ 0 \end{bmatrix} = \begin{bmatrix} 2 \\ -1 \\ 0 \end{bmatrix} = 2\begin{bmatrix} 1 \\ -0.5 \\ 0 \end{bmatrix}$$

$\Rightarrow X^{(1)} = [1\ -0.5\ 0]^T$ is the first approximate eigen vector and 2 is eigen value.

Repeating it, we get

$$AX^{(1)} = \begin{bmatrix} 2 & -1 & 0 \\ -1 & 2 & -1 \\ 0 & -1 & 2 \end{bmatrix}\begin{bmatrix} 1 \\ -0.5 \\ 0 \end{bmatrix} = \begin{bmatrix} 2.5 \\ -2 \\ 0.5 \end{bmatrix} = 2.5\begin{bmatrix} 1 \\ -0.8 \\ 0.2 \end{bmatrix}$$

$\Rightarrow X^{(2)} = [1 -0.8\ 0.2]^T$ and eigen value = 2.5.

$$AX^{(2)} = \begin{bmatrix} 2 & -1 & 0 \\ -1 & 2 & -1 \\ 0 & -1 & 2 \end{bmatrix} \begin{bmatrix} 1 \\ -0.8 \\ 0.2 \end{bmatrix} = \begin{bmatrix} 2.8 \\ -2.8 \\ 1.2 \end{bmatrix} = 2.8 \begin{bmatrix} 1 \\ -1 \\ 0.428571 \end{bmatrix}$$

$\Rightarrow X^{(3)} = [1\ \ -1\ \ 0.428571]^T$ and eigen value = 2.8.

$$AX^{(3)} = \begin{bmatrix} 2 & -1 & 0 \\ -1 & 2 & -1 \\ 0 & -1 & 2 \end{bmatrix} \begin{bmatrix} 1 \\ -1 \\ 0.428571 \end{bmatrix} = \begin{bmatrix} 3 \\ -3.428571 \\ 1.857142 \end{bmatrix}$$

$$= 3.428571 \begin{bmatrix} 0.875000 \\ -1 \\ 0.541666 \end{bmatrix}$$

$\Rightarrow X^{(4)} = [0.875000\ \ -1\ \ 0.541666]$ and eigen value = 3.428571.

$$AX^{(4)} = \begin{bmatrix} 2 & -1 & 0 \\ -1 & 2 & -1 \\ 0 & -1 & 2 \end{bmatrix} \begin{bmatrix} 0.875000 \\ -1 \\ 0.541666 \end{bmatrix} = \begin{bmatrix} 2.750000 \\ -3.416666 \\ 2.083332 \end{bmatrix}$$

$$= 3.416666 \begin{bmatrix} 0.804878 \\ -1 \\ 0.609756 \end{bmatrix}$$

$\Rightarrow X^{(5)} = [0.804878\ \ -1\ \ 0.609756]^T$ and eigen value = 3.416666.

$$AX^{(5)} = \begin{bmatrix} 2 & -1 & 0 \\ -1 & 2 & -1 \\ 0 & -1 & 2 \end{bmatrix} \begin{bmatrix} 0.804878 \\ -1 \\ 0.609756 \end{bmatrix} = \begin{bmatrix} 2.609756 \\ -3.414634 \\ 2.219512 \end{bmatrix}$$

$$= 3.414634 \begin{bmatrix} 0.764286 \\ -1 \\ 0.649999 \end{bmatrix}$$

$\Rightarrow X^{(6)} = [0.764286\ \ -1\ \ 0.649999]^T$ and eigen value = 3.414634.

$$AX^{(6)} = \begin{bmatrix} 2 & -1 & 0 \\ -1 & 2 & -1 \\ 0 & -1 & 2 \end{bmatrix} \begin{bmatrix} 0.764286 \\ -1 \\ 0.649999 \end{bmatrix} = \begin{bmatrix} 2.528572 \\ -3.414285 \\ 2.299998 \end{bmatrix}$$

$$= 3.414285 \begin{bmatrix} 0.740586 \\ -1 \\ 0.673640 \end{bmatrix}$$

$\Rightarrow X^{(7)} = [0.740586 \ -1 \ 0.673640]^T$ and eigen value = 3.414285.

$$AX^{(7)} = \begin{bmatrix} 2 & -1 & 0 \\ -1 & 2 & -1 \\ 0 & -1 & 2 \end{bmatrix} \begin{bmatrix} 0.740586 \\ -1 \\ 0.673640 \end{bmatrix} = \begin{bmatrix} 2.481172 \\ -3.414226 \\ 2.347280 \end{bmatrix}$$

$$= 3.414226 \begin{bmatrix} 0.726716 \\ -1 \\ 0.687500 \end{bmatrix}$$

$\Rightarrow X^{(8)} = [0.726716 \ -1 \ 0.6875]^T$ and eigen value = 3.414226.

$$AX^{(8)} = \begin{bmatrix} 2 & -1 & 0 \\ -1 & 2 & -1 \\ 0 & -1 & 2 \end{bmatrix} \begin{bmatrix} 0.726716 \\ -1 \\ 0.6875 \end{bmatrix} = \begin{bmatrix} 2.453432 \\ -3.414216 \\ 2.375000 \end{bmatrix}$$

$$= 3.414216 \begin{bmatrix} 0.718593 \\ -1 \\ 0.695621 \end{bmatrix}$$

$\Rightarrow X^{(9)} = [0.718593 \ -1 \ 0.695621]^T$ and eigen value = 3.414216.

$$AX^{(9)} = \begin{bmatrix} 2 & -1 & 0 \\ -1 & 2 & -1 \\ 0 & -1 & 2 \end{bmatrix} \begin{bmatrix} 0.718593 \\ -1 \\ 0.695621 \end{bmatrix} = \begin{bmatrix} 2.437186 \\ -3.414214 \\ 2.391242 \end{bmatrix}$$

$$= 3.414214 \begin{bmatrix} 0.713835 \\ -1 \\ 0.700378 \end{bmatrix}$$

$\Rightarrow X^{(10)} = [0.713835 \ -1 \ 0.700378]^T$ and eigen value = 3.414214.

After tenth iteration, we follow that the largest eigen value is 3.414214 and the corresponding eigen vector is $[0.713835 \ -1 \ 0.700378]^T$.

Example 10: Determine the largest eigen value and corresponding eigen vector of the matrix

$$A = \begin{bmatrix} 1 & -3 & 2 \\ 4 & 4 & -1 \\ 6 & 3 & 5 \end{bmatrix} \text{ by taking } X^0 = \begin{bmatrix} 1 \\ 1 \\ 1 \end{bmatrix}.$$

Solution: Here

$$AX^0 = \begin{bmatrix} 1 & -3 & 2 \\ 4 & 4 & -1 \\ 6 & 3 & 5 \end{bmatrix} \begin{bmatrix} 1 \\ 1 \\ 1 \end{bmatrix} = \begin{bmatrix} 0 \\ 7 \\ 14 \end{bmatrix} = 14 \begin{bmatrix} 0 \\ 0.5 \\ 1 \end{bmatrix}$$

$\Rightarrow X^{(1)} = [0 \quad 0.5 \quad 1]^T$ is eigen value = 14.

$$AX^{(1)} = \begin{bmatrix} 1 & -3 & 2 \\ 4 & 4 & -1 \\ 6 & 3 & 5 \end{bmatrix} \begin{bmatrix} 0 \\ 0.5 \\ 1 \end{bmatrix} = \begin{bmatrix} 0.5 \\ 1 \\ 6.5 \end{bmatrix} = 6.5 \begin{bmatrix} 0.076923 \\ 0.153846 \\ 1 \end{bmatrix}$$

$\Rightarrow X^{(2)} = [0.076923 \quad 0.153846 \quad 1]^T$ and eigen value = 6.5.

$$AX^{(2)} = \begin{bmatrix} 1 & -3 & 2 \\ 4 & 4 & -1 \\ 6 & 3 & 5 \end{bmatrix} \begin{bmatrix} 0.076923 \\ 0.153846 \\ 1 \end{bmatrix} = \begin{bmatrix} 1.615385 \\ -0.076924 \\ 5.923076 \end{bmatrix} = 5.923076 \begin{bmatrix} 0.272727 \\ -0.012987 \\ 1 \end{bmatrix}$$

$\Rightarrow X^{(3)} = [0.272727 \quad -0.012987 \quad 1]^T$ and eigen value = 5.923076.

$$AX^{(3)} = \begin{bmatrix} 1 & -3 & 2 \\ 4 & 4 & -1 \\ 6 & 3 & 5 \end{bmatrix} \begin{bmatrix} 0.272727 \\ -0.012987 \\ 1 \end{bmatrix} = \begin{bmatrix} 2.2311688 \\ 0.03896 \\ 6.597401 \end{bmatrix}$$

$$= 6.597401 \begin{bmatrix} 0.3381890 \\ 0.0059054 \\ 1 \end{bmatrix}$$

$\Rightarrow X^{(4)} = [0.338189 \quad 0.0059054 \quad 1]^T$ and eigen value = 6.597401.

$$AX^{(4)} = \begin{bmatrix} 1 & -3 & 2 \\ 4 & 4 & -1 \\ 6 & 3 & 5 \end{bmatrix} \begin{bmatrix} 0.338189 \\ 0.0059054 \\ 1 \end{bmatrix} = \begin{bmatrix} 2.3204728 \\ 0.3763776 \\ 7.0468502 \end{bmatrix}$$

$$= 7.0468502 \begin{bmatrix} 0.3292921 \\ 0.05341075 \\ 1 \end{bmatrix}$$

$\Rightarrow X^{(5)} = [0.3292921 \quad 0.05341075 \quad 1]^T$ and eigen value = 7.0468502.

$$AX^{(5)} = \begin{bmatrix} 1 & -3 & 2 \\ 4 & 4 & -1 \\ 6 & 3 & 5 \end{bmatrix} \begin{bmatrix} 0.3292921 \\ 0.05341075 \\ 1 \end{bmatrix} = \begin{bmatrix} 2.16905985 \\ 0.5308114 \\ 7.03780751 \end{bmatrix}$$

$$= 7.03780751 \begin{bmatrix} 0.30820110 \\ 0.0754228 \\ 1 \end{bmatrix}$$

$\Rightarrow X^{(6)} = [0.30820110 \quad 0.0754228 \quad 1]^T$ and eigen value = 7.03780751.

$$AX^{(6)} = \begin{bmatrix} 1 & -3 & 2 \\ 4 & 4 & -1 \\ 6 & 3 & 5 \end{bmatrix} \begin{bmatrix} 0.30820110 \\ 0.0754228 \\ 1 \end{bmatrix} = \begin{bmatrix} 2.0819327 \\ 0.5344956 \\ 7.075475 \end{bmatrix}$$

$$= 7.075475 \begin{bmatrix} 0.294246 \\ 0.075542 \\ 1 \end{bmatrix}$$

$\Rightarrow X^{(7)} = [0.294246 \quad 0.075542 \quad 1]^T$ and eigen value = 7.075475.

$$AX^{(7)} = \begin{bmatrix} 1 & -3 & 2 \\ 4 & 4 & -1 \\ 6 & 3 & 5 \end{bmatrix} \begin{bmatrix} 0.294246 \\ 0.075542 \\ 1 \end{bmatrix} = \begin{bmatrix} 2.06762 \\ 0.479152 \\ 6.992102 \end{bmatrix}$$

$$= 6.992102 \begin{bmatrix} 0.295708 \\ 0.0685276 \\ 1 \end{bmatrix}$$

$\Rightarrow X^{(8)} = [0.295708 \quad 0.0685276 \quad 1]^T$ and eigen value = 6.992102.

$$AX^{(8)} = \begin{bmatrix} 1 & -3 & 2 \\ 4 & 4 & -1 \\ 6 & 3 & 5 \end{bmatrix} \begin{bmatrix} 0.295708 \\ 0.0685276 \\ 1 \end{bmatrix} = \begin{bmatrix} 2.0901252 \\ 0.4569424 \\ 6.9798308 \end{bmatrix}$$

$$= 6.9798308 \begin{bmatrix} 0.299452 \\ 0.065466 \\ 1 \end{bmatrix}$$

$\Rightarrow X^{(9)} = [0.299452 \quad 0.065466 \quad 1]^T$ and eigen value = 6.9798308.

$$AX^{(9)} = \begin{bmatrix} 1 & -3 & 2 \\ 4 & 4 & -1 \\ 6 & 3 & 5 \end{bmatrix} \begin{bmatrix} 0.299452 \\ 0.065466 \\ 1 \end{bmatrix} = \begin{bmatrix} 2.103054 \\ 0.459672 \\ 6.99311 \end{bmatrix}$$

$$= 6.99311 \begin{bmatrix} 0.300732 \\ 0.657321 \\ 1 \end{bmatrix}$$

$\Rightarrow X^{(10)} [0.300732 \quad 0.0657321 \quad 1]^T$ and eigen value = 6.99311.

After tenth iteration, we follow that the largest eigen value is 6.99311 and the

corresponding eigen vector is $X = \begin{bmatrix} 0.300732 \\ 0.0657321 \\ 1 \end{bmatrix}$.

Example 11: Using QR method, find the eigen value of the following matrix $A = \begin{bmatrix} 3 & 2 \\ 1 & 4 \end{bmatrix}$ correct up to one decimal place.

Solution: We have $\tan \alpha = -\dfrac{a_{21}}{a_{11}} = -\dfrac{1}{3}$

$$\sin \alpha = \dfrac{-1}{\sqrt{10}} = -0.3162$$

$$\cos \alpha = \dfrac{3}{\sqrt{10}} = 0.9487$$

$h_{11} = a_{11}\cos \alpha - a_{21}\sin \alpha = 3 \times (0.9487) - 1 \times (-0.3162) = 3.1623$
$h_{12} = a_{12}\cos \alpha - a_{22}\sin \alpha = 2 \times (0.9487) - 4 \times (-0.3162) = 3.1622$
$h_{21} = 0$
$h_{22} = a_{12}\sin \alpha + a_{22}\cos \alpha = 2 \times (-0.3162) + 4 \times (0.9487) = 3.1624$

Now $A_1 = Q_1 R_1 = \begin{bmatrix} 0.9487 & -0.3162 \\ 0.3162 & 0.9487 \end{bmatrix} \begin{bmatrix} 3.1623 & 3.1622 \\ 0 & 3.1624 \end{bmatrix} = \begin{bmatrix} 3 & 2 \\ 0.9999 & 4 \end{bmatrix}$

and $A_2 = R_1 Q_1 = \begin{bmatrix} 3.1623 & 3.1622 \\ 0 & 3.1624 \end{bmatrix} \begin{bmatrix} 0.9487 & -0.3162 \\ 0.3162 & 0.9487 \end{bmatrix} = \begin{bmatrix} 4 & 2 \\ 1 & 3 \end{bmatrix}$

Again $\tan \alpha = -\dfrac{1}{4}, \sin \alpha = -\dfrac{1}{\sqrt{17}} = -0.2425, \cos \alpha = \dfrac{4}{\sqrt{17}} = 0.97$

$h_{11} = 4 \times 0.97 - 1 \times (-0.2425) = 4.1225$
$h_{12} = 2 \times 0.97 - 3 \times (-0.2425) = 2.6675$
$h_{21} = 0, \ h_{22} = 2.4250$

$$A_3 = R_2 Q_2 = \begin{bmatrix} 4.1225 & 2.6675 \\ 0 & 2.425 \end{bmatrix} \begin{bmatrix} 0.97 & -0.2425 \\ 0.2425 & 0.97 \end{bmatrix} = \begin{bmatrix} 4.6457 & 1.5878 \\ 0.5881 & 2.3522 \end{bmatrix}$$

$$\tan \alpha = -\dfrac{0.5881}{4.6457} = -0.1256, \sin \alpha = 0.1256, \cos \alpha = 0.9921$$

$h_{11} = 4.6457 \times 0.9921 - 0.5881 \times (0.1256) = 4.6829$
$h_{12} = 1.8707, h_{21} = 0, h_{22} = 2.1342$

$$A_4 = R_3 Q_3 = \begin{bmatrix} 4.6829 & 1.8707 \\ 0 & 2.1342 \end{bmatrix} \begin{bmatrix} 0.9921 & -0.1256 \\ 0.1256 & 0.9921 \end{bmatrix} = \begin{bmatrix} 4.8808 & 1.2677 \\ 0.2681 & 2.1173 \end{bmatrix}$$

$$\tan \alpha = -\dfrac{0.2681}{4.8808} \sin \alpha = -0.05485, \cos \alpha = 0.9985$$

$h_{11} = 4.8808 \times 0.9985 - 0.2681 \times (-0.0548) = 4.8882$
$h_{12} = 1.3814, h_{21} = 0, h_{22} = 2.0446$

$$A_5 = R_4 Q_4 = \begin{bmatrix} 4.8882 & 1.3814 \\ 0 & 2.0446 \end{bmatrix} \begin{bmatrix} 0.9985 & -0.0548 \\ 0.0548 & 0.9985 \end{bmatrix} = \begin{bmatrix} 4.9566 & 1.0938 \\ 0.1120 & 2.0415 \end{bmatrix}$$

$$\tan \alpha = -\frac{0.1120}{4.9566}, \sin \alpha = -0.0226, \cos \alpha = 0.9997$$

$$h_{11} = 4.9566 \times 0.9997 - 0.1120 \times (-0.0226) = 4.9576$$

$$h_{12} = 1.1396, h_{21} = 0, h_{22} = 2.0162$$

$$A_6 = R_5 Q_5 = \begin{bmatrix} 4.9576 & 1.1396 \\ 0 & 2.0162 \end{bmatrix} \begin{bmatrix} 0.9997 & -0.0226 \\ 0.0226 & 0.9997 \end{bmatrix}$$

Comparing the diagonal elements of A_5 and A_6, we observe that

$$|4.9819 - 4.9566| = 0.0253$$
$$|2.0156 - 2.0415| = 0.0259$$

which is less than 0.05. Hence eigen values are 5 and 2.

EXERCISE 11.1

1. Using Jacobi's method find the eigen values of the matrix $A = \begin{bmatrix} 2 & 1 & 0 \\ 1 & 4 & 1 \\ 0 & 1 & 4 \end{bmatrix}$.

[Ans. 1.52, 5.17 and 3.31]

2. Using House-Holder's method reduce the following matrix to the tridiagonal form

$$A = \begin{bmatrix} 2 & 1 & 1 \\ 1 & 1 & 0 \\ 1 & 0 & 1 \end{bmatrix}. \qquad \text{Ans.} \begin{bmatrix} 2 & \sqrt{2} & 0 \\ \sqrt{2} & 1 & 0 \\ 0 & 0 & 1 \end{bmatrix}$$

3. Using House-Holder's method reduce the matrix $A = \begin{bmatrix} 2 & 0 & 1 \\ 0 & 3 & -2 \\ 1 & -2 & -1 \end{bmatrix}$ to the tridiagonal form.

$$\text{Ans. } W_2^T = \begin{bmatrix} 0, \dfrac{1}{\sqrt{2}}, \dfrac{1}{\sqrt{2}} \end{bmatrix}, O = \begin{bmatrix} 1 & 0 & 0 \\ 0 & 0 & -1 \\ 0 & -1 & 0 \end{bmatrix}, OAO = \begin{bmatrix} 2 & -1 & 0 \\ -1 & 1 & -2 \\ 0 & -2 & 3 \end{bmatrix}$$

4. Transform the matrix $\begin{bmatrix} 1 & 2 & 2 \\ 2 & 1 & 2 \\ 2 & 2 & 1 \end{bmatrix}$ to the tridiagonal form using Given's method.

Hence find largest eigen value and the corresponding eigen vector of the tridiagonal matrix.

$$\text{Ans.} \begin{bmatrix} 1 & 2\sqrt{2} & 0 \\ 2\sqrt{2} & 3 & 0 \\ 0 & 0 & 1 \end{bmatrix}; 5; \begin{bmatrix} 1 \\ 1 \\ 1 \end{bmatrix}$$

5. Determine the largest eigen value and the corresponding eigen vector of the following matrices:

(a) $\begin{bmatrix} 1 & 6 & 1 \\ 1 & 2 & 0 \\ 0 & 0 & 3 \end{bmatrix}$

Ans. 4.028669, $X = \begin{bmatrix} 1 \\ 0.4887 \\ 0.0910 \end{bmatrix}$

(b) $\begin{bmatrix} 1 & 3 & -1 \\ 3 & 2 & 4 \\ -1 & 4 & 10 \end{bmatrix}$

Ans. 11.66433, $X = \begin{bmatrix} 0.0251409 \\ 0.4219855 \\ 1 \end{bmatrix}$

(c) $\begin{bmatrix} 1 & 2 & 3 \\ 0 & -4 & 2 \\ 0 & 0 & 7 \end{bmatrix}$

Ans. 7, $X = \begin{bmatrix} 0.560209 \\ 0.182809 \\ 1 \end{bmatrix}$

(d) $\begin{bmatrix} 10 & 7 & 8 & 7 \\ 7 & 5 & 6 & 5 \\ 8 & 6 & 10 & 9 \\ 7 & 5 & 9 & 10 \end{bmatrix}$

Ans. 30.29, $X = \begin{bmatrix} 0.96 \\ 0.69 \\ 1.00 \\ 0.94 \end{bmatrix}$

6. Obtain the largest eigen value and the corresponding eigen vector for the equations

$$(2 - \lambda) x_1 - x_2 = 0$$
$$-x_1 + (2 - \lambda) x_2 - x_3 = 0$$
$$-x_2 + (2 - \lambda) x_3 = 0$$

by using Rayleigh power method.

[Ans. 3.41, $X = [0.74 \quad -1 \quad 0.67]^T$]

7. Find the eigen values of the matrix $\begin{bmatrix} 2 & i & 0 \\ i & 2 & 0 \\ 0 & 0 & 3 \end{bmatrix}$.

[Ans. 3, 1, 3]

Chapter
12   Statistical Techniques

12.1 INTRODUCTION

Statistical inference is that branch of statistics which is concerned with using probability concept to deal with uncertainty in decision making.

Statistical inference refers to the process of selecting and using a sample statistic to draw inference about a population parameter based on a subset of it—the sample drawn from the population. Statistical inference treats two different classes of problems:

i. **Hypothesis testing**, i.e. to test some hypothesis about parent population from which the sample is drawn.

ii. **Estimation**, i.e. to use the 'statistics' find from the sample as estimate of the unknown 'parameter' of the population from which the population is drawn.

In both the above cases the particular problem at hand is designed in such a way that inference about relevant population values can be made from sample data.

Here we are giving some definitions which will be used throughout the chapter.

12.2 POPULATION OR UNIVERSE

The collection of all objects of a specified type in a given region at a particular point or period of time is named as a population or universe. Thus, we may assume a population of persons or a population of snake or birds in a forest, etc. depending on the nature of data required.

A universe consisting a finite number of objects is called a finite universe. A universe with infinite number of objects is called infinite universe.

The collection of all possible ways in which a specified event can happen is called a hypothetical universe.

12.3 SAMPLE

A finite subset of a universe is called a sample. The number of objects in a sample is called the sample size. The process of selecting a sample from a larger population is called sampling.

12.4 STATISTICAL HYPOTHESIS

A statistical hypothesis is a consideration about a population parameter. This assumption may or may not be true.

12.5 PARAMETERS OF STATISTICS

The statistical constants of the population such as mean, the variance etc. are known as the parameters. The statistical concepts of the sample from the members of sample to estimate the parameters of the population from which the sample has been drawn is known as statistic. μ is population mean whereas $\bar{x}$ is sample mean.

12.6 HYPOTHESIS TESTING

A hypothesis test is a statistical test which is used to decide that there is enough evidence in a sample of data to infer that a certain condition is true for the entire population. In other words, a hypothesis in statistics is simply a quantitative statement about a population, e.g. a coin may be tosses 200 times and we get heads 60 times and tails 140 times. We may now be interested in testing the hypothesis that the coin is unbiased.

12.6.1 Procedure for Testing Hypothesis

a. **Set up a hypothesis:** The first thing in hypothesis testing is to set up a hypothesis about a population parameter. The two hypothesis in a statistical test are generally mentioned:

   i. *Null hypothesis*: According to RA Fisher, null hypothesis is the hypothesis which is tested for possible rejection under the assumption that it is true. Null hypothsis is denoted by H_0, e.g. in case of single statistic, null hypothesis (H_0) will be that the sample statistic does not differ significantly from the hypothetical parameter value and in case of double statistic, H_0 will be that the sample statistic do not differ significantly.

   ii. *Alternative hypothesis*: Any hypothesis which is complementary to null hypothesis is called an alternative hypothesis. It is denoted by H_1, e.g. if we want to test the null hypothesis that the problem has a specified mean μ_0 (say), then the null hypothesis is $H_0: \mu = \mu_0$ and the alternative hypothesis are:
   - $H_1: \mu \neq \mu_0$ (i.e. $\mu > \mu_0$ or $\mu < \mu_0$) (Two tailed alternative hypothesis)
   - $H_1: \mu > \mu_0$ (Right tailed alternative hypothesis)
   - $H_1: \mu < \mu_0$ (Left tailed alternative hypothesis)

b. **Set up a suitable significance level:** Having set up the hypothesis, the next step is to test the validity of H_0 against that of H_1 at as a certain level of significance.

   In testing a given hypothesis, the maximum probability with which we would be willing to take risk is called level of significance of the test. The level of significance is also known as the size of rejection region or the size of critical region. It is denoted by α. The level of significance which are generally used in test of significance are 1%, 5% and 10%.

   For example, if 10% level of significance is choosen in deriving a test of hypothesis, then there are about 10% chances in 100 that we would reject the hypothesis when it should be accepted, i.e. we are about 9.5% confident that we made the right decision. In such a case we say that the hypothesis has been rejected at 5% level of significance which means that we could be wrong with probability 0.05.

   Figure 12.1 illustrates the region in which we would accept or reject the null hypothesis. When it is being tested at 5% level of significance and a two-tailed test is applied.

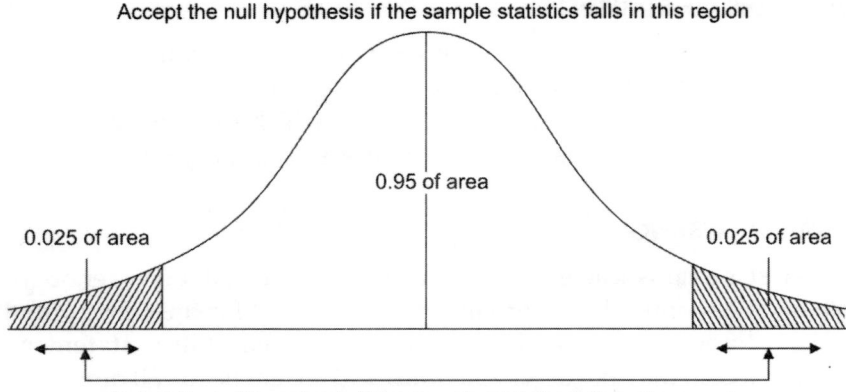

Fig. 12.1

Setting a test criterion or statistic: In this step, we construct a test criterion or statistic. The value of test statistic is calculated from the distribution of sample statistic by using the following formula

$$\text{Test statistic} = \frac{\text{Value of sample statistic} - \text{value of hypotherized population parameters}}{\text{Standard error of sample statistic}}$$

or
$$z = \frac{t - E(t)}{SE(t)}$$

where t is sample statistic.

Note:
1. Select the test statistic and determine its value from the sample data. This value is called the observed value of the test statistic.
2. If only small sample information is available, the use of the normal distribution would be inappropriate.

c. **Decision making:** In last step, we may draw statistical conclusions and take decisions. A statistical conclusion or statistical decision is a decision either to accept or to reject the null hypothesis. The decision will depend on whether the computed value of the test statistic falls in the region of rejection or the region of acceptance.
   i. If $|z| < z_\alpha$, i.e. if the calculated value of z is less than z_α, then H_0 is accepted.
   ii. If $|z| > z_\alpha$, i.e. if the calculated value of z is greater than z_α, then H_0 is rejected.

12.7 ERROR IN TESTING OF HYPOTHESIS

When a statistical hypothesis is tested there are two types of errors possible:
   i. The hypothesis is true but our test rejects it. (Type-I error)
   ii. The hypothesis if false but our test accept it. (Type-II error)

In a statistical hypothesis testing experiment, a type-I error is committed by rejecting the null hypothesis when it is true. The probability of committing a type-I error is denoted by α, where

$$\alpha = \text{Probability (Type-I error)}$$
$$= \text{Probability (Rejecting } H_0/H_1 \text{ is true)}$$

On the other hand, a type-II error is committed by not rejecting (i.e. accepting) the null hypothesis when it is false. The probability of committing a type-II error is denoted by β, where

$$\beta = \text{Probability (Type II error)}$$
$$= \text{Probability (Not rejecting or accepting } H_0/H_1 \text{ is false)}$$

12.8 TWO-TAILED AND ONE-TAILED TESTS OF HYPOTHESIS

In any test, the critical region is represented by a portion of the area under the probability curve of the sampling distribution of the test statistic.

A test of any statistical hypothesis where the alternative hypothesis is one-tailed (right tailed or left tailed) is called a one-tailed test.

In two tailed test, the critical region is given by the portion of the area lying in both the tails of the probability curve of the test statistic.

Note: In a particular problem, whether one-tailed or two-tailed test is to be applied depends entirely on the nature of the alternative hypothesis. If the alternative hypothesis is two-tailed, we apply two-tailed test and if alternative hypothesis is one-tailed, we apply one-tailed test.

12.9 STANDARD ERROR

The standard deviation of the sampling distribution is called the standard error (SE). It plays an important role in the theory of large samples and it forms a basis of the testing of hypothesis. If t is any statistic, for large sample, $z = \dfrac{t - E(t)}{\text{SE}(t)}$ is normally distributed with mean 0 and variance unity.

Having discussed the above concepts, let us now discuss the various situations where we have to apply different tests of significance. For the sake of convenience and clarity, these situations may be summed up under the following three heads:

  i. Test of significance for attributes
  ii. Test of significance for variables (large samples)
 iii. Test of significance for variables (small samples)

Here, we will study only the test of significance for large samples.

12.10 TEST OF SIGNIFICANCE FOR LARGE SAMPLING

Hypothesis testing involving large samples ($n > 30$) is based on the assumption that the population from which the sample is drawn has a normal distribution. Consequently, the sampling distribution of mean $\bar{x}$ is also normal. Even if the population does not have a normal distribution, the sampling distribution of mean $\bar{x}$ is assumed to be normal.

Thus, in all the large sample tests, we compute the test statistic z under the null hypothesis, where z is a normal distribution with mean 0 and variance one.

Based on the above area property of the standard normal distribution, the critical values for z statistic at 10%, 5% and 1% level of significance can be seen from Table 12.1.

Table 12.1: Critical values for z-statistics

Level of significance	Right tailed test	Left tailed test	Two tailed test
$\alpha = 0.10$	$z > 1.28$	$z < -1.28$	$|z| > 1.645$
$\alpha = 0.05$	$z > 1.645$	$z < -1.645$	$|z| > 1.96$
$\alpha = 0.01$	$z > 2.33$	$z < -2.33$	$|z| > 2.58$

12.1 Z-TEST

We shall now discuss the following four large sample tests:
1. Testing the significance of population proportion
2. Testing the significance of the difference in proportion
3. Testing the significance of population mean
4. Testing the significance of the difference between two sample means or two population means

12.11.1 Testing the Significance of Population Proportion

This test is used to find the significant difference between proportion of the sample and population.

Let x be the number of success in n independent trials with constant probability p_0 of success of each trial

$$E(x) = np_0; \; V(x) = np_0q_0; \; q_0 = 1 - p_0$$

where q_0 is the probability of failure.

Let $\bar{p} = \dfrac{x}{n}$ called the observed proportion of success.

$$E(\bar{p}) = E\left(\frac{x}{n}\right) = \frac{1}{n}E(x) = \frac{np_0}{n} = p_0$$

$$V(\bar{p}) = V\left(\frac{x}{n}\right) = \frac{1}{n^2}V(x) = \frac{1}{n^2}np_0q_0 = \frac{p_0q_0}{n}$$

Standard error $\qquad (\bar{p}) = \sqrt{\dfrac{p_0q_0}{n}}$

Now test statistic $\qquad z = \dfrac{\bar{p} - E(\bar{p})}{SE(\bar{p})} = \dfrac{\bar{p} - p_0}{\sqrt{\dfrac{p_0q_0}{n}}}$

Note: The probable limit for the observed proportion of success are

$$\bar{p} + z_\alpha \cdot SE(\bar{p})$$

where z_a is the significant value at level of significance α.

In this case, the probable limit will be

$$\bar{p} \pm z_\alpha \sqrt{\frac{p_0q_0}{n}}$$

1. If we take 95% confidence limits, then the above formula becomes

$$\bar{p} \pm 1.96 \sqrt{\frac{p_0q_0}{n}}$$

2. If we take 90% confidence limit, then the above formula becomes

$$\bar{p} \pm 1.645 \sqrt{\frac{p_0 q_0}{n}}$$

Example 1: An auditor claims that 8% of customers ledger accounts are carrying mistakes of posting and balancing. A random sample of 500 was taken to test the accuracy of posting and balancing and 20 mistakes were found. Are these sample results consistent with the claim of auditor? Use 5% level of significance.

Solution: Null hyypothesis: Let us take the null hypothesis that the claim of the auditor is valid, i.e.

$$H_0 : p_0 = 0.08$$

Alternative hypothesis: $H_1 : p_0 \neq 0.08$ (Two tailed test)

Now given that

$$\bar{p} = \frac{x}{n} = \frac{20}{500} = 0.04, n = 500 \text{ and } \alpha = 0.05$$

$$q_0 = 1 - p_0 = 1 - 0.08 = 0.92$$

Test statistic:
$$z = \frac{\bar{p} - p_0}{\sqrt{\dfrac{p_0 q_0}{n}}} = \frac{0.04 - 0.05}{\sqrt{\dfrac{0.08 \times 0.92}{500}}} = \frac{-0.01}{\sqrt{\dfrac{0.0736}{500}}}$$

$$= \frac{-0.01}{\sqrt{0.0001472}} = -0.8264$$

$$|z| = 0.8264$$

Comparison: The critical value of z at 5% level of significance for two tailed test is 1.96.

Thus
$$|z_{cal}| < z_{tab}$$

So, the null hypothesis is accepted. Hence we conclude that the claim of auditor is valid.

Example 2: A coin was tossed 400 times and the head turned up 212 times. Test the hypothesis that the coin is unbiased.

Solution: Null hypothesis: Let us take that the coin is unbiased, i.e.

$$H_0 : p_0 = 0.5$$

Alternative hypothesis: $H_1 : p_0 \neq 0.5$ (Two tailed test)

$$\bar{p} = \frac{x}{n} = \frac{212}{400} = 0.53$$

Population proportion $p_0 = 0.5$

$\therefore$
$$q_0 = 1 - p_0 = 1 - 0.5 = 0.5$$

Test statistic
$$z = \frac{\bar{p} - p_0}{\sqrt{\dfrac{p_0 q_0}{n}}} = \frac{0.53 - 0.5}{\sqrt{\dfrac{0.5 \times 0.5}{400}}} = 1.2$$

$$|z| = 1.2$$

Comparison: The critical value of z at 5% level of significance for two tailed test is 1.96.

$\therefore \ |z_{cal}| < z_\alpha$. Hence, the null hypothesis is accepted and hence we conclude that the coin in unbiased.

Remark: If level of significance is not given in the question, then generally, we use 5% level of significance.

Example 3: A manufacturer claims that at least 90% of the equipments which he supplied to a factory conformed to the specification. An examination of the sample of 200 pieces of equipment revealed that 16 were faulty. Test the claim of the manufacturer.

Solution: Null hypothesis: Let us take that at 90% of the equipments supplied conformed to the specification, i.e.

$$H_0 = p_0 \geq 0.9$$

Alternative hypothesis: $H_1 < 0.9$ (Left tailed test)

Given $\bar{p}$ = percent of pieces conforming to the specification

$$= 1 - \frac{16}{200} = 0.84, n = 200, \alpha = 0.05$$

Also $q_0 = 1 - p_0 = 1 - 0.9 = 0.1$

Test statistic $z = \dfrac{\bar{p} - p_0}{\sqrt{\dfrac{p_0 q_0}{n}}} = \dfrac{0.84 - 0.90}{\sqrt{\dfrac{0.9 \times 0.1}{200}}} = \dfrac{-0.06}{0.0212}$

$$|z| = 2.83$$

Since the critical value of z at 5% level of significance for left tailed test $= -1.645$

$$|z_\alpha| = 1.645$$

Conclusion: $|z_{cal}| > |z_\alpha|$. So, the null hypothesis is rejected. Hence we conclude that the proportion of conforming to specification is not 90%.

Example 4: A machine is producing bolts of which a certain fraction is defective. A random sample of 400 is taken from a large batch and is found to contain 20 defective bolts. Does this indicate that the proportion of defectiveness is more than that claimed by manufacturer where the manufacturer claims that only 4% of his product is defective. Find 95% confidence limits of the proportion of defective bolts in batch.

Solution: Null hypothesis: H_0: The manufacturer claim is accepted, i.e.

$$p_0 = \frac{4}{100} = 0.04, q_0 = 1 - p_0 = 1 - 0.04 = 0.96$$

Alternative hypothesis: H_1: $p_0 > 0.05$ (Right tailed test)

Here $\bar{p} = \dfrac{x}{n} = \dfrac{20}{400} = 0.05$

Level of significance $\alpha = 0.05$

Test statistic $z = \dfrac{\bar{p} - p_0}{\sqrt{\dfrac{p_0 q_0}{n}}} = \dfrac{0.05 - 0.04}{\sqrt{\dfrac{0.04 \times 0.96}{400}}}$

$$|z| = 1.0416$$

Since the critical value of z at 5% level of significance for right tailed test $= 1.645$.

Conclusion: $|z_{cal}| > |z_\alpha|$. So, the null hypothesis is accepted. Hence we conclude that the manufacturer claim is accepted.

Confidence limits: We know that it is given by the formula

$$\bar{p} \pm z_\alpha \sqrt{\frac{p_0 q_0}{n}} = 0.05 \pm 1.645 \sqrt{\frac{0.04 \times 0.96}{400}}$$

$$= 0.05 \pm 1.645 \times 0.0096$$

$$= 0.05 \pm 0.0158$$

$$= 0.0658, 0.0342$$

Example 5: A manufacturer claims that only 4% of his products supplied by him are defective. A random sample of 600 products contained 36 defectives. Tese the claim of the manufacturer.

Solution: Null hypothesis: H_0: The manufacturer claim is accepted, i.e.

$$p_0 = \frac{4}{100} = 0.04, q_0 = 1 - p_0 = 1 - 0.04 = 0.06$$

Alternative hypothesis: H_1: $p_0 \neq 0.04$ (Two tailed test)

Here
$$\bar{p} = \frac{x}{n} = \frac{36}{600} = 0.06$$

Level of significance: Let us take $\alpha = 0.05$

Test statistic
$$z = \frac{\bar{p} - p_0}{\sqrt{\frac{p_0 q_0}{n}}} = \frac{0.06 - 0.04}{\sqrt{\frac{0.04 \times 0.06}{600}}} = \frac{0.02}{0.002} = 10$$

$$|z| = 10$$

Since the critical value of z at 5% level of significance for two tailed test = 1.96.

Conclusion: $|z_{cal}| > |z_\alpha|$. So, H_0 is rejected. Hence, we conclude that the manufacturer claim is not valid.

EXERCISE 12.1

1. A company manufacturing a certain type of breakfast cereal claims that 60 percent of all housewives prefer that type of cereal. A random sample of 300 housewives contains 165 who do prefer that type of cereal. At 5% level of significance, test the claim of the company.   [Ans. $|z| = 1.77, H_0$ is rejected]

2. A sales clerk in the departmental store claims that 60% of the shoppers entering the store leave without making a purchase. A random sample of 50 shoppers showed that 35 of them left without buying anything. Are these sample results consistent with the claim of the sales clerk? Use 5% level of significance.

     [Ans. $|z| = 1.44, H_0$ is accepted]

3. A bag contains defective articles, the exact number of which is not known. A sample of 100 from the bag gives 10 defective articles. Find the limit for the proportion of defective articles in the bag.   [Ans. 0.1588, 0.0412]

4. A sample of size of 600 persons selected at random from a large city shows that the percentage of males in the sample is 53. It is believed that the ratio of males to

the total population in the city is 0.5. Test whether the belief is confirmed by the observation. [Ans. H_0 is accepted]

5. 400 apples are taken at random from a large basket and 40 are found to be bad. Estimate the proportion of bad apples in the basket and assign limits within which the percentage most probably lies. [Ans. 0.1294, 0.0706]

12.11.2 Testing the Significance of the Difference in Proportion

Consider two samples x_1 and x_2 of sizes n_1 and n_2 respectively taken from two different populations. To test the significance of the difference between the sample proportions $\hat{p}_1$ and $\hat{p}_2$, the test statistic under the null hypothesis H_0: that there is no difference between the two sample proportions, is given by

$$z = \frac{\hat{p}_1 - \hat{p}_2}{SE(\hat{p}_1 - \hat{p}_2)} = \frac{\hat{p}_1 - \hat{p}_2}{\sqrt{\hat{p}\hat{q}\left(\dfrac{1}{n_1} + \dfrac{1}{n_2}\right)}}$$

where

$$\hat{p}_1 = \frac{x_1}{n_1}; \ \hat{p}_2 = \frac{x_2}{n_2} \text{ and } \hat{p} = \frac{n_1\hat{p}_1 + n_2\hat{p}_2}{n_1 + n_2}$$

or

$$\hat{p} = \frac{x_1 + x_2}{n_1 + n_2}, \hat{q} = 1 - \hat{p}$$

Example 6: A machine produced 15 defective articles in a batch of 500. After overhauling it produced 2 defectives in a batch of 100. Has the machine improved?

Solution: Null hypothesis: There is no significant difference between two population proportion i.e.

$$H_0 : \hat{p}_1 = \hat{p}_2$$

Alternative hypothesis: $H_1 : \hat{p}_1 > \hat{p}_2$ (Right tailed test)

Level of significance: Let us assume 5% level of significance.

Test statistic

$$z = \frac{\hat{p}_1 - \hat{p}_2}{\sqrt{\hat{p}\hat{q}\left(\dfrac{1}{n_1} + \dfrac{1}{n_2}\right)}}$$

$$\hat{p}_1 = \frac{x_1}{n_1} = \frac{15}{500} = 0.03$$

$$\hat{p}_2 = \frac{x_2}{n_2} = \frac{2}{100} = 0.02$$

$$\hat{p} = \frac{x_1 + x_2}{n_1 + n_2} = \frac{15 + 2}{500 + 100} = \frac{17}{600} = 0.0283$$

$$\hat{q} = 1 - \hat{p} = 1 - 0.0283 = 0.9717 \text{ (approx.)}$$

$\therefore$

$$z = \frac{0.03 - 0.02}{\sqrt{0.0283 \times 0.9717\left(\dfrac{1}{500} + \dfrac{1}{100}\right)}} = \frac{0.01}{0.0182} = 0.5494$$

Conclusion: The critical value of z at 5% level of significance for right tailed test is 1.645.

$\therefore$ $|z_{cal}| < z_\alpha$. H_0 is accepted. Hence, there is no significant difference between two population proportions.

Example 7: Random sample of 400 men and 700 women were asked whether they would like to have a flyover near their residence. 200 men and 250 women were in favour of the proposal. Test the hypothesis that proportions of men and women in favour of the proposal, are same against that they are not, at 5% level.

Solution: Null hypothesis: Consider that there is no significant difference between the opinions of men and women as far as proposal of flyover is concerned, i.e.

$$H_0 : \hat{p}_1 = \hat{p}_2$$

Alternative hypothesis: $H_1 := \hat{p}_1 \neq \hat{p}_2$ (Two tailed test)

Level of significance: 5% level of significance is given.

Test statistic
$$z = \frac{\hat{p}_1 - \hat{p}_2}{\sqrt{\hat{p}\hat{q}\left(\frac{1}{n_1} + \frac{1}{n_2}\right)}}$$

$$\hat{p}_1 = \frac{x_1}{n_1} = \frac{200}{400} = 0.5$$

$$\hat{p}_2 = \frac{x_2}{n_2} = \frac{250}{700} = 0.357$$

$$\hat{p} = \frac{x_1 + x_2}{n_1 + n_2} = \frac{200 + 250}{400 + 700} = \frac{450}{1100} = 0.409$$

$$\hat{q} = 1 - \hat{p} = 1 - 0.409 = 0.591$$

$\therefore$
$$z = \frac{0.5 - 0.357}{\sqrt{0.409 \times 0.591\left(\frac{1}{400} + \frac{1}{700}\right)}}$$

$$= \frac{0.143}{\sqrt{0.2417 \times 0.0039}} = \frac{0.143}{0.0307}$$

$$|z| = 4.66$$

Conclusion: The critical value of z at 5% level of significance for two tailed test is 1.96. $\therefore$ $|z_{cal}| > |z_\alpha|$. H_0 is rejected. Hence, we conclude that men and women differ significantly as regards proposal of flyover is concerned.

Example 8: In a simple random sample of 600 men taken from a big city, 400 are found to be drinkers. In another simple radom sample of 900 men taken from another city, 450 are drinkers. Do the data indicate that there is a significant difference in the habit of drinking alcohol in the two cities?

Solution: Null hypothesis: Consider that there is no significant difference in the habit of drinking alcohol in two cities, i.e.

$$H_0 : \hat{p}_1 = \hat{p}_2$$

Alternative hypothesis: $H_1 := \hat{p}_1 \neq \hat{p}_2$ (Two tailed test)

Level of significance: We will take 5% level of significance if it is not given in problem.

Test statistic
$$z = \frac{\hat{p}_1 - \hat{p}_2}{\sqrt{\hat{p}\hat{q}\left(\dfrac{1}{n_1} + \dfrac{1}{n_2}\right)}}$$

$$\hat{p}_1 = \frac{x_1}{n_1} = \frac{400}{600} = 0.667$$

$$\hat{p}_2 = \frac{x_2}{n_2} = \frac{450}{900} = 0.5$$

$$\hat{p} = \frac{x_1 + x_2}{n_1 + n_2} = \frac{400 + 450}{600 + 900} = \frac{850}{1500} = 0.567$$

$$\hat{q} = 1 - 0.567 = 0.433$$

$$\therefore \quad z = \frac{0.667 - 0.5}{\sqrt{0.567 \times 0.433\left(\dfrac{1}{600} + \dfrac{1}{900}\right)}} = 6.423$$

$$|z| = 6.423$$

Conclusion: The critical value of z at 5% level of significance for two tailed test is 1.96.

$\therefore |z_{cal}| > z_\alpha$. Since H_0 is rejected. Hence, we conclude that there is significant difference in the habit of drinking alcohol in two cities.

Example 9: In a referendum submitted to the students body at a university, 850 men and 560 women voted. 500 men and 320 women voted yes. Does this indicate a significant difference of opinion between men and women on this matter at 1% level of significance.

Solution: Null hypothesis: Consider that there is no significant difference between men and women in the referendum, i.e.

$$H_0 : \hat{p}_1 = \hat{p}_2$$

Alternative hypothesis: $H_1 := \hat{p}_1 \neq \hat{p}_2$ (Two tailed test)

Test statistic
$$z = \frac{\hat{p}_1 - \hat{p}_2}{\sqrt{\hat{p}\hat{q}\left(\dfrac{1}{n_1} + \dfrac{1}{n_2}\right)}}$$

$$\hat{p}_1 = \frac{x_1}{n_1} = \frac{500}{850} = 0.588$$

$$\hat{p}_2 = \frac{x_2}{n_2} = \frac{320}{560} = 0.571$$

$$\hat{p} = \frac{x_1 + x_2}{n_1 + n_2} = \frac{500 + 320}{850 + 560} = \frac{820}{1410} = 0.58$$

$$\hat{q} = 1 - \hat{p} = 1 - 0.58 = 0.42$$

$\therefore$

$$z = \frac{0.588 - 0.571}{\sqrt{0.58 \times 0.42 \left(\frac{1}{850} + \frac{1}{560} \right)}} = \frac{0.017}{0.026} = 0.654$$

$|z| = 0.654$

Conclusion: The critical value of z at 1% level of significance is 2.58.

$\therefore |z_{cal}| < |z_\alpha|$. Since H_0 is accepted. Hence, we conclude that there is no significant difference between men and women in the referendum.

EXERCISE 12.2

1. Before an increase in excise duty on tea, 400 people out of a sample of 500 persons were found to be tea drinkers. After an increase in the excise duty, 400 persons were known to be tea drinkers in a sample of 600 people. Do you think that there has been a significant decrease in the consumption of tea after the increase in the excise duty? [Ans. $|z| = 4.93$, H_0 is rejected]

2. 500 units from a factory are inspected and 12 are found to be defective. 800 units from another factory are inspected and 12 are found to be defective. Can it be concluded that at 5% level of significance production at the second factory is better than in first factory? [Ans. $|z| = 1.184$, H_0 is accepted]

3. In two large population, there are 30 and 25 percent respectively of blue eyed people. Is this difference likely to be hidden in samples of 1200 and 900 respectively from the two population? [Ans. $|z| = 2.538$, H_0 is rejected]

4. In a year there are 956 births in a town A, of which 52.5% were males, while in towns A and B combined, this proportion in a total of 1405 births was 0.496. Is there any significant difference in the proportion of male births in the two towns? [Ans. $|z| = 3.368$, H_0 is rejected]

5. A cigarette manufacturing firm claims that its brand A of the cigarette outsells its brand B by 8%. If it is found that 42 of a sample of 200 smokers prefer brand A and 18 out of another random sample of 100 smokers prefer brand B, test whether the 8% difference is valid claim (use 5% level of significance). [Ans. $|z| = 1.02$, H_0 is accepted]

12.11.3 Testing the Significance of Population Mean

Suppose there is a population with unknown mean μ and known standard deviation σ. The z statistic test is given by

$$z = \frac{\bar{x} - \mu}{\frac{\sigma}{\sqrt{n}}}$$

where $\bar{x} = \dfrac{\Sigma x}{n}$.

Remarks:

1. If standard deviation σ is unknown, then a sample standard deviation s is used to estimate σ. The value of the z-test statistic is given by $z = \dfrac{\bar{x} - \mu}{s / \sqrt{n}}$.

2. Confidence limit will be $\bar{x} \pm \dfrac{\sigma}{\sqrt{n}} z_\alpha$.

Example 10: The mean weight obtained from a random sample of size 100 is 60 gms. The standard deviation of the weight distribution of the population is 3 gms. Test the statement that the mean weight of the population is 65 gms at 5% level of significance. Also set up 99% confidence limits of the mean weight of the population.

Solution: Null hypothesis: There is no significant difference between sample and population mean, i.e.

$$\mu = 65$$

Alternative hypothesis: H_1: $\mu \neq 65$ (Two tailed test).

Level of significance: 5% level of significance.

Test statistics $$z = \frac{\bar{x} - \mu}{\sigma/\sqrt{n}} = \frac{60 - 65}{3/\sqrt{100}} = -16.67$$

$$|z| = 16.67$$

Conclusion: The critical value of $|z|$ at 5% level of significance for two tailed test is 1.96.

$\therefore$ $|z_{cal}| > z_\alpha$. Since H_0 is rejected. Hence, we conclude that the sample is not drawn from the population with mean 65.

To find 99% condifence limit: It is given by

$$\bar{x} \pm 2.58 \frac{\sigma}{\sqrt{n}} = 60 \pm 2.58 \left(\frac{3}{\sqrt{100}}\right) = 60 \pm 0.774$$

The limits are (60.774, 59.226).

Example 11: The mean life time of a sample of 400 flourescent light bulbs produced by a company is found to be 1600 hours with a standard deviation of 150 hours. Test the hypothesis that the mean life time of the bulbs produced in general is higher than the mean life of 1570 hours at $\alpha = 0.01$ level of significance.

Solution: Null hypothesis: Consider that mean life time of bulbs is not more than 1570 hours, i.e.

$$H_0: \mu \leq 1570$$

Alternative hypothesis: H_1: $\mu > 1570$ (Right tailed test).

Level of significance: 1% level of significance.

Test statistics $$z = \frac{\bar{x} - \mu}{s/\sqrt{n}} = \frac{1600 - 1570}{150/\sqrt{400}} = 4$$

$$|z| = 4$$

Conclusion: The critical value of z at 1% level of significance for right tailed test is 2.33.

$\therefore$ $|z_{cal}| > |z_{tab}|$. Since H_0 is rejected. Hence, we conclude that the mean life time of bulbs produced by the company may be higher than 1570 hours.

Example 12: A random sample of 400 flowers stems has an average length of 9.6 cm. Can this be regarded as a sample from a large population with mean of 9.8 cm and a standard deviation of 2.25 cm?

Solution: Null hypothesis: There is no significant difference between sample and population mean, i.e.

$$H_0: \mu = 9.8$$

Alternative hypothesis: $H_1: \mu \neq 9.8$ (Two tailed test).

Level of significance: Consider 5% level of significance.

Test statistics
$$z = \frac{\bar{x} - \mu}{\sigma/\sqrt{n}} = \frac{9.6 - 9.8}{2.25/\sqrt{400}} = \frac{-0.2}{2.25/20} = -1.78$$

$$|z| = 1.78$$

Conclusion: The critical value of z at 5% level of significance for two tailed test is 1.96. $\therefore |z_{cal}| < |z_{tab}|$. Since H_0 is accepted. Hence, we conclude that the sample has been drawn from the population whose mean is 9.8 cm and standard deviation of 2.25 cm.

EXERCISE 12.3

1. A sample of 100 measurements of breaking strength of cotton threads gave a mean of 7.4 ounces and standard deviation of 1.2 ounces. Find 95% confidence limits for the mean breaking strength. [Ans. 7.6352, 7.1648]

2. A packaging device is set to fill detergent powder packets with a mean weight of 5 kg with a standard deviation of 0.21 kg. The weight of packets can be assumed to be normally distributed. The weight of packets is known to drift upwards over a period of time due to machine fault, which is not tolerable. A random sample of 100 packets is taken and weighed. This sample has a mean weight of 5.03 kg. Can we conclude that the mean weight produced by the machine has increased? Use a 5% level of significance. [Ans. $|z| = 1.428$, H_0 is accepted]

3. An ambulance service claims that it takes, on an average, 8.9 minutes to reach its destination in emergency calls. To check on this claim, the agency which licenses ambulance services had them timed on 50 emergency calls, getting a mean of 9.3 minutes with a standard deviation of 1.8 minutes. At the level of significance of 0.05, does this constitute evidence that the figure claimed is too low? [Ans. $|z| = 1.574$, H_0 is accepted]

4. An auto company decided to introduce a new six cylinder car whose mean petrol consumption is claimed to be lower than that of the existing auto engine. It was found that the mean petrol consumption for 50 cars was 10 km per litre with a standard deviation of 3.5 km per litre. Test for the company at 5% level of significance, the claim that in the new car petrol consumption is 9.5 km per litre on the average. [Ans. $|z| = 1.010$, H_0 is accepted]

5. A sample of 100 households in a village was taken and the average income was found to be Rs. 628 per month with a standard deviation of Rs. 60 per month. Find the standard error of mean and determine 99% confidence limit within which the income of all the people in this village are expected to lie. Also test the claim that the average income was Rs. 640 per month. [Ans. 615.1, 640.9]

12.11.4 Testing the Significance of the Difference Between Two Sample Mean or Population Mean

Let the independent random samples of large size n_1 and n_2 be drawn from the first and second population respectively. Let the sample means so calculated be $\bar{x}_1$ and $\bar{x}_2$.

Let σ_1^2 and σ_2^2 be the variances of these samples respectively. In this case, we shall use the z-test statistics.

$$z = \frac{\bar{x}_1 - \bar{x}_2}{SE(\bar{x}_1 - \bar{x}_2)} = \frac{\bar{x}_1 - \bar{x}_2}{\sqrt{\dfrac{\sigma_1^2}{n_1} + \dfrac{\sigma_2^2}{n_2}}}$$

Remarks:

1. If $\sigma^2 = \sigma_1^2 = \sigma_2^2$, then under H_0: $\mu_1 = \mu_2$

$$z = \frac{\bar{x}_1 - \bar{x}_2}{\sigma\sqrt{\dfrac{1}{n_1} + \dfrac{1}{n_2}}}$$

2. If $\sigma_1^2 \neq \sigma_2^2$ and σ_1^2 and σ_2^2 are not known, then they are estimated from sample values.

This results in some error, which is practical immaterial, if samples are large. These estimates for large samples are given by

$$\left.\begin{array}{l} \sigma_1^2 = s_1^2 \\ \sigma_2^2 = s_2^2 \end{array}\right| \text{ Since samples are large}$$

Then
$$z = \frac{\bar{x}_1 - \bar{x}_2}{\sqrt{\dfrac{s_1^2}{n_1} + \dfrac{s_2^2}{n_2}}}$$

Example 13: The average hourly wage of a sample of 150 workers in a plant A was Rs. 2.56 with a standard deviation of Rs. 1.08. The average wage of a sample of 200 workers in plant B was Rs. 2.87 with a standard deviation of 1.28. Can an applicant safely assume that the hourly wages paid by plant B are higher than those paid by Plant A.

Solution: Null hypothesis: There is no significant difference between the mean level of wages of workers in plant A and plant B, i.e.

$$H_0: \mu_1 = \mu_2$$

Alternative hypothesis: H_1: $\mu_1 < \mu_2$ (Left tailed test)

Level of significance: Consider 5% level of significance.

Test statistics
$$z = \frac{\bar{x}_1 - \bar{x}_2}{\sqrt{\dfrac{\sigma_1^2}{n_1} + \dfrac{\sigma_2^2}{n_2}}} = \frac{2.56 - 2.87}{\sqrt{\dfrac{(1.08)^2}{150} + \dfrac{(1.28)^2}{200}}} = -2.46$$

$$\therefore \qquad |z| = 2.46$$

Conclusion: The critical value of z at 5% level of significance for left tailed test is 1.645. $\therefore$ $|z_{cal}| > |z_{tab}|$. Since H_0 is rejected. Hence, we conclude that the average hourly wages paid by plant B are certainly higher than those paid by plant A.

Example 14: From the data given in table, compute the standard error of the differences of the two sample means and find out if the two means significantly differ at 5% level of significance.

	No. of items	Mean	SD
Group I	45	175	2.9
Group II	70	172.1	3.2

Solution: Null hypothesis: Consider that there is no significant difference between the two population means, i.e.

$$H_0: \mu_1 = \mu_2$$

Alternative hypothesis: $H_1: \mu_1 \neq \mu_2$ (Two tailed test)

Level of significance: Given that 5% level of significance.

Test statistics $\qquad z = \dfrac{\bar{x}_1 - \bar{x}_2}{\sqrt{\dfrac{s_1^2}{n_1} + \dfrac{s_2^2}{n_2}}} = \dfrac{175 - 172.1}{\sqrt{\dfrac{(2.9)^2}{45} + \dfrac{(3.2)^2}{70}}} = 5.01$

Conclusion: The critical value of z at 5% level of significance for two tailed test is 1.96.
∴ $|z_{cal}| > |z_{tab}|$. Since H_0 is rejected. Hence, we conclude that there is no significant difference between samples.

Example 15: A product is manufactured in two ways. A pilot test on 64 items from each method indicates that the products of method 1 have a sample mean tensible strength of 106 lbs and a standard deviation of 12 lbs, whereas in method 2 corresponding values of mean and standard deviation are 100 lbs and 10 lbs, respectively. Greater tensile strength in the product is preferable. Use an appropriate large sample test of 5% level of significance to test whether or not method 1 is better for processing the product. State clearly the null hypothesis.

Solution: Null hypothesis: Consider that there is no significant difference between the two methods for processing the product, i.e.

$$H_0: \mu_1 = \mu_2$$

Alternative hypothesis: $H_1: \mu_1 > \mu_2$ (Right tailed test)

Level of significance: Given that 5% level of significance.

Test statistics $\qquad z = \dfrac{\bar{x}_1 - \bar{x}_2}{\sqrt{\dfrac{s_1^2}{n_1} + \dfrac{s_2^2}{n_2}}} = \dfrac{106 - 100}{\sqrt{\dfrac{(12)^2}{64} + \dfrac{(10)^2}{64}}} = 3.07$

Conclusion: The critical value of z at 5% level of significance for right tailed test is 1.645.
∴ $|z_{cal}| > |z_{tab}|$. Since H_0 is rejected. Hence, we conclude that the method 1 is better than method 2.

Example 16: A firm believes that the tyres produced by process A on an average last longer than tyres produced by process B. To test this belief, random sample of tyres produced by the two processes were tested and the results are:

Process	Sample size	Average life time (in km)	SD (in km)
A	40	20400	900
B	40	19800	900

Is there evidence at a 5% level of significance that the firm is correct in its belief?

Solution: Null hypothesis: Consider that there is no significant difference in the average life of tyres produced by processes A and B, i.e.

$$H_0: \mu_1 = \mu_2$$

Alternative hypothesis: $H_1: \mu_1 \neq \mu_2$ (Two tailed test)

Level of significance: 5% level of significance is given.

Test statistics

$$z = \frac{\bar{x}_1 - \bar{x}_2}{\sigma\sqrt{\dfrac{1}{n_1} + \dfrac{1}{n_2}}} = \frac{20400 - 19800}{900\sqrt{\dfrac{1}{40} + \dfrac{1}{40}}} = \frac{600}{201.25} = 2.98$$

$$|z| = 2.98$$

Conclusion: The critical value of z at 5% level of significance for two tailed test is 1.96.

$\therefore |z_{cal}| > |z_{tab}|$. Since H_0 is rejected. Hence, we conclude that the tyres produced by process A last longer than those produced by process B.

EXERCISE 12.4

1. In a certain factory there are two independent processes manufacturing the same item. The average weight in a sample of 250 items produced from on process is found to be 120 ozs, with a standard deviation of 12 ozs, while the corresponding figure in a sample of 400 items from the other process are 124 and 14. Is this difference significant. [Ans. $|z| = 3.87, H_0$ is rejected]

2. A sample of heights of 6400 soldiers has a mean of 67.85 inches and a SD of 2.56 inches. While another sample of heights of 1600 sailors has a mean of 68.55 inches with SD of 2.52 inches. Do the data indicate that the sailors are on the average taller than soldiers? [Ans. H_0 is rejected]

3. The following table presents data on the values of a harvested crop stored in the open and inside a godown.

	Sample size	Mean	Variance
Outside	40	117	8685
India	100	132	27315

Assuming that the two samples are random and they have been drawn from normal population with equal variance, examine if the mean value of the harvested crop is affected by weather conditions.

[Ans. $|z| = 0.342, H_0$ is accepted]

4. Two types of new car produced in India are tested for petrol mileage. One group consisting of 36 cars averaged 14 km per litre, while the other group consisting of 72 cars averaged 12.5 kms per litre.

   i. What test statistic is appropriate if $\sigma_1^2 = 1.5$ and $\sigma_2^2 = 2.0$?

   ii. Test whether there exists a significant difference in petrol consumption of these two types of cars. (Use $\alpha = 0.01$) [Ans. $|z| = 5.703, H_0$ is rejected]

5. An experiment was conducted to compare the mean time in days required to recover from a common cold for person given daily dose of 4 mg of vitamin C versus those who were not given a vitamin supplement. Suppose that 35 adults were randomly selected for each treatment category and that the mean recovery times and standard deviations for the two groups were as follows:

	Vitamin C	No vitamin supplement
Sample size	35	35
Sample mean	5.8	6.9
Sample standard deviation	1.2	2.9

Test the hypothesis that the use of vitamin C reduces the mean time required to recover from a common cold and its complications at the level of significance $\alpha = 0.05$. [Ans. $|z| = 2.605$, H_0 is rejected]

12.11.5 Testing the Significance of Two Sample Standard Deviations

In this case, we shall use the test statistic as

$$z = \frac{s_1 - s_2}{\sqrt{\dfrac{\sigma_1^2}{2n_1} + \dfrac{\sigma_2^2}{2n_2}}}$$

where σ_1 and σ_2 are population standard deviations.

Remark: When population standard deviations are not known, then

$$z = \frac{s_1 - s_2}{\sqrt{\dfrac{s_1^2}{n_1} + \dfrac{s_2^2}{n_2}}}$$

Example 17: Random samples drawn from two countries gave the following data relating to the height of adult male:

	Country A	Country B
Mean height (in inches)	67.42	67.25
Standard deviation	2.50	2.25
Number of samples	1000	1200

Is the difference between the standard deviation significant?

Solution: Null hypothesis: H_0: $\sigma_1 = \sigma_2$

Alternative hypothesis: H_1: $\sigma_1 \neq \sigma_2$

Level of significance: Consider 5% level of significance

Test statistic $\quad z = \dfrac{s_1 - s_2}{\sqrt{\dfrac{s_1^2}{2n_1} + \dfrac{s_2^2}{2n_2}}}$

$$= \frac{2.50 - 2.25}{\sqrt{\dfrac{(2.50)^2}{2 \times 1000} + \dfrac{(2.25)^2}{2 \times 1200}}} = \frac{0.25}{0.072} = 0.35 \,(\text{approx.})$$

Conclusion: The critical value of z at 5% level of significance for two tailed test is 1.96. $\therefore$ $|z_{cal}| < |z_{tab}|$. Since H_0 is accepted. Hence, we conclude that the difference in two standard deviations is not significant.

Appendix

Short Answer Type Questions

1. Describe briefly the floating point representation of numbers.
2. Explain underflow and overflow conditions of error in floating point's addition and subtraction.
3. If true value = 10/3, approximate value = 3.33, then find the absolute and relative errors.   [Ans. $e_a = 0.003333$, $e_r = 0.000999$]
4. Write difference between the truncation error and round off error.
5. What do you understand by machine epsilon of a computer? Explain.
6. Two numbers are given as 2.5 and 48.289, both of which being correct to the significant figure given. Find their product.   [Ans. 1.2×10^2]
7. Represent 44.85×10^6 in normalized floating point mode.   [Ans. $0.4485\,E\,8$]
8. Show that the following rearrangement of equation $x^3 + 6x^2 + 10x - 20 = 0$ does not yield a convergent sequence of successive approximation by iteration method near $x = 1$.
9. What is the difference between Regula–Falsi method and Secant method?
10. What is the condition for the convergence of iteration method. [Ans. $|\phi'(x)| < 1$]
11. Write the criterion for the convergence of Newton–Raphson method.
$$\text{[Ans. } |f(x) f''(x)| < |f'(x)|^2]$$
12. Write the formula for Aitken's Δ^2 method.
13. How can the rate of convergence of two methods be compared? Explain by taking an example.
14. What is the order of convergence of Newton–Raphson method?   [Ans. 2]
15. What is deflated polynomial? Explain.
16. Define μ and δ.
17. Prove $\Delta^3 y_0 = y_3 - 3y_2 + 3y_1 - y_0$.
18. State Newton divided difference formula for interpolation.
19. Write a formula for cubic spline.
20. Write the condition of natural spline.   [Ans. $M_0 = M_n = 0$]
21. Express $1 + x - x^2 + x^3$ as sum of Chebyshev polynomials.
$$\left[\text{Ans. } \frac{T_0}{2} + \frac{7}{4}T_1 - \frac{T_2}{2} + \frac{T_3}{4}\right]$$

22. Express $T_0(x) + 2T_1(x) + T_2(x)$ as polynomial in x. [Ans. $2x + 2x^2$]

23. What is the principle of least square?

24. State Lagrange's interpolation formula.

25. Write the normal equation of the curve $y = \dfrac{C_0}{x} + C_1\sqrt{x}$.

$$\left[\text{Ans. } \Sigma\frac{y}{x} = C_0\Sigma\frac{1}{x^2} + C_1\Sigma\frac{1}{\sqrt{x}}, \Sigma y\sqrt{x} = C_0\Sigma\frac{1}{\sqrt{x}} + C_1\Sigma x\right]$$

26. Define truncation error in Lagrange interpolation.

27. Write the formula for piecewise linear interpolation.

28. Define Chebyshev polynomial.

29. Using Gram–Schmidt orthogonalization process, compute the first two orthogonal polynomials $P_0(x)$, $P_1(x)$ which are orthogonal on the interval $[0, 1]$ with respect to weight function $w(x) = 1$. [Ans. $P_0(x) = 1, P_1(x) = x - 1/2$]

30. Prove $E = 1 + \Delta$.

31. What are the errors in trapezoidal and Simpson's rules of numerical integration?

32. Find the error in the derivative of $f(x) = \cos x$ by computing directly and using the approximation $f'(x) = \dfrac{f(x + h) - f(x - h)}{2h}$ at $x = 0.8$ choosing $h = 0.01$.

[Ans. Error $= 1.414050$]

33. Distinguish between interpolation and extrapolation.

34. When does Simpson's rule given exact value?

35. Write the formula for first derivative of Newton's forward formula for interpolation.

36. Write the error terms in first derivative of Lagrange interpolation formula.

[Ans. $-1/2\,hf''(\xi), x_0 < \xi < x_1$]

37. Define numerical differentiation.

38. Write the truncation error in Simpson's 3/8 rule.

39. Compute the error in the evaluation of $\int_4^{5.2} e^x dx$ by Simpson's one-third rule.

[Ans. 1.93×10^{-3}].

40. Write two point formula for Gauss–Legendre integration method.

41. Write the error term in three point formula for Gauss–Legendre integration method.

42. Obtain an approximate value of $I = \int_{-1}^{1} e^x \, dx$ by using Lobatto three point formula. [Ans. 2.36205]

43. Write the Radau three point formula for integration.

44. Write the error term in Radau three point formula for integration.

45. Differentiate ill conditioned and well conditioned methods.

46. Write the formula for second order Runge–Kutta method.

47. Write the formula for third order Runge–Kutta method.

48. Write the formula for fourth order Runge–Kutta method.

49. Write the condition of rate of convergence of relaxation method.

50. Differentiate the Type-I and Type-II error.

51. Define level of significance.

52. Define test of hypothesis.

53. Determine the condition number of matrix $A = \begin{bmatrix} 1 & 7 & -4 \\ 4 & -3 & 8 \\ 12 & -1 & 3 \end{bmatrix}$.   [Ans. 16]

54. Define condition number.

55. Write the Adam–Bashforth predictor–corrector formula.

56. By using Runge–Kutta method of second order, find $y(0.1)$.
    Given $y' = -y$, $y(0) = 1$.   [Ans. 0.905]

57. Define statistical hypothesis.

58. Define partial pivoting and complete pivoting.

59. A coin was tossed 400 times and the head turned up 216 times. Test the hypothesis that the coin is unbiased.   [Ans. Accepted]

Index